PLANTS
AND PEOPLE

Origin and Development
of Human–Plant Science
Relationships

PLANTS
AND PEOPLE

Origin and Development of Human–Plant Science Relationships

Christopher Cumo

CRC Press
Taylor & Francis Group
Boca Raton London New York

CRC Press is an imprint of the
Taylor & Francis Group, an **informa** business

CRC Press
Taylor & Francis Group
6000 Broken Sound Parkway NW, Suite 300
Boca Raton, FL 33487-2742

International Standard Book Number-13: 978-1-4987-0708-4 (Paperback)

Library of Congress Cataloging-in-Publication Data

Cumo, Christopher, author.
    Plants and people : origin and development of human--plant science relationships / author: Christopher Cumo.
        pages cm
    Includes bibliographical references and index.
    ISBN 978-1-4987-0708-4
    1. Agriculture--History. 2. Human-plant relationships--History. I. Title. II. Title: Origin and development of human--plant science relationships.

S419.C78 2015
635--dc23   2015011903

Visit the Taylor & Francis Web site at
http://www.taylorandfrancis.com

and the CRC Press Web site at
http://www.crcpress.com

Contents

Preface

This book aims to fill gaps in the literature available to the scholar and casual reader alike. At the same time, it is a synthesis of what is known about the sciences, history, and agriculture. To turn first to the sciences, the reader may note my preference for the word "sciences" rather than "science." This preference is not absolute, but stems from the recognition that the sciences are not a single discipline but rather a multiplicity. The traditional high school quartet of earth science, biology, chemistry, and physics barely scratches the surface. One might add entomology, virology, bacteriology, plant physiology, zoology, microbiology, molecular biology, cellular biology, agronomy, soil microbiology, nutrition, and biological anthropology to gain a glimpse of the breadth of the sciences. Even history of science and philosophy of science impinge on these disciplines. Of course, there is a place for the word "science," where one wishes to denote those branches of knowledge that use the scientific method. Earlier uses of the term are likewise important. To take one example, Columella, the first-century CE Roman agricultural writer and stylist, listed agriculture as a "science" (see Chapter 5). He did not elaborate on this linkage, but he must have meant something like an organized body of knowledge, which is what he delivered in his 12 volumes *On Agriculture.*

If my book comes to terms with the sciences, it must break through the thicket of the scientific lexicon, a formidable language that is a kind of insider's speak. The problem, at least in part, is the fact that the sciences produce so much information so rapidly that one is pressed to take stock of it all. This problem mushrooms when one realizes that all branches of knowledge produce a voluminous amount of literature that grows day by day. No one anymore can aspire to be a polymath. Not every reader has the time or desire to conquer the technical literature of the sciences. In this context, I have attempted to write plain, unadorned prose to cut through the code language that takes up so much of the sciences. I am attempting to fill the gap between technicality and the desire for simplicity. For this reason, my book may hold special appeal for the interested layperson or the student who aspires to be a scientist but has not climbed the ladder very far.

The second field under scrutiny is history, a discipline that is in flux. The older fields—American history, European history, and military history, for example—appear to have lost much of their appeal. Today, the emphasis is on gender studies, courses that treat racial and social issues, the study of ordinary people, Islamic studies, Asian studies, Latin American history, and African studies. Particularly hard hit has been the history of science. It never commanded the allegiance of a large number of scholars and now appears in a tenuous position. It may appear to be too much of the same old story about the lives and careers of educated white men, most of them Europeans or their descendants who have populated the Americas. Is such a focus too elitist for today's historians? As a historian of science by training, I think it is too early to consign the field to the morgue. My intent has been to use this field to map the study of plants and of the plant sciences and their relation to human needs and wants over a vast time. One might glimpse in this book how this study has changed since the days of Theophrastus, pupil of Greek philosopher Aristotle and arguably the founder of the science of botany. I also want to stress items of continuity. The works of Roman writers Varro and Columella on the use of manures and legumes came down to the early twentieth century. It was still possible for American soil scientist Charles Thorne to write about manures in the early twentieth century just as Varro and Columella had 2000 years earlier (see Chapter 5). Ultimately, these are the insights that make history of science relevant today. In indulging in history, I have attempted to treat all the iterations of the plant sciences from their inception to the present, with the understanding that this is a tall order.

Third is my focus on agriculture, the discipline that potentially unites the study of the plant sciences and history. From its inception about 10,000 years ago (perhaps even earlier), agriculture has been a prehistoric and then a historic process. Like the sciences, and history for that matter, agriculture is incredibly diverse. The origin and development of agriculture differed in essential aspects in Papua New Guinea, southwestern Asia, the Americas, and several other places. Indeed, there was no uniformity in agricultural developments throughout the Americas. It may be difficult to pinpoint the moment when the sciences and agriculture intersected. In some sense, the first farmers created a kind of proto science that predated the origin of the sciences, if one is right to trace the latter contribution to the Greeks more than 2000 years ago. The very attempt to select the best seeds for planting next season was an attempt to alter the genome of an edible plant. Simple as this activity may have been, it is difficult to deny its scientific character. As with my attempt to unify the sciences and history, this study introduces agriculture as the third leg of this triangle, a leg that cannot stand without the other two. This attempt to create a triangle among the sciences, history, and agriculture contributes fresh insights to all three fields. My microscope has focused on plants and the plant sciences as the means to build this triangle.

Acknowledgments

Writing a book or even an article can be a solitary endeavor but one that never occurs in a vacuum. I am grateful to many people. From the outset, Randy Brehm, senior editor at CRC Press, took my proposal for this book seriously. Without her assistance and persistence, I doubt that this project could have lifted from the ground. The Stark County District Library in the township where I reside offered me access to innumerable resources. When I needed a book that the library did not have, the librarians swiftly borrowed it from another library. They never complained even when my requests numbered in the hundreds. The library also maintains an online database of peer-reviewed journals, from which I accessed hundreds of articles. I could not have written this book without access to these services, all of them free to the public. Finally, my family deserves special thanks. My wife Gerianne and my daughters Francesca and Alexandra have been a great comfort to me.

Introduction

The preface indicated my intent to unify the study of the sciences—the focus being on the plant sciences—history and agriculture. The sciences and agriculture are the prime movers, but because they occurred over time, they have a natural historical element. My intent has been to give the reader an overview of the relationships among plants, people, and the sciences. It is at once a scientific work and a book that offers an olive branch to the humanities. C. P. Snow noted the difficulty of bridging the sciences and the humanities.[1] I am a bit more optimistic and believe that books with an essentially scientific character can appeal to a broad readership if they have historical depth to them. We all want to know how humans created the present out of the past. This book attempts to satisfy this curiosity in examining how humans used plants, the sciences, and, of course, agriculture to shape a world of some 7 billion people.

This book has two natural starting points. The biology of plants informs Chapter 1 and evolutionary biology Chapter 2. Chapter 1 introduces the basic concepts and insights that biology has gained from the study of plants. The chapter covers much ground but with simple prose. One need not use the excuse of having a superficial knowledge of biology as justification for skipping this chapter. In fact, the cumulative nature of this book and of the sciences, in general, should discourage the desire to select chapters of interest while ignoring others. Chapter 2 is in some ways the most historical because the study of life is an evolutionary process that imparts change over an immense time. In this sense, history is not enough because most of this chapter takes place deep in prehistory. Plants and people did not appear suddenly on earth, as a literal reading of Genesis might imply. Both took millions of years to assume their present form. Of the two, plants are by far the more ancient. In evolutionary terms, humans are latecomers to the planet.

This sense of change over time is essential for an understanding of Chapter 3, which considers the early uses of plants. My treatment here is confined to the era before the Neolithic Revolution. Here, the reader finds that plants have always been essential to the primates, a type of mammal that has given rise to several species, including humans. Among our lineage, it is clear that the Australopithecines ate primarily plants. Meat appears to have been a rare item of nourishment. The members of our genus, *Homo*, also ate a great diversity of plants, though meat appears to have grown in importance. We also used plants early in our evolution as the foundation for making shelters. The earliest use of sugarcane plants may have been to thatch simple huts in Papua New Guinea, again well before the onset of agriculture.

The constancy of change over time reappears in Chapter 4, which overviews the Agricultural or Neolithic Revolution. Perhaps, most startling is the fact that humans went from eating many species of wild plants to domesticating a comparatively small number of edible plants, a trend that accelerated over the millennia. This narrowing of the diet may have scarred humans with diseases of dietary deficiency. This chapter attempts to do some justice to the enormous diversity of agricultural systems that emerged, many of them fairly quickly, between about 10,000 and 5000 years ago.

It is not possible to simplify this process, though one might note the prevalence of seed agriculture in several areas of the temperate zones. The seeds of choice were usually grains or legumes. One might think of the soybean in China, the pea in southwestern Asia and throughout the Mediterranean Basin, and *Phaseolus* beans in the Americas. In the tropics, different systems emerged, based largely on roots and tubers. Papua New Guinea is an example, but a counterexample may be the Andean highlands, which, though in the tropics, shared the climate of the temperate zone. There people domesticated five tubers, one of which, the potato, became a world staple. Because roots and tubers are not preserved well in the hot, damp tropics, it is difficult to know how old agriculture really is in Southeast Asia and Papua New Guinea. The answer must be many millennia. The same is not true of Australia despite its proximity to Southeast Asia. Only the arrival of the British led to the establishment of agriculture, which has been quite successful in that continent.

Chapter 5 reexamines ancient agriculture through the lens of the first stirrings of the sciences among the Greeks and Romans. Space deterred me from including the Chinese. The focus is on the empirical nature of ancient science and on the durability of this vision into the present. The Greeks and Romans all wrote treatises on plants, particularly on their use in agriculture. The Chinese writings may be the earliest, but they tend to be fragmentary. Those that have come down to the present from Greece and Rome are complete. Several writers, the Romans Cato, Varro, and Columella are all good examples, as they recounted their own observations of what worked and what did not. Columella, when speaking about grape varieties, commented that he would write about only those that he had grown. This empirical stance is now taken for granted. Is it possible to conceive of the sciences outside an empirical framework? Although Varro used the term "art" to refer to the study and practice of agriculture, the Romans, as a rule, deemed agriculture a science (see Chapter 5). Here an important and inseparable linkage was forged.

Chapter 6 examines one of the pivotal events of the early modern era: the Columbian Exchange. A torrent of activity swept aside old ways. Old World diseases depopulated the Americas, and plants and animals crossed between eastern and western hemispheres. It is not possible in a book of this sort to trace the entirety of this exchange, but attention focuses on a handful of New World plants that crossed to the Old World, and Old World plants that entered the Americas. Consider sugarcane alone. A native of the Old World tropics, almost certainly Papua New Guinea, sugarcane came to the notice of Europeans during their trade with the Arabs during the Middle Ages. A tropical crop, sugarcane could not tolerate European winters. Matters changed with the discovery of the Americas. Spanish Italian Christopher Columbus planted sugarcane in the Caribbean in 1493. From such humble beginnings arose a sugar empire in tropical America. With the near extirpation of the Amerindians, Europeans looked for a new source of labor, enslaving Africans. This action reshaped the definition of slavery. It was now a racial category infused with animus, prejudice, and violence. Sugarcane stoked the American economy but cost African Americans their lives and dignity.

Chapter 7 explores another vast topic, the rise of publicly funded agricultural education, research, and extension, with the primary focus on the United States, though British and German forerunners receive treatment. The sources of this movement came from many directions. The Enlightenment focused attention on scientific rationalism and its power to improve the world. Americans, perhaps best represented by third U.S. president Thomas Jefferson, emphasized the importance of practical knowledge and were unshaken in their belief that the sciences benefited agriculture. By benefiting agriculture, the sciences benefited all Americans. Another factor emerged from an egalitarian sentiment. The emphasis on higher education as a bastion of blue bloods did not endear the common person. Education must open its doors to the children of farmers, teaching them to be better farmers for the benefit of all Americans. The land grant universities attempted to do just that. At the same time, Americans wanted these universities to conduct research on agriculture, a mandate that extended to the U.S. Department of Agriculture. The emphasis on the agricultural sciences reached fruition in the nineteenth century in the creation of the agricultural experiment stations.

Chapter 8 overviews one of the great achievements of this public research system in the development of hybrid corn. The science itself goes back to the work of Austrian monk Gregor Mendel, whose paper on pea hybridization in 1866 laid in obscurity until its rediscovery in 1900. Within the decade, American geneticists and plant breeders were applying Mendel's ideas to the breeding of corn. By 1917, this work came to fruition in the first practical method to produce lots of corn seeds through hybridization. The 1930s and 1940s witnessed spectacular growth in the planting of these hybrids in the Corn Belt and the West. At the same time arose, through hybridization, the possibility of deriving pest-resistant varieties of corn. This chapter examines one such success: the breeding of hybrids resistant to the European Corn Borer.

If hybrid corn was a revolutionary development, so was the attempt to export American plant sciences to the developing world in what has become known as the Green Revolution, the subject of Chapter 9. Here again, the land grant universities played an important role as the University of Minnesota educated Norman Borlaug, perhaps the most influential figure in the Green Revolution. Even now, it is difficult to assess the movement. Proponents point to the unprecedented growth in the yields of wheat and rice, with wheat in Mexico and northern India and Pakistan and rice throughout southern Asia, notably in the Indian peninsula. Detractors point to the fact that the poorest farmers did not benefit from the Green Revolution and that it, by increasing the use of agrochemicals and water, is unsustainable and even dangerous. Where will we turn for potable water when farmers take so much through irrigation?

Chapter 10 attempts to overview what may be the revolution of the future. If the hybrid corn revolution and the Green Revolution defined the past, biotechnology may revolutionize the future. My aim has been to lay bare the essential sciences behind the breakthroughs of biotechnology. It is clear that biotechnology can produce crops, Bt corn, for example, resistant to pests and others, namely Roundup Ready soybeans, corn, and sugar beets resistant to the herbicide glyphosate. These achievements are important though one must be cautious. Biotechnology has so far guarded plants from biotic and abiotic stresses, protecting yield stability. From the vantage of the sciences, though, it is not clear whether this new generation of plants has markedly higher yields than the current generation of elite cultivars derived by the breeding techniques of the Green Revolution, at least under optimal conditions. One must also consider the fact that biotech crops are not popular everywhere.

Chapter 11 tries, as difficult as it may be, to peek into the future. The question it poses is simple: Can agriculture and the agricultural sciences feed the world as population continues to mushroom? The starting point to such an inquiry is *An Essay on the Principle of Population,* British cleric Thomas Malthus' 1798 contribution to the emerging social science of demography. Malthus was clear: Human population always outruns the food supply so that famine, disease, and warfare limit population to what agriculture can sustain. The history of the twentieth century appears to have disproved Malthus. Global population exploded, yet the hybrid corn revolution and the Green Revolution kept pace with this growth and even exceeded it, producing a worldwide net surplus of food. Yet, in the long run, Malthus must be right. The planet is finite and so can support only a fixed number of people. Moreover, the heady days of the Green Revolution are behind us. Yield gains have slowed and funding for the agricultural sciences has ebbed. How can agriculture and the sciences feed perhaps 10–12 billion people in 2050 given these trends? Future generations of scientists must answer this question.

REFERENCE

1. C. P. Snow, *The Two Cultures* (Cambridge, UK: Cambridge University Press, 1998), 2.

1 The Biology of Plants

THE RISE OF THE SCIENCES

To comprehend the ways in which biology has shaped the study of plants, one must, by prelude, grasp something of the role of biology in the sciences. The term "sciences" is purposeful because many disciplines comprise them. All sciences share the same method, so there is value, too, in the word "science" to encompass this method. I mean, of course, the scientific method, though it is not my concern at the moment. I wish to construct a simple notion of science as the systematic study of the natural world. It is fashionable in some circles to conceive of science as the outgrowth of the labors that spanned Polish astronomer and mathematician Nicholas Copernicus to British polymath Isaac Newton, but science is much older, tracing its roots to classical Greece.[1] The pre-Socratic philosophers, Greek philosopher Plato (Figure 1.1) noted in the *Apology*, aimed to identify the elements upon which the world was built.[2] According to Plato, his mentor Socrates showed an interest in this inquiry but finally came to crisis. Socrates determined that scientific inquiry made no difference in how he lived his life. He henceforth devoted his energies to exploring the issue of how one should create a purposeful life. The concern seems almost existential, but it is possible to exaggerate the Socratic break with science, as British classicist Francis Cornford did.[3] Although Plato shared Socrates' desire to craft a meaningful life, Plato did not ignore science. Indeed, in the *Timaeus*, a late dialog, Plato concerned himself with describing how the universe came into being.[4] It is enough to say that Plato had scientific interests that Greek intellectuals shared. In this context, the Harvard University biologist Ernst Mayr may have been too hasty in depicting Plato's corpus as an impediment to scientific advancement.[5] As a generalization, the Romans were more engineers than scientists, though the demise of the Greco-Roman world, surely by the fifth century CE, appears to have put the sciences into eclipse. Faith rather than science characterized the medieval world, although the dazzling achievements of the early modern world heightened the status of the sciences.

PLANT BIOLOGY: THE CELL

In the plant and animal kingdoms, the basic unit of organization is the cell. Keep in mind that there are two types of cells: prokaryote and eukaryote. The prokaryote is older and is distinguished by the absence of a nucleus.[6] Archaebacteria are part of this group. The prokaryotic cell typifies the simplest bacteria and must have arisen not long after the origin of life. The eukaryote, a later development, represents an advancement because it organizes its genetic material in a nucleus, the command center of the eukaryote. The eukaryote is the building block of the plant cell and so is our focus. It is astonishing to think that although plants do not have a skeleton as

FIGURE 1.1   Plato: Arguably the most influential philosopher in history, Plato was also a keen student of cosmogony and mathematics. (From Shutterstock.)

vertebrates do to make them erect, trees are perfectly erect and may weigh several hundred pounds. In addition, a tree trunk may support branches that themselves weigh hundreds of pounds. What accounts for this strength and relative rigidity? The answer lies in the cell wall. Although animal cells have a flexible, thin membrane, a plant cell has a fibrous wall.[7] To be sure, this wall may impart flexibility to a plant, but it is also a source of strength, snapping a plant erect as it grows. The cell wall was the part of a plant cell that English scientist Robert Hooke discovered in 1665.[8] A type of sugar, cellulose, is the principal constituent of the cell wall. Cellulose is an arrangement of glucose that may be 15,000 molecules in length.[9] A grazing[10] animal such as a cow derives much of its energy from the cellulose it eats. In turn, humans consume beef. Whether eating a food plant, pork, beef, or another meat or poultry, humans directly or indirectly depend on cellulose for energy. One might wonder what binds together all the cellulose molecules in the cell wall. The answer is hemicellulose, which one might liken to glue or another adhesive to hold together a cell wall. Glycoproteins, a type of protein bound to sugar molecules, are also present in cell walls. In sum, a cell wall may be 80% cellulose, with the rest being hemicellulose and glycoproteins.

Despite the importance of the nucleus, one might be surprised that most of the chemical activity in a plant cell occurs outside of the nucleus, in the cytoplasm. In this context, the nucleus directs chemical reactions in the cytoplasm. An important part of this activity is the synthesis of proteins. The production of proteins is important not only to plants but also to herbivores. When humans, an omnivore, eat a food plant, legumes such as peas, chickpeas, beans, lentils, or peanuts, for example, they derive protein. In fact, the consumption of a whole grain and a legume, brown rice and beans, for example, provides all essential amino acids that humans need in their diet. Once a plant cell has manufactured a protein or another chemical, it may share the molecule with other cells through the plasma membrane, a remarkably thin and porous structure at the boundary between cells. The membrane is a regulator of sorts, impeding the passage of some chemicals but abetting the transfer of others. By being selectively permeable, the membrane may permit the buildup of a molecule, often an ion, in a cell. Carbohydrates and proteins make up a portion of the cell membrane.

Given all the fanfare devoted to the nucleus, some attention must focus on it. Some scientists have compared the nucleus to a computer. Certainly, it has the instructions and blueprints from which the rest of the cell is built. The deoxyribose nucleic acid (DNA) in the nucleus is the blueprint, or template if you like, out of which the cell assembles proteins and other molecules. This process may be complex, but its omission here will not impair our understanding of the cell. It suffices that DNA provides the information that the rest of the cell uses to grow, to differentiate one structure from another, and ultimately when to die. It is important to note that a somatic cell can double the content of DNA in its nucleus and then divide, forming two daughter cells, each with the correct amount of DNA. This process is essential for growth. As a rule, the nucleus, not a large structure, varies from 2 to 15 μm in diameter.[11] Plants differ from some algae and fungi in having a single nucleus per cell. A nuclear envelope cordons off the nucleus, differentiating it from the rest of the cell. This structure gives the nucleus its shape. The envelope has pores that allow chemicals to exit and enter the nucleus. Messenger ribonucleic acid (mRNA) is a good example of a molecule that exits a nucleus, whereas proteins may enter the nucleus. Chapter 10 provides greater detail about the organization and role of DNAs and RNAs in the cell.

Beyond the cell wall, beyond even the cell membrane, is the endoplasmic reticulum. Through it, ions and other types of chemicals transit from one cell to another. One might think of the endoplasmic reticulum as a series of tubes through which cells exchange chemicals. On the outer surface of the endoplasmic reticulum are ribosomes, which contain RNAs and proteins. Also not large, ribosomes are only about 20 nm in diameter.[12] Ribosomes may also populate a cell apart from the endoplasmic reticulum. They may aggregate in groups of 5–100. They play an important role in connecting amino acids to form long chains of protein. In this way, the ribosomes aid RNA by initiating the assembly of amino acids. In addition, one may find ribosomes in chloroplasts.

Plastids, the chloroplasts, are perhaps the emblematic structure in a plant cell. They are large relative to other plant structures within the cytoplasm. They may vary in size and shape, with some resembling certain types of algae. The most characteristic chloroplasts tend to resemble frisbees. Others have been likened to a rugby ball. The simplest plants may have only one or two chloroplasts, although most plants tend to aggregate 75–125 chloroplasts.[13] Obviously, the disparity is large. Two membranes encase each chloroplast. Of the two, the outer appears to be a variant of the endoplasmic reticulum. The inner appears to be exceedingly old, perhaps deriving from the cyanobacteria early in the evolution of life. A chloroplast contains stacks of grana, which resemble checkers or coins or some other discs. Each stand may accumulate 40–60 grana, each linked to its lower and upper granum. These contain the thylakoid membrane, a key structure in the production of chlorophyll, the pigment that gives plants their characteristic greenness. One may determine a plant's health at a glance. A green plant is usually healthy, although the incidence of the water mold *Phytophthora infestans* in the 1840s occasionally left a potato plant green; the tuber of such plants often appeared acceptable but quickly rotted. The incidence of this disease caused the last subsistence crisis in Europe (Chapter 6). Moreover, it seemed to validate British cleric Thomas Malthus who feared that populations tend to outrun their food supplies.[14] A green plant that has yellowed is not healthy. Disease or nutritional deficiency has impaired its ability to manufacture chlorophyll. A chloroplast is a liquid that contains enzymes, a type of protein, essential in the process of photosynthesis. Photosynthesis is the most important process on earth. Through photosynthesis, the chloroplasts convert the energy from sunlight into sugars, a point made later in this chapter. The plant stores this energy as biomass. Anyone who has eaten an apple or another food has consumed the energy that an apple tree, for example, has stored from sunlight. Without photosynthesis, earth would contain no animals, including humans. Plants have made the first civilizations to the most recent technologies possible. One hopes that solar panels will become ubiquitous, capturing the sun's energy by a different mechanism than plants have done for eons. In addition, the chloroplasts harbor RNAs and ribosomes to direct the assemblage of proteins. Chloroplasts are not the end of the story. Although chloroplasts and chlorophyll are responsible for the greenness of plants, chloroplasts produce yellow, red, and orange pigments. They account for the red or yellow color of a ripe tomato. One might note in this context that the red or yellow ripe tomatoes are vastly superior in nutrients and flavor than the putatively red tomatoes at the supermarket. Because a ripe tomato is subject to spoilage during transit, the food industry does not favor it. Instead, growers harvest tomatoes green, well before they have reached the peak of nutrients and flavor. Shippers spray these tomatoes with ethylene gas to simulate redness. Although such a tomato looks red, it never had a chance to ripen because of early picking. The consumer is stuck with a solid, insipid sphere. The attempt of genetic engineering to solve this problem is a subject for Chapter 10.

Mitochondria are also important structures. Known simply as the powerhouse of the cell, they release energy through cellular respiration. This energy sustains a plant, allowing it to grow and reproduce. In flowering plants, reproduction yields seeds. In several agronomically important plants—rice, corn, wheat, barley, rye, and oats, for example—humans derive energy directly from the seeds, known as grains

FIGURE 1.2 Barbara McClintock: Part of a long line of scientists who chose corn as their research plant, Barbara McClintock received a Nobel Prize in physiology or medicine in 1983.

or kernels (Figure 1.2). This is also true of seed legumes such as peas, beans, lentils, chickpeas, and peanuts. In fact, human populations have grown so large because of a diet of grains. In other cases, a fruit surrounds the seeds. Apples, oranges, grapes, and tomatoes are good examples, although scientists have bred citrus fruits and grapes with few, if any, seeds. Potatoes, sweet potatoes, cassava, turnips, and carrots, among others, are in a different class, although they are also valuable additions to the diet. In some places, northern Europe, for example, potatoes rivaled bread as a staple of the masses. Potatoes are a tuber. That is, they are swollen underground stems. Sweet potatoes, in contrast, are swollen roots and are important, in contrast to potatoes, in warm environs.

THE PLANT GOES UNDERGROUND

As potatoes and sweet potatoes suggest, the underground components of a plant, whether stem or root, are immensely important to humans and to the plants themselves. Roots may be particularly conspicuous in large trees. As a tree grows, its roots thicken. In some cases, this process causes problems, for example, the elm. City planners know not to plant elms near sidewalks because their shallow roots, snaking under a sidewalk, enlarge to the point of fracturing or otherwise marring sidewalks. Roots perform functions vital to a plant's survival and maturation. They absorb water and nutrients in the form of ions from the soil. The roots absorb several ions. In the 1840s, the German chemist Justus von Liebig identified the ions of nitrogen, potassium, and phosphorus as the big three nutrients.[15] A plant cannot survive without the big three and several ions in small quantities. von Liebig even calculated a law of the minimum. One might, for example, calculate the amount of nitrogen that the roots absorb from the soil. Suppose, for simplicity, that 1 ha of potatoes absorb through their roots one metric ton of nitrogen from the soil; after the harvest, the farmer must restore not less than one metric ton of nitrogen to that acre. For centuries, the use of animal and even human excrement served this purpose. Indeed, well into the twentieth century, the American scientist Charles Embree Thorne counseled farmers to apply manure to their soils liberally.[16] Since Thorne's day, the transition to inorganic fertilizers, probably at our peril, has been rapid. These do nothing to improve soil structure, enhance water-holding capacity, or add organic matter.

Another essential feature of roots is to anchor plants to the soil. This means, of course, that plants cannot flee from adverse conditions. Through seed dispersal, however, the next generation may find more hospitable conditions. Although many roots may be shallow, others penetrate the soil to a depth greater than 5 m.[17] There they find water and nutrients that may not be available to shallow roots. The long and large taproot of the sugar beet, for example, is renowned for its sugar content. Rice displays an adaptive feature by being able to tolerate saturated soils. Strictly speaking, rice roots do not gain special advantage from inundation, which serves to suppress weeds whose roots cannot penetrate waterlogged soils. In this respect, rice plants benefit from the minimization of competition from weeds.

The roots are the first structure to emerge from a seed. Through evolution, roots have learned to penetrate the soil rather than migrate above ground. Doubtless gravity helps roots orient themselves in the soil. The first root is the radicle. The radicle, as in the sugar beet, may form a large, conspicuous taproot. Alternatively, a large number of secondary roots may branch off from the radicle and the emerging stem. The number of roots may be staggering. A single ryegrass plant may have 15 million roots, a number twice the population of humans.[18] Put end to end, these roots would travel 400 miles. The roots of some plants are highly specialized. The roots of soybeans, peas, peanuts, beans, or other legumes contain nodules in which reside symbiotic bacteria. They benefit by having a place to live. In turn, they benefit legume roots by converting atmospheric nitrogen, hydrogen, and oxygen in the soil, which roots cannot absorb, into nitrate or ammonium ions, which plant roots readily absorb. These nitrifying bacteria are obviously essential for the health of legumes and thus benefit humans by making these plants a good source of protein. As a rule, monocots, plants with a single embryo leaf, possess a more fibrous root system than dicots, plants with two embryo leaves. Corn and rice are examples of monocots, whereas peas and peanuts are dicots. The fibrous root systems of monocots are adept at holding soil in place, lessening the danger of erosion. Among food plants, dicots are more numerous than monocots, but the latter probably provides the majority of the body's calories.

As a matter of structure, one may divide roots into zones. First is the root cap, the tip of root surrounded by a number of parenchyma cells. The root cap itself may be large, although in other plants it is diminutive. The root cap protects the root as it burrows through the soil. This is important because the interior of the root cap is delicate. The root cap is sure to earn its keep in rocky soils that impede the progress of roots. The outermost cells in the root cap exude a slime of sorts that apparently eases a root's passage through the soil. Interestingly, this slime also attracts nitrifying bacteria, which we have seen enable the roots to absorb nitrogenous ions. Perhaps, because the root cap bears the brunt of a root's passage through the soil, its cells do not wear well, living less than 1 week before new cells replace them. In essence, a plant is continuously remaking its root caps. The root cap is also important because it apparently senses gravity, pulling a root deeper into the soil. Starchy grains on the tip of a root cap appear to be the gravity sensors. The roots whose root caps are damaged tend to grow horizontally rather than downward.

Immediately behind the root cap is the region of the root responsible for cell division. This region produces new cells for the cap. A single cell in this region produces two new daughter cells every 12–36 h.[19] This is much faster than the norm for a plant cell, which may divide only every 200–500 h. In the region of cell division, the production of new cells tends to peak once or twice every 24 h so that the output of new cells is not constant over time but rather fluctuates. Curiously, noon and midnight tend to correspond with the peaks of cell production.

About 1 cm behind the root cap and behind the region of cell division is the region of elongation. As the name suggests, in this region, root cells may expand several fold as they mature. As these cells lengthen, they become slightly stouter. As the cells elongate, they produce one or two vacuoles. Largely, these vacuoles may account for as much as 90% of a cell's length.[20] Once a root cell in this region expands to maximum length, it maintains its size until death.

Next to the region of elongation is the region of maturation. This latter region is responsible for causing what was a homogeneous mass of root cells to differentiate into specialists. The region of differentiation is sometimes called the root hair zone. Roots have hair-like structures that resemble short threads. They absorb water and nutrients from the soil and thus have the most direct contact with the soil. With many root hairs, a root is able to maximize its surface area in contact with the soil, assuring the maximum absorption of water and nutrients. Indeed, a root may have nearly 40,000 root hairs per square centimeter.[21] Corn is an example of a plant with abundant root hairs. Ryegrass has an even larger number of root hairs. It is easy to disturb or damage root hairs upon transplanting a plant, although the Chinese followed this practice with rice seedlings for millennia without apparent ill effect. The root cap is active in the formation of new root hairs. Surely, evolutionary adaptation root hairs are just thin enough to impede the entrance of bacteria or fungi, although this barrier can be breached.

The roots of sweet potatoes, yams, and other root crops efficiently store nutrients, converting them into starch. As the plant ages, these roots swell, providing humans a vital source of nourishment. Carrots, beets, turnips, and radishes are curiosities, having both swollen roots and stems. In this sense, they are not strictly root crops. We have seen that the potato is still another type, being an enlarged underground stem and so is not a root crop. The common name potato applies to both the potato and the sweet potato, but this is due to Spanish confusion upon encountering these crops in the Americas. The two are not closely related. Other plants have roots particularly adept at accumulating water. The pumpkin and watermelon are good examples of plants that develop large water-abundant fruits due to the efficiency of their roots at absorbing water. Tomato and citrus fruits also display this capacity. As a rule, many of the plants that humans domesticated millennia ago have roots that are exceptional in storing large quantities of water. Just as an example, a watermelon is roughly 95% water. This capacity to absorb water is an evolutionary adaptation against lean times.

In modernity, the propagation of plants depends largely on the production and dispersal of seeds. Other plants, however, have roots capable of generating a new plant, a clone of the parent, for example, pineapple. The new offshoot of the root is known as a sucker. Cherry, apple, and pear trees, among others, propagate new trees from suckers. Other plants produce so-called aerial roots that do not contact the soil. The orchid is an example. Some old conifers bear roots that are partly above ground. In the case of aerial roots, vanilla is an interesting plant whose roots, exposed to sunlight, photosynthesize. Some plants, several dicots, are examples and develop roots capable of pulling a plant deeper into the soil, anchoring it more securely. Dandelion is an example. Perhaps because of this phenomenon, the dandelion root and lower stem are difficult to extract. It is well known that even a small portion of the dandelion root is capable of regenerating a new plant, a frustration to homeowners nearly everywhere, although it is worth noting that the American author Henry David Thoreau was not alone in eating dandelion leaves as a salad green.[22]

In all this diversity roots are far from trivial. They were surely part of the diet of early humans, and several have attained the status of world staples. Many such crops are biennials, flowering only in the second year. Because these plants need long durations of warmth, many are tropical. In the first year, these plants accumulate water and nutrients. Humans harvest the roots at this stage, not letting them complete their life cycle. Among the root crops, it is difficult to overstate the importance of sweet potatoes, yams, and cassava to the people of the tropics and even warm regions of the temperate zone. Cassava yields more calories per acre than any other root crop. Cassava may even rival other world crops in its caloric density. The masses in the tropics depend on cassava in much the way that India, China, and Southeast Asia have depended on rice. One cannot, however, regard cassava as nutrient-dense. Fish and other sources of protein must complement it. Cassava is also deficient in beta-carotene, the precursor of vitamin A. Sweet potatoes, in contrast, do not suffer this deficiency. Roots are also a source of spices and medicines. The Japanese even make alcohol from sweet potatoes. Tobacco roots are responsible for the manufacture of nicotine, which circulates through the plant and has been the bane of humanity.

STEM

A plant of course has more than an underground component. Above ground are stems, foliage, and, in many cases, flowers and seeds, the latter often encased in fruit. Some stems may have an underground component. We have seen the importance of the underground stem in the formation of the potato. Humans have long recognized the malleability of stems. It has long been possible to graft the stems of one plant onto the rootstock of another plant, often a closely related species.

Stems are of two types. Determinate stems cease growth when a plant flowers. Indeterminate stems take no note of flowering but grow throughout a plant's life. Tomato, pea, and soybean stems are all examples of indeterminate growth, although scientists have managed to breed varieties of each species with determinate stems.

The result is a compact plant, often called a dwarf, whose foliage is easier to manage and whose ratio of edible biomass is greater than that of their indeterminate counterparts. In many respects, the phenomenon of determinate growth had much to do with the success of the Green Revolution, discussed in Chapter 9.

The tip of a stem, such as a root cap, is a region of active growth. The robust activity of a stem tip determines how long the stem will be. In the case of trees, long stems have produced the magnificent redwoods of California. The buds from which leaves emerge form along the length of a stem. Of special importance to humans are the rhizomes, that part of a stem that develops underground. The stolen is a type of underground stem from which tubers develop, the outstanding example being, as we have seen, the potato. What we think of as wood is simply the stem of a tree. Wood has long been important because of the diversity of its uses. Because some stems are less dense than water, they float. Wood thus gave humans their first watercraft. Boats gave humans mobility. No longer were they confined to land. Watercraft may have been important in the early migrations to parts of Southeast Asia and Australia, although they played no role in the colonization of the Americas, at least not until the arrival of Christopher Columbus. Although brick, concrete, and steel are important, the construction of buildings has long depended on wood. Wood is also a source of paper, although not of the papyrus that was so important to ancient Egypt. Papyrus too is the product of a stem, in this case, the stem of the papyrus plant that was ubiquitous along the Nile River, particularly in the delta. Ancient papyri are important sources of historical and religious information. The stems of some plants may vary by season. In spring, the Venus's flytrap, for example, produces long, erect stems, keeping the traps well above ground. This description is relative because compared to many other plants, the Venus's flytrap is diminutive. As summer progresses, the stem shortens and its growth begins to incline toward horizontal, positioning the traps closer to the ground. This progression is interesting and must in some way influence the type of organisms on which the plant preys.

LEAVES

The importance of plant leaves is difficult to overstate. Leaves contain the cells and chloroplasts that covert sunlight into biomass. Some scientists have compared leaves with solar panels, although the method of energy storage is not identical. Whatever its size or shape, a leaf begins as a primordium in the bud. As a bud first forms in spring, the primordium may contain only about 200 cells.[23] The onset of warm weather stimulates cell division, so that in only a few weeks a primordium may grow to millions of cells. When a leaf becomes mature, it has a short, thin stalk known as a petiole. The leaf blade, the lamina, has a dense network of veins. The leaves of some plants may lack a petiole. The term for such a leaf is sessile. Climate determines the longevity of leaves. In the tropics and subtropics warmth, sunlight, and rain are generally abundant. Leaves photosynthesize year-round and so remain intact longer than is the case for a deciduous plant. A tropical plant produces new leaves year-round. In the temperate zone, the situation is different. The warm and lengthening days of spring and the warmth of summer allow leaves to photosynthesize during these seasons. In autumn, however, as daylight wanes and temperatures cool, leaves become a liability because they continue to consume water while producing less energy. A temperate plant typically compensates by shedding its leaves in autumn. Such plants are known as deciduous. This time is special for the inhabitants of the temperate zone. Leaves bereft of their chlorophyll display their red, orange, and yellow underlying colors. People revel in touring the countryside decorated in such a profusion of colors. Even evergreens do not retain all their leaves, known as needles. Rather, their leaves seldom live longer than 2–7 years, when they are replaced. This principle holds true for evergreens in the tropics and subtropics.

Leaves vary from plant to plant. The diminutive leaves of a soybean plant in the temperate zone are not as large and spectacular as the leaves of a palm tree in subtropical Florida. All leaves form from auxiliary buds on the stems. A simple leaf, the leaves of monocots are examples, originates from a single leaf blade. A compound leaf has multiple parts. Pinnate compound leaves pair their leaflets along a part of the petiole, known as the rachis. In contrast, palmate compound leaves group all leaflets at the petiole's end. Further subdivision is possible, creating an array of small leaflets. A transparent, protective layer of cells envelops each leaf, whatever its classification. The layer is transparent because it must allow the passage of sunlight to the rest of the leaf. This envelope is known as the epidermis. So important is sunlight that during the course of a day, some leaves may twist to receive maximum sunlight as the sun appears to move across the sky. Because of earth's rotation on its axis, the sun's apparent motion is an illusion, though one that most humans retained into the fifteenth century. The underside of leaves and sometimes even leaf surfaces contain stomata. Their importance is difficult to overstate because when they open, the plant leaf emits oxygen necessary for humans and many other organisms. The stomata also engage in transpiration, a process by which plant leaves emit water vapor. This too is an important process because the atmosphere needs water vapor to form clouds. These clouds produce the rainwater that sustains life. Transpiration, however, lowers the amount of water present in a plant. If unregulated, transpiration would be a fatal process. Even desert plants, cacti are good examples, have stomata. How then do they escape dehydration and death? The answer lies in their evolutionary adaptation to the desert. Cacti open their stomata for only brief periods to conserve water. This process of regulation makes possible the existence of vegetation in the desert, although the population of plants in such environs is necessarily sparse. Stomata are also important because when open they take in carbon dioxide. In many ways, then, due to the stoma (the singular of stomata), plants are the great carbon sinks, although the point is subject to debate. Some agronomists hold to the belief that an increase in carbon dioxide in the atmosphere benefits plants because the gas is essential to photosynthesis. In this context, global warming and climate change might benefit plants. It is worth noting, however, that the grasses evolved in an era of comparatively low carbon dioxide levels. How will they fare as the concentration of carbon dioxide continues to increase? The question is not academic. Civilizations have for many centuries depended on the great grasses: rice, corn, wheat, barley, oats, rye, and other species. Could further increases in carbon dioxide levels curb yield increases? It is also worth noting that the planet does not appear to be oscillating between low and high carbon dioxide concentrations. Humans appear to be the culprits as automobiles, factories, and power plants emit enormous quantities of carbon dioxide and other greenhouse gases.

Because leaves are variable by species, one can identify plants by examining their leaves. Even a nascent gardener can distinguish between potato, tomato, and pepper plants, even though all belong to the same family. Indeed, the ability to recognize plants by sight is one of gardening's pleasures. In this regard, the diversity of plants is astonishing. A water lily, for example, may have a leaf with a diameter of about 2 m and a weight of more than 220 kg.[24] Yet, the leaf will not sink. In contrast, try to float a rock of only one-tenth of this weight. Plants, this example makes clear, have properties that transcend what is possible in the inorganic world. The shapes and colors that leaves may take is another point of diversity. Plant leaves may be flattened, papery, scaly, feathery, cup-shaped, and needle-like, and not every plant is green. A beautiful species of Venus's flytrap is red (Figure 1.3).[25] A leaf may be smooth or hairy, slick or sticky, waxy or glossy, aromatic or producing an unpleasant odor, and edible or poisonous. Some plants have edible and poisonous organs. Tomato and potato leaves are toxic, yet a tomato or potato is not only edible but nutritious as well. The arrangement of leaves is not random, but appears to aggregate in one of the three types. Many species arrange their leaves alternately or in a spiral along a stem. Other plants may produce opposite arrangements of leaves, with two of them attached to each node. Still, other plants have leaf whorls, producing at least three leaves at a node. Corn leaves are a good example of this phenomenon.

Whatever the shape or grouping, a leaf contains three principal regions: the epidermis, the mexophyll, and the vein. The epidermis, as mentioned earlier, is noted, in almost all cases, for its lack of chloroplasts. That is, the epidermis is metabolically inactive. The epidermis contains a waxy coating known as the cuticle. The epidermis, depending on the plant leaf, may produce other waxy substances, again with a view toward protecting the leaf. Pollution appears to degrade these substances. Given the importance of plants to other lives, human carelessness toward them is problematic. Some plant leaves secrete additional wax as defense against predatory insects. In fact, the act of feeding on leaves prompts this production. The mesophyll is an important structure because it is the region of photosynthesis. The uppermost portion of the

FIGURE 1.3 Red Venus's Fly Trap: An attractive specimen, this red Venus's Fly Trap demonstrates that a plant need not be green.

mesophyll, the palisade mesophyll, contains more than 80% of a leaf's chloroplasts.[26] The lower region, the spongy mesophyll, has a comparatively loose network of cells and the remaining chloroplasts. If parenchyma tissue has chloroplasts, it is known as chlorenchyma tissue. This tissue may occur in other parts of a plant, not merely in the leaves. Veins, known as vascular bundles, run throughout the mesophyll. Xyla and phloem tissue comprise veins. Thick parenchyma cells envelop each vein, probably to protect a vein in the manner that the epidermis protects the rest of a leaf. The parenchyma cells are known as the bundle sheath. One might think of the veins as a skeleton of sorts, given their function in supporting leaves. The phylum transmits sugars and various other carbohydrates through the leaves. The xylem conducts water from root to leaf, a vital function. The xyla are conspicuous in trees. The veins in dicots appear to branch in several directions; in monocots, veins are parallel to one another, running the length of a leaf. Leaves have adapted to special functions. In a forest, for example, the upper canopy of leaves harvests the most sunlight and therefore need not be large. The understory of trees does not receive abundant sunlight and so has evolved large leaves to capture as much sunlight as possible. The effect of shading has been mischievous. On several occasions, researchers have reported lower than expected yields from semidwarf varieties of rice and soybeans. There was no cause for vexation. Taller test plants had simply shaded the semidwarfs, causing an artificial loss in yields. Grown among other semidwarfs, these rice and soybeans lived up to their yield potential. Other leaves take on the appearance and function of tendrils, winding around various structures as they climb upward. The pea is an example of this phenomenon. The leaves of desert plants assume the shape of pines, exposing less surface area to the air and thus reducing water loss. The spine is also a formidable defense against herbivores. Spines are sometimes confused with thorns and prickles. In fact, all three are confused. One reads about the thorns of a rosebush when the structures are really prickles. Whereas spines are modified leaves, thorns arise from modifications to the stem and prickles are outgrowths of the epidermis or cortex. All are defensive structures. Perhaps, most striking of all are the leaves that natural selection has modified into traps. The traps of the Venus's flytrap snap shut to devour a hapless insect. The leaves of pitcher plants converge to form a kind of basin in which insects drown. Large pitcher plants can even trap small rodents and frogs. Contrary to television commercials, these plants, especially the Venus's flytrap, do not grow large enough to eat humans. These capture mechanisms delighted Charles Darwin, who realized that they had evolved in response to the nutrient-poor soils in which these plants lived. Because the carnivorous plants could not derive sufficient nourishment from the soil, they evolved traps that captured a meal.

Humans use plants with leaves in a variety of settings. The shade trees enhance recreation and leisure and capture the human need for aesthetic appeal. Shade trees planted near a home may reduce the homeowner's electric bill by lessening the need to use air conditioning in summer. Leaves of the true vegetables—cabbage, lettuce, spinach, Swiss chard, and Brussels sprouts—nourish people. Leaves provide a variety of herbs: thyme, marjoram, oregano, tarragon, peppermint, spearmint, wintergreen, basil, dill, sage, cilantro, and savory. Important as they are, they were not the spices of the fabled spice trade. Humans also extract drugs from various leaves. Leaves are also the source of fiber and textiles, although leaves do not rival cotton fibers in this regard. In the Andes Mountains, the Amerindians burned yareta leaves for heat long before the arrival of Spanish conquistador Francisco Pizarro. Leaf oils scent soaps, lotions, perfume, and even repel mosquitoes. Although undesirable, cocaine is derived from coca leaves. Central and South American cocaine has flooded the U.S. narcotics market with dangerous results. The tobacco leaf is another undesirable product. Its ill effects on humans have been the subject of many books. Given all the harm it does, one might express surprise that tobacco remains legal. Marijuana, a controversial drug, is nothing other than the dried leaves of cannabis plants. The medical community, particularly in the United States, appears to be divided over the effects of marijuana, although some states have legalized it in recent years. Beyond these uses, the tealeaf has emerged as a pillar of civilizations. After water, tea appears to be the most widely consumed beverage. Before the advent of DDT (dichlorodiphenyltrichloroethane) in the 1940s, scientists had derived several pesticides from the extracts of leaves. The leaves of a Mexican plant are lethal to cockroaches. The potential of plant-based pesticides appears to hold promise against a variety of insects and may be more appealing than the synthetic pesticides that degrade the environment. Because some plants yield leaves with up to 50% protein, their potential as nourishment should be explored.

FLOWERS

Although flowering plants, known as angiosperms, are ubiquitous today, this was not always so. The first plants lacked flowers. One may, for example, scrutinize a fern without discovering any flowers. Measured over the eons of geological time, flowering plants are relative newcomers to the biota. Because of their omnipresence, humans take flowering plants for granted. There is even a species of orchid that develops flowers and the rest of its anatomy underground. There appears to be no limit to where flowering plants may grow, excepting the poles. The angiosperms contain some 250,000 species, of which hollycock pictured in Figure 1.4 is an example, and there may be many more awaiting discoveries.[27] Scientists believe that the tropics contain species of angiosperms never catalogued by humans. Of the diversity of species of flowering plants, only 11 provide 80% of the world's food. The grasses contain 10 of these species, and the 11th, ranking fifth, is the Andean potato. Flowers may attain a multiplicity of sizes, shapes, and colors. The dainty but beautiful potato flower sparked a revolution of sorts in eighteenth century France. Curiously, it is possible to conceive of a potato flower as an afterthought of sorts. Although it is pollinated and yields seeds, these are seldom planted. Rather, farmers divide a potato into subsections, each with at least one "eye." Each eye yields a new plant. This method of propagation has serious consequences because each new plant is a clone of the parent, causing the potato to be a genetically uniform crop, a circumstance that played a horrible role in the Irish Potato Famine of the 1840s. Other flowers are not dainty. An Indonesian plant in the genus *Rafflesia* yields flowers more than 1 m in diameter. Each weighs about 45 kg.[28]

Some flowers are exceedingly rare. Historians who have surveyed the written record of the last 5000 years find fewer than 100 instances of the corpse flower actually blooming. As a rule, the more showy and spectacular a flower, the greater the

FIGURE 1.4   Flower: These hollyhock flowers are both beautiful and functional in making reproduction possible.

likelihood that it attracts insects or birds to pollinate it. Such flowers may also emit an odor to attract pollinators. The plants with unremarkable flowers, for example, corn, usually depend on the wind to spread pollen from one flower to the stigma of another. Plants lead sexual lives in the sense that pollen is the male gamete and the ovum the female sex cell. As in humans and other sexually reproducing species, the pollen and ovum are haploids. That is, they contain half of the full complement of chromosomes in a somatic cell. The union of pollen grain and ovum restores the full complement of chromosomes, creating in a seed the embryo of a new plant. A seed, once planted, may proceed from germination to full maturity in a month, although the agronomically important crops need months to produce edible fruits, seeds, tubers, or roots. Even the interior of a stem, in the case of sugarcane, may yield calories, although one might note that given sugar's ubiquity, it provides no nutrients. In this sense, sugar is referred to as providing empty calories.

An annual plant flowers and completes its life cycle in a single year. A biennial plant produces foliage in its first year but flowers and reproduces only in its second year. The carrot is an example of a biennial. Fortunately, for humans, it fills its taproot in its first year so that the farmer treats it as an annual. A perennial plant, for example, sugarcane, may require several growing seasons to flower. Because sugarcane is harvested before it flowers, the belief persisted for centuries that sugarcane did not flower. This is untrue, but it is worth noting that sugarcane does not yield abundant blooms. Rather, similar to the potato, sugarcane is propagated vegetatively. A section of stem contains nodes from which a new plant is germinated. Again, the danger of genetic uniformity is present. Perhaps, this risk is acceptable because the loss of a sugarcane crop would not deprive humans of nourishment. Biennials and perennials need long growing seasons. Many, sugarcane is again the exemplar, reside in the tropics and subtropics. Citrus fruits are another example of a perennial that must have tropical or subtropical conditions if they are to survive, reproduce, and yield fruits. This is not the place for a detailed study of citrus fruits, but their use as an anti-scurvy agent long predated the discovery of vitamin C.

ANATOMY OF FLOWERS

At the end of a stalk is a peduncle, which may have still smaller stalks known as pedicels. One or both structures swell into a small pad called a receptacle, from which the parts of a flower emerge. The sepals of a flower, known as the calyx, may fuse together; the calyx protects the parts of a flower while it remains in the bud. The most conspicuous part of a flower may be the petals, known as the corolla. The petals may not be fused into a single entity, as is true of peach flowers. Petunia flowers, in contrast, are fused. Many grasses, for example, corn, lack petals altogether. The term perianth refers to both calyx and corolla. Around the base of a flower are the stamens, which contain the male gametes. At the center of a flower is the pistil, containing the female gametes. This unity applies to perfect flowers. Other flowers may contain only male or female parts. In such cases, cross-pollination is essential. The stamen contains the anthers, which in turn have the pollen. Atop the pistil is the stigma, which connects to the base of the ovary. The ovary contains the ovum, the female gamete. A flower, as in a peach tree, may arise as an autonomous entity, but in many plants, such as lilac, grapefruit, and grapes, flowers group into inflorescences. An inflorescence may contain hundreds of flowers. In some cases, these flowers open simultaneously, although in other instances the opening of flowers is staggered.

DICOTS AND MONOCOTS

Of the two, dicots are older. About 75% of all plants are dicots.[29] The majority of the flowering plants are dicots. Remember that a new dicot produces two embryo leaves rather than the one that monocots yield. Monocots must have evolved from a species of dicot, although the history of this transformation has not been traced. Over the millennia, monocots have assumed enormous economic importance. Remember that the grasses are monocots, among which are rice, wheat, corn, barley, oats, rye, sugarcane, and other agronomically important crops. For millennia, rice, for example, has sustained the people of China, Southeast Asia, India, and Africa. In fact, African rice is its own species distinct from the more familiar Asian species. Corn played a similar role in pre-Columbian America, although it now feeds pigs and cattle in the developed world. Yet, one cannot ignore the agronomically important dicots, among which are the legumes. In this class are soybeans, peas, lentils, beans, chickpea, peanuts, and other valuable crops. The domestication of the legumes came at a fortunate moment, yielding proteins superior to what could be obtained by eating grains alone. So important were soybeans to China that one ancient text listed them among the five sacred grains.[30] This is not true because soybeans are a legume, not a grain. The chickpea and lentil nourished India, and the pea, chickpea, and lentil were staples of Roman agriculture. That is, the ancients built their civilizations on a foundation of dicots and monocots.

FRUITS

There has long been confusion about the designations fruit and vegetable. In 1893, the U.S. Supreme Court shunted aside science in ruling that a tomato is a vegetable rather than a fruit.[31] Rather than astonishment, U.S. tomato growers were elated. The ruling gave tomatoes the protection of a tariff to keep cheap Mexican "vegetables" out of the United States. The court had acted for economic rather than scientific gains. In its decision, the court even recognized the botanical classification of the tomato as a fruit. That is, the tomato and other fruits are mature ovaries, swollen in many cases. A fruit usually contains seeds. The flesh of a fruit may attract herbivores to eat them. The seeds pass through the body so that, in effect, the fruit is a means of seed dispersal. In some cases, the seeds are abundant, as is true for tomatoes, strawberries, apples, and several other fruits. Other fruits contain a single, usually large seed known as a stone. The peach, cherry, nectarine, apricot, and plum all contain a single seed.

It is amazing to think how many putative vegetables—tomatoes, cucumbers, and squashes, for example—are really fruits. Because a fruit develops from an ovary, only angiosperms can yield fruits. As they develop in a fruit, seeds exude hormones that cause fruits to grow. In some instances, the fruit may become large. A single beefsteak tomato may weigh 1 kg. The banana is conspicuous in lacking seeds. The proto banana had seeds but over millennia humans selected for seedlessness. Instead, the banana develops by a process known as parthenocarpy.

The skin of a fruit is known as the exocarp and is usually thin compared with the rest of the fruit. The endocarp, another layer, encases the seeds. The endocarp may be hard, as is true in the peach, apricot, nectarine, plum, and cherry. Alternatively, the endocarp may be comparatively fragile. The endocarp that surrounds apple seeds is fibrous but can be easily eaten. Between endocarp and exocarp is the mesocarp, the flesh that herbivores and humans often find irresistible. The pericarp comprises exocarp, mesocarp, and endocarp and properly speaking is the entire fruit minus the seeds. The stone fruits are known as drupes. Not only the peach and its relatives fall into this category, but also nuts. The berry is another prominent type of fruit. It develops from a compound ovary and is similar to the tomato; it usually bears many seeds. The entire paricarp is fleshy and edible. The seeds themselves are edible, a single tomato contains hundreds of seeds, yet not one will harm the stomach or intestines. Berries include the tomato, grape, persimmons, peppers, and eggplant. Given their informal status as berries, one might express surprise that strawberries, raspberries, and blackberries are not true berries. The vernacular does not always coincide with the scientific. Blueberries, gooseberries, and cranberries are true to their names. A specialized type of berry, the pepo, includes pumpkin and other squashes, cucumber, watermelon, and cantaloupe. Berries known as hesperidia include citrus fruits, among them, the orange, lemon, lime, grapefruit, and tangerine.

SEEDS

Seeds are precious, each containing the embryo of a new plant. Wild species of flowering plants have many ways to disperse their seeds. Dispersal is important in allowing a plant to distribute its progeny over space. Domesticated plants are noted by an inability to disperse seeds on their own. Corn is a perfect example. Unable to disperse seeds, it depends wholly on humans for planting the next generation of seeds. Wheat and barley are other examples. Before domestication, these plants shattered. That is, they dropped their seeds to the ground when ripe. According to the canonical gospels, Jesus may have had this phenomenon in mind when he claimed that unless a wheat plant died it could not yield a new generation of plants.[32] During domestication, humans selected for wheat and barley plants that did not shatter but held their seeds (grains or kernels) until human harvest.

To return to wild plants, one notes the ease of seed dispersal. In many cases, wind plays a prominent role. Wind disperses dandelion seeds, often over great distances and in profusion. It is no wonder that the dandelion is a worldwide weed. Humans have put men on the moon, but they have little prospect of eradicating dandelions. Maple seeds are another example; they twirl in the wind like maladapted helicopter blades, also spreading over distance. In fact, a pair of maple seeds may spread as far as 6 miles from the parent tree. In addition to wind, animals disperse seeds. Birds, for example, feed on the fruit of cherry trees, swallowing the seed. Perhaps miles away, the bird defecates, depositing seed and dung on the ground. The cherry seed thus has a built-in source of fertilizer. These examples, and others like them, demonstrate the role that fruits play in attracting animals and thus dispersing seeds. It is also possible for mammals to rub their fur against seeds, acquiring them in the process. Over the course of a mammal's trek, it unwittingly sheds these seeds. Some plants even produce fruit that acts as a laxative, speeding seeds' progress through the digestive tract. In some cases, blackbirds retain such seeds no longer than 15 min, although most mammals will hold seeds for about 1 day before excretion.[33] The giant tortoises of the Galapagos Islands retain seeds as long as 2 weeks. This length of time appears necessary to stimulate the seeds to germination upon excretion. Squirrels, known to gather acorns, typically do not eat all, allowing some to germinate. Some seeds adhere to the feet of birds and mammals and in this way disperse. Ants likewise disperse seeds in their search for food. Given these phenomena, it seems evident that flowering plants coevolved with insects, birds, and mammals. The important role of bees and other insects in pollination strengthens the case for coevolution. Some fruits and seeds float, allowing water currents to disperse them. There has been debate over whether ocean currents might have carried coconuts to several Polynesian islands. Transit by ocean current probably is not correct because brine that penetrates a seed will kill the embryo. Rain may cause seeds to dislodge from a plant, dispersing them. Among the prime movers of seed dispersal, humans must rank high. During the Columbian Exchange, for example, humans spread corn from the Americas to Europe, Asia, Africa, and Australia, making it a world staple. Other plants, as Chapter 6 makes clear, shared a similar global expansion. The crops that have become world staples owe their status to human agency.

Seeds contain a hilium, the point of attachment to the ovary wall. The hilium is conspicuous on many true beans. The careful opening of a bean reveals the two embryo leaves of the new plant. The two leaves are known as cotyledons. The rest of the bean possesses a reservoir of carbohydrates and nutrients to sustain the new plant during germination. Of course, only viable seeds can yield a new plant. Most seeds require a period of inactivity, dormancy, to germinate. This appears particularly true of the seeds of plants of the temperate zone. A legume seed, a bean, for example, has a thick, tough outer coat that does not readily absorb the water or air that would trigger immediate germination. One may overcome dormancy by cracking the seed coat or dipping it in an acid for perhaps 1 min. A seed may also contain chemicals that postpone germination. Desert plants are renowned in this regard. Only abundant rains dissipate these chemicals, allowing germination at those rare moments when a storm breaks over the desert. A brief shower will not suffice, protecting a seed from germinating when there is insufficient water to sustain the new plant's growth. Apple, pear, citrus, and tomato fruits likewise have chemical inhibitors to prevent premature germination. Once the seeds are free from the fleshy fruit, they germinate easily. In temperate zones, many seeds must pass through a prolonged cold spell before they will germinate. As in the life of a new plant, water is critical. Many seeds will not germinate unless they have absorbed 10 times their weight in water. The absorption of water drives the action of enzymes capable of breaking proteins into their constituent amino acids and fat into soluble lipids for uptake by the radicle. Other enzymes convert starch into simple sugars, again for uptake. In the early stages of germination, the roots must have sufficient oxygen. The presence of too much water eliminates this oxygen, causing a seedling to die. Too much water may also encourage the growth of fungi that may likewise kill a seedling. This phenomenon is known as damping off disease. Soil temperature plays an important role in germination. Although ideal temperatures vary by species, many plants as a rule germinate between roughly 1°C and about 42°C.[34] The upper range does not characterize the germination of temperate plants. Also, as a rule, the germination temperatures of crops are more narrow. The role of light in germination is not easy to generalize because some seeds need light to germinate, whereas others must have darkness.

The longevity of seeds is also not easy to quantify, given reports that wheat seeds recovered from a pharaoh's tomb in Egypt or from a cave or from an Amerindian grave have germinated after thousands of years of dormancy. Most scientists doubt these reports. In any case, wheat was unknown in pre-Columbian America. Yet, seeds from the tundra have germinated after 10,000 years. Lotus seeds may germinate after 1200 years, both impressive achievements. The seeds of willow, cottonwood, orchid, and tea, in contrast, may remain viable only a few weeks. Stored at low temperature and humidity, seeds stand the best chance of longevity.

WATER

Water is essential to all organisms, including plants. Absorbed through the roots, water tends to move freely from cell to cell in a process known as osmosis. Osmosis is a type of diffusion, which in turn is the movement of molecules from an area of greater to lesser concentration. This process is readily observed in air, and liquids are no different. The membrane of a plant cell tends to be semipermeable, controlling the entrance and exit of water and other chemicals. A membrane may even bar a chemical from entrance to or exit from a cell. An important consequence of osmosis is that it ceases to occur when the concentration of water becomes equal in adjoining cells. Stasis is the result. Pressure may act to thwart osmosis as well. Osmotic potential is the amount of pressure within a cell to stop osmosis. A cell reaches osmotic potential when the cell walls will no longer expand to allow more water into a cell. Water that enters a cell exerts a pressure of its own, turgor pressure, which enhances the structural rigidity of a cell. The combination of osmotic potential and turgor pressure is a cell's water potential. Water flows from an area of high potential to that of low potential until equilibrium is reached. Cells also absorb and retain water through imbibition. The wetting of starch and other colloids creates an electrical potential that attracts water. Remember that water is a weakly ionic compound with the negative charge at the oxygen atom and the positive charge at the hydrogen atoms.

In addition to osmotic and ionic pressures, active transport causes chemicals to enter or exit a cell. In this case, a cell must expend energy to move a molecule from outside its wall to the inside or stop the exit of a molecule. Because this process requires energy, it is expensive but necessary. Active transport prevents ions from moving from an area of higher concentration in a cell to an area of lower concentration in the soil, exactly what is in the plant's interest. Without active transport, a plant that leaked nitrate ions would never have enough nitrogen to grow. Through active transport, mangrove trees can survive in brine because although they absorb the salt water, they eject the salt through specialized glands.

Remarkably, more than 90% of the water a plant absorbs passes through it, exiting the stomata as water vapor.[35] Various experiments have demonstrated that, covered by a jar, a plant will mist the jar in only an hour or two. A corn plant can transpire 4 gallons of water per week, an impressive figure. A single acre of corn may emit 350,000 gallons of water in just one growing season. A single beech tree may transpire 1000 gallons of water per day. If humans were so profligate, each person would need to drink more than 10 gallons of water per day. It is rare for humans, except those engaged in strenuous activity, to drink more than one-tenth this amount. Similar to humans, a plant is about 90% water. Part of this water evaporates from the leaves to cool the plant on hot days.

Many factors account for the movement of water from roots to the leaves. Because water is weakly polar, the negative end forms a hydrogen bond with the positive end of another water molecule. This allows water molecules to adhere to one another. Moreover, the xyla are akin to narrow tubes through which water raises in a process known as capillary action. The cells of the stomata, releasing water, develop a low water potential, and so osmosis transfers water to this area of low water potential. In a sense, the stomata cells pull water into them from adjoining cells, creating a pressure that lifts water throughout a plant.

Representing just 1% of the surface area of a leaf, stomata are important regulators of water, carbon dioxide, and oxygen. A stoma has two guard cells and what one might call a pore capable of opening and shutting. The guard cells have thick walls that by movement open or shut a stoma. The turgor pressure in the guard cells regulates this movement. When turgor pressure is low, a stoma opens. It closes as turgor pressure increases. Capable of photosynthesis, the guard cells expend energy to accrue potassium from nearby cells. The potassium regulates turgor pressure. The presence of sufficient potassium opens a stoma. When a leaf is not photosynthesizing, potassium exits the guard cells, causing stomata to close. Potassium accomplishes these processes because it brings water with it into the guard cells, thereby increasing turgor pressure. The exit of potassium coincides with a reduction in water in the guard cells, lessening turgor pressure. As a rule, stomata are open during the day and closed at night. A number of desert plants, however, operate in reverse, opening their stomata only at night when water stress is a minimum. This is a peculiarity, given that the stomata do not acquire carbon dioxide during the day when leaves are photosynthetically active. The plant instead stores carbon dioxide that it has acquired during the night for use during the day. Humidity plays an important role in transpiration. Transpiration slows as humidity rises and quickens as humidity lowers. Increasing temperatures likewise accelerate transpiration.

Water is crucial to a plant for many reasons, not the least of which is because it transports important ions into and out of cells. These ions are essential for plant life and so for virtually all life on earth. Among the important elements that a plant needs are nitrogen, potassium, phosphorus, carbon, hydrogen, oxygen, sulfur, calcium, iron, magnesium, sodium, chlorine, copper, manganese, cobalt, zinc, molybdenum, and boron. One might note that the elements sodium and chlorine in edible plants suffice for human needs. Nonetheless, humans tend to crave table salt, adding more than enough crystals to food, provoking debate among nutritionists, some of whom link a salt-laden diet with high blood pressure. Recent research however suggests that an ultralow amount of salt in the diet may not benefit people. Scientists, beginning in the nineteenth century, classified plant nutrients as macronutrients and micronutrients. As these designations suggest, plants need significantly more macro- than micronutrients, although this fact does not discount the importance of micronutrients. The big three—nitrogen, potassium, and phosphorus—were early recognized as macronutrients, but calcium, magnesium, and sulfur are also in this group. In fact, nitrogen, potassium, phosphorus, and calcium constitute 99% of a plant's nutrients. There is some variance in the requirement for micronutrients. Ferns, for example, must have aluminum.[36]

PHOTOSYNTHESIS

Photosynthesis (Figure 1.5), the process by which plants convert sunlight, carbon dioxide, and water into sugars, is basic to life. Although humans are omnivores, they could not have survived without plant nourishment and calories. The animals on which early humans hunted or scavenged were often herbivores. But recent research suggests that even meat-eating Neanderthals (sometimes rendered Neandertals) still consumed about 15% of their calories from plants.[37] Photosynthesis is also crucial to a modern world in which fossil fuels play a large—many would say too large—role in the global economy. Coal and oil today provide about 90% of the energy humans use to power their lives. Coal is nothing but the fossilized remains of plants that converted sunlight into sugars millions of years ago. What the plant captured from

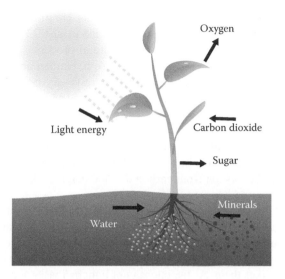

Oxygen

Light energy

Carbon dioxide

Sugar

Minerals

Water

FIGURE 1.5 Photosynthesis diagram: Photosynthesis is basic to all life. Through it, plants manufacture their own energy and animals derive their nourishment from plants. (From Shutterstock.)

the sun over millions of years is what humans are squandering in a few centuries. Photosynthesis is also important because it provides the oxygen all higher animals must have for respiration. The amount of sugars that plants yield from photosynthesis is astonishing. In just 1 year, plants form as many as 220 billion tons of sugar molecules. Plants and the energy they provide must at some point limit human population growth in a Malthusian scenario, the subject of Chapter 11. As it has for millennia, starvation targets humans in food-scarce environments, whereas many people in the West are awash in calories. Obesity ranks as a leading pathology in the United States and other developed nations. This is not the place for an analysis of the causes of obesity, but one cannot help but feel distress at the ubiquity of junk and fast foods.

Using light, plant cells manufacture an energy-dense molecule known as adenosine triphosphate (ATP). During the process of manufacturing, ATP and ultimately a type of sugar known as glucose occurs in the chloroplasts. Using the energy from light, a chloroplast combines the carbon dioxide, the cell acquired from the air, and the water acquired from the soil to manufacture glucose ($C_6H_{12}O_6$). The leftover oxygen exits the stomata into the atmosphere to fuel cellular respiration. Although many steps are involved in this process, one may simplify the basic equation as $6CO_2 + 12H_2O + \text{light} \rightarrow C_6H_{12}O_6 + 6O_2 + 6H_2O$. Transpiration releases the gaseous by-products into the atmosphere as oxygen and water vapor.[38]

Given the necessity of carbon dioxide to photosynthesis, it is striking that it is not especially abundant in the atmosphere, comprising less than 1% of the gases that constitute the atmosphere. This carbon dioxide, absorbed through the stomata, reaches the chloroplasts after a series of transformations. In this sense, plants are carbon sinks. In a single growing season, an acre of corn consumes more than 5000 pounds of carbon through the absorption of carbon dioxide. This calculates to an absorption of 11 tons of carbon dioxide. Were it not continually replaced, plants would consume all the atmosphere's carbon dioxide in just 22 years. The amount of carbon dioxide should be static, given that natural processes such as volcanism emit carbon dioxide at the rate at which plants use it. Human activity—the wanton burning of fossil fuels and the cutting down of forests—is, however, beginning to increase carbon dioxide levels, leading to debate over whether a carbon dioxide-enriched atmosphere will benefit plants.[39]

Given the importance of water to photosynthesis, it is striking that a plant cell uses less than 1% of the available water in photosynthesis. Most of the oxygen from these water molecules exits the stomata without having participated in photosynthesis. In contrast, the oxygen from carbon dioxide is more likely to be used in the manufacture of glucose.

The light available to plants comes in various wavelengths. About 40% of the light is visible to humans. This light, passed through a prism, differentiates into its component colors. A plant uses all colors with violet, blue, orange, and red light being the most important in photosynthesis. Leaves are efficient, absorbing about 80% of the visible light that strikes them. As the intensity of light increases, a plant photosynthesizes at a higher rate and therefore must have extra carbon dioxide and water. If one ranks a sunny day at noon on the equator as 100% of available light, most plants need to harness at least 30% of this light. Obviously, the real world does not conform to the ideal. Clouds, pollution, and shade all reduce the availability of light.

One might think of photosynthesis as a process with two phases, one that occurs in the presence of light and one that can occur in the absence of light. The phase that depends on light is commonly known as the light-dependent reactions. These reactions occur in the chloroplast (Figure 1.6), using light to split water molecules to yield electrons, hydrogen ions, and oxygen. An electron transport system moves electrons to other cells. The chloroplasts use light and electrons to produce ATP. The

Chloroplast anatomy

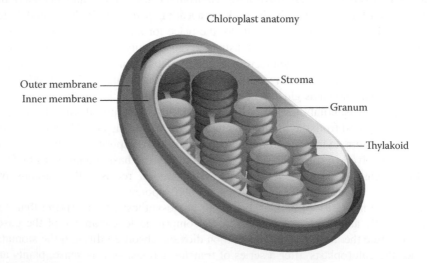

FIGURE 1.6   Chloroplast diagram: A chloroplast is the structure in which photosynthesis occurs. (From Shutterstock.)

chloroplasts, using hydrogen ions from water molecules, make nicotinamide adenine dinucleotide phosphate into reduced nicotinamide adenine dinucleotide phosphate (NADPH). NADPH is important in the second phase, the light-independent reactions. These reactions complete the transformation of the energy in light into biochemical energy in the form of glucose. It was once fashionable to designate this phase as the dark reactions, but because darkness is not necessary, the phrase has fallen from favor. Indeed, the reactions may occur in the presence of light, but they need not. In fact, these reactions ordinarily do occur in the presence of light and in the stroma of the chloroplasts. The main process is known as the Calvin cycle to honor its discoverer Nobel laureate Melvin Calvin at the University of California. In this cycle, carbon dioxide joins a 5-carbon sugar, ributose bisphosphate. Over a series of reactions, glucose is the product. ATP and NADPH furnish the electrons necessary for these reactions. During this phase, a plant may store this sugar as starch.

From an agronomic standpoint, one must draw a distinction between types of photosynthesis.[40] One may classify photosynthesis by the C3 or C4 pathway, depending on whether the process produces three or four carbons. From the perspective of evolution, the C3 pathway evolved first. The C4 pathway was a comparative latecomer, an evolutionary novelty of sorts, that arose at least 62 times in 18 families of plants, several of agronomic value. The distinction between the C3 and C4 pathways is important because, particularly in warm climates, the C4 pathway is much more efficient than the older C3 pathway. Plants with the C4 pathway must have had a selective advantage. C4 plants may grow 50% more rapidly than their C3 forbearers. Enzymes catalyze both processes. Curiously, C3 plants possess all the enzymes necessary for a transition to the C4 pathway. One might suppose in this context that the transition from the C3 to C4 pathway should be relatively straightforward, but efforts have yet to produce results. This circumstance is particularly frustrating because rice, a grass that feeds more humans than any other crop, possesses the C3 pathway and so is not as efficient as one might expect a variety of rice engineered with the C4 pathway to be. These efficiencies function at several levels. The C4 pathway makes a plant more efficient at metabolizing nitrogenous compounds and water. A C4 plant should accordingly be more drought-tolerant than an otherwise comparable C3 plant. Because the leaves of plants conduct photosynthesis, these efficiencies all accrue to these structures. In this context, scientists in several parts of the world must work toward these improvements in rice and other C3 crops.

NOTES

1. John P. McKay, et al., *Understanding Western Society: A Brief History* (Boston and New York: Bedford/St. Martin's, 2012), 490–497.
2. Plato, *The Apology*, in *Great Dialogues of Plato*, trans. W. H. D. Rouse (New York: New American Library, 1956), 425.
3. Francis M. Cornford, *Before and after Socrates* (Cambridge, UK: Cambridge University Press, 1932), 29–30.
4. Plato, *Timaeus*, in *The Collected Dialogues of Plato including the Letters*, eds. Edith Hamilton and Huntington Cairns (Princeton: Princeton University Press, 1989), 1162.
5. Ernst Mayr, *The Growth of Biological Thought: Diversity, Evolution, and Inheritance* (Cambridge, MA: The Belknap Press of Harvard University Press, 1982), 38–39, 87.

6. T. Clason, et al., "The Structure of Eukaryotic and Prokaryotic Complex I," *Journal of Structural Biology* 169 (2010): 81–88; Wanjie, ed. *The Basics of Plant Structures*, 9–10.

7. G. Agoda-Tanjawa, et al., "Properties of Cellulose/Pectins Composites: Implication for Structural and Mechanical Properties of Cell Wall," *Carbohydrate Polymers* 90 (2012): 1081–1091.

8. Howard Gest, "Homage to Robert Hooke (1635–1703): New Insights from the Recently Discovered Hooke Folio," *Perspectives in Biology & Medicine* 52 (2009): 392.

9. Kingsley Stern, *Introductory Plant Biology* (New York: McGraw-Hill, 2003), 35.

10. Ibid, 36.

11. "Molecular Expressions: Cell Biology and Microscopy, Structure and Function of Cells and Viruses," accessed December 5, 2014, http://micro.magnet.fsu.edu/cells/plants/nucleus.html.

12. Stern, *Introductory Plant Biology*, 37.

13. Ibid.

14. Thomas Malthus, *An Essay on the Principle of Population* (Oxford: Oxford University Press, 2000): 13.

15. Fritz W. Went, *The Plants* (New York: Time-Life Books, 1965): 163; Justus Liebig, *Chemistry in Its Applications to Agriculture and Physiology* (Philadelphia: T. B. Peterson, 1847), 30–37.

16. Charles E. Thorne, "The Maintenance of Fertility," *Circular 40 of the Ohio Agricultural Experiment Station* (1905): 17–21.

17. Stern, *Introductory Plant Biology*, 68.

18. Ibid.

19. Stern, *Introductory Plant Biology*, 67; Peter H. Raven, Ray E. Evert, and Susan E. Eichhorn, *Biology of Plants*, 6th edition (San Francisco: W. H. Freeman, 2013), 591.

20. "Vacuole," accessed December 5, 2014, http://www.princeton.edu/~achenay/tmve/wiki100k/docs/vacuole.html.

21. Raven, *Biology of Plants*, 604–605.

22. Anita Sanchez, *The Teeth of the Lion: The Story of the Beloved and Despised Dandelion* (Blacksburg, VA: McDonald and Woodward, 2006), 22–26.

23. Stern, *Introductory Plant Biology*, 107.

24. Ibid, 108.

25. Peter D'Amato, *The Savage Garden: Cultivating Carnivorous Plants*, revised (Berkeley, CA: Ten Speed Press, 2013), 81–82.

26. Stern, *Introductory Plant Biology*, 109.

27. Ibid, 128.

28. Ibid, 107.

29. Ibid, 108.

30. Ibid.

31. Ibid, 109.

32. Ibid, 128.

33. Ibid.

34. Ibid, 129.

35. Ibid.

36. Ibid.

37. Michael P. Richards and Ralf W. Schmitz, "Isotope Evidence for the Diet of the Neanderthal Type Specimen," *Antiquity* 82 (2008): 553.

38. Stern, Introductory Plant Biology, 168; Manetas, *Alice in the Land of Plants*, 19–20.

39. Sylvia Aubry, Naomi J. Brown, and Julian M. Hibbard, "The Role of C3 Proteins Prior to Their Recruitment into the C4 Pathway," *Journal of Experimental Botany* 62 (2011): 3047–3051.

40. Ibid.

2 Evolution of Plants and People

TOWARD NATURAL SELECTION

A chapter on the evolution of plants and people must begin with some understanding of the term "evolution." Scientists can trace the origin of change over time to the Greeks, but a formal understanding had to await modernity. The nineteenth century was a time of intellectual flowering in many fields, including natural history, the discipline that would become biology in the twentieth century. A number of people proposed theories of evolution. By theory, I do not mean, as is the case in popular culture, speculation or guess. I mean a logical, integrated construct of ideas that scientists may test with the aim of strengthening or refuting it. A theory is thus a powerful tool, not a wild assertion. To consider a brief example, creationists often emphasize that evolution is a theory with the expectation that the public will not take it seriously because most people erroneously believe that a theory is nothing more than a crackpot idea that scientists gin up in their spare time. Creationists thus prey upon popular misunderstandings to score points with those either hostile to or with very little training in the sciences.

In the case of evolution, French naturalist Jean Baptiste Lamarck and British naturalists Charles Darwin and Alfred Russel Wallace published theories of evolution in the nineteenth century.[1] Lamarck was the first of this trio, and his ideas survived into the early twentieth century. Even American amateur botanist Luther Burbank held Lamarckian views.[2] In simple terms, Lamarck believed in the inheritance of acquired characteristics. Consider a slender man who, through the dint of strenuous exertion, builds a muscular physique. Lamarck believed that such a man could father a child, boy or girl, with naturally large biceps or quadriceps or other muscle groups, even without the same physically demanding upbringing that the father had endured. In this way, evolution progresses from slight to robust humans, an idea that is incorrect, with special capacities for strength and conditioning. The difficulty with Lamarck's ideas was that they were not easily testable. That is, they lay on the border of science rather than squarely within it.

Nevertheless, even Darwin, to an extent, subscribed to Lamarckism.[3] This may seem strange to anyone familiar with the notion of Darwinism as foreign to Lamarckism. There is reason for this contrast because Darwin developed a new mechanism, natural selection, to explain how evolution occurred. Darwin came to his conclusions in the 1830s but hesitated to publish because of the religious uproar he feared. Remarkably, Wallace came to the same ideas in 1858, galvanizing Darwin (Figure 2.1) to publish the next year his landmark *On the Origin of Species*.[4] Darwin in term derived much from British theologian Thomas Malthus, whose *Essay on the Principle of Population*, Darwin had read. In the late eighteenth century, Malthus

FIGURE 2.1   Charles Darwin: the architect of the theory of evolution by natural selection, Charles Darwin remains a titan in Western thought.

conceived of the dangers by which populations of organisms tended to outrun their food supply.[5] The Irish Potato Famine of the 1840s would validate Malthus' dire forecast. Given this situation, Darwin raised two premises and drew one conclusion. First, there is competition for scarce resources, an obvious idea that derived from his reading of Malthus.[6] One may state Darwin's second premise bluntly: organisms vary in their attributes. Even a cursory glance at the plant and animal kingdoms reveals astonishing variation. Humans, for example, run through a gradation of heights, weights, complexions, and many more traits. Our noses are not all the same length. Hair color varies. The timber of the human voice modulates. Some people cannot play the simplest tune on a piano. Others are virtuosos. Given competition and diversity, Darwin concluded that an organism, with any advantage, however slight, would on average be more likely to survive long enough to leave offspring.[7] The progeny might inherit the selective advantages of the parents, although Darwin did not have much idea of how organisms inherited traits. Darwin understood that survival and reproductive advantages, over millennia and even longer, radiated throughout a population, so that over time a population became better adapted to its environment.

British naturalist J. B. S. Haldane demonstrated how natural selection worked, at least on a small scale, in the early twentieth century.[8] Moths of a particular species are either white or dark. In the era before the Industrial Revolution, light-colored lichens covered trees and white moths prevailed because when they alighted on a tree, they were inconspicuous and so not eaten by birds. White moths were then more common than black moths. With the Industrial Revolution, however, pollution covered the trees, killing the lichens. Now, the bare bark and the shoot protected black moths from detection and they, rather than the white moths, became predominant. The important point about Darwinism is that it applies to all life. This book concerns plants, people, and the sciences. Accordingly, the evolution of plants and people is the focus of this chapter.

PUNCTUATED EQUILIBRIUM: A MODIFICATION OF DARWINISM

Darwin had posited gradual evolution, with each new species a modification of its precursor. In 1972, American scientists Niles Eldridge and Stephen Jay Gould (Figure 2.2) proposed that evolution was not at all gradual.[9] One does not find gradual changes in the fossil record. Rather, one finds long periods when species change little if at all. These long periods of stasis occasionally give way to sudden evolutionary events: the appearance of a new species fully formed. With rare exceptions, the fossil record does not show the evolution of transitional forms. Eldridge and Gould took the fossil record at face value. That is, little happens for long stretches

FIGURE 2.2 Stephen Jay Gould: a popular professor at Harvard University, Stephen Jay Gould was one of the creators of the theory of punctuated equilibrium, a modification of Darwinian gradualism. (Photo Services: Harvard Public Affairs and Communication.)

of time. Evolutionary novelty is a sudden, rapid burst of speciation. The fossil record appears to be fragmentary because new species arise quickly in small founder populations that the fossil record seldom preserves. Only when the new species increases in number does the fossil record pinpoint its existence. Mass extinctions may be the punctuating event, wiping the slate clean so to speak, opening niches that new species populate. We will come to understand that the evidence for the evolution of plants corresponds more neatly to the theory of punctuated equilibrium rather than to Darwinian gradualism.

EVOLUTION OF PLANTS

Scientists have written many books about the evolution of plants. This chapter cannot hope to replicate their comprehensiveness. Rather, the aim is to help the reader see the scaffolding upon which our understanding of plant evolution is built. This chapter aims to provide generalizations in which warranted and sufficient details help the reader see this scaffolding. The number of plant species is vast, and this chapter cannot hope to treat even a fraction of them. With so much information, I had to choose what to include and exclude. Because of their marked importance to humans, I had to mention the evolution of the grasses, but I stopped short of treating the evolution of root and tuber crops. Sweet potatoes and potatoes are, for example, hardly less consequential than the grasses, but the treatment of roots and tubers must await Chapter 4. Because they evolved long before humans, plants deserve our immediate attention. It is astonishing to think that plants have occupied eons of earth's history.

EARLIEST LIFE

The universe is about 13.7 billion years old, an antiquity that we cannot really comprehend.[10] Earth is about 4.6 billion years old, still a vast span of time. Earth was initially very hot, perhaps too hot to permit the evolution of life in the initial stages of earth's history. When it cooled sufficiently, life seems to have arisen rapidly. Life may have arisen about 3.5 billion years ago. The first life, whether based on carbon or not, must have been a molecule, probably a small one at first, capable of replicating itself. Such a molecule would never have left any trace in the fossil record, and it seems reasonable to suppose that life might be even older than 3.5 billion years. Eventually, a carbon-based chain, something perhaps akin to RNA or DNA (Chapter 1), must have appeared in the primeval ocean. It is important to note that both RNA and DNA have the ability to replicate, thus meeting the minimal requirements of life.

The initial environment in which life arose must have still been quite warm. Volcanoes must have been very active, spewing large quantities of carbon dioxide and methane, both greenhouse gases. One must suppose that this early atmosphere was hotter than our present environment. The sun was less luminous than it is today, meaning that it emitted less heat. Yet, this fact probably did not overcome the greenhouse effect that was so prevalent on the early earth. About 3.5 billion years ago, earth may have been as hot as 50°C. This period of intense heat may have persisted for the first 300 million years of life's tenure on earth. The early atmosphere had little oxygen, although the subsequent evolution and proliferation of photosynthetic organisms would raise the concentration of oxygen. As late as 2.2 billion years ago, earth's atmosphere had only about 1% oxygen. As it is today, nitrogen may have been the most abundant gas in the atmosphere of the early earth. Without much oxygen, the atmosphere cannot have had a stout ozone layer to screen the earth from solar radiation of all wavelengths. This radiation likely imperiled the formation of organic molecules necessary for a robust increase in the number and type of organisms.

TOWARD THE FIRST PLANTS

Among the early forms of life, prokaryotic cells (Chapter 1) must have been numerous. All fungi, plants, and animals, however, derive from eukaryotic cells (Chapter 1). Because of their importance to the evolution of subsequent life, the evolution of the eukaryote must rank as among the great achievements of prehistory. The first photosynthetic microbe arose approximately 3.3 billion years ago.[11] Fossil evidence from Michigan in the United States puts the origin of algae, which may have evolved from these microbes, about 2.1 billion years ago, another milestone in prehistory because algae photosynthesize sunlight, carbon dioxide, and water into the sugar glucose, oxygen, and water vapor (Chapter 1). Algae and later plants contributed the oxygen that is part of the modern atmosphere. Without sufficient oxygen, the evolution of aerobic organisms, including humankind, would never have been possible. Rocks in Canada dated to 1.2 billion years ago contain fossils of organisms that appear similar to red algae in the genus *Bangia*.

Green algae (Figure 2.3), the precursors of plants, may have arisen about 800 million years ago.[12] Today, there are more than 7000 species of green algae, which scientists have classified into five large groups, each of which appears to trace its descent to a separate founder species. The first true plants were aquatic, a necessary condition for these simple organisms that had no structures to conduct water to differentiated parts. The first terrestrial plants, another milestone, arose roughly 430 million

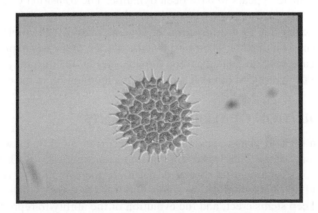

FIGURE 2.3 Green algae; all plants trace their lineage to green algae, an ancient photosynthesizer.

years ago. Measured on the vast geological scale that spans the formation of earth to the present, the origin of land plants is a comparatively recent event. Alternatively, measured on the scale of the human lifespan, plants are indeed ancient. The comparative recentness of land plants probably stems from the relatively recent formation of soil. A land plant cannot grow on nothing more than bare rock. It needs a medium of pulverized rock, organic matter to retain moisture and nutrients, and many kinds of microorganisms. We know this medium as soil. One needed as well a climate of relative constancy and moderation.

By about 500 million years ago, a shallow sea covered about two-thirds of what is today North America.[13] About 60 million years later, the climate cooled enough for glaciers to form near the poles, locking up ocean water as ice. Sea levels diminished accordingly. These events provided the environment in which evolved the first land plants. As a precursor, the first soils appear to have formed about 440 million years ago. The weathering of rocks supplied these prehistoric soils with iron and phosphorus, both essential to plants. The acids exuded from microbes helped break down the surface of rocks, liberating a number of minerals that the rocks had previously locked up. Lichens, the pairing of fungi and green algae, also contributed to the weathering of these rocks. Yet, the rate of weathering must not have produced minerals at the rate plants would have used them. Other processes must have been at work. It seems possible that the decay of aquatic life might have contributed sulfur and nitrogen to these early soils. The decay of phytoplankton, for example, releases sulfur in the form of sulfur dioxide for incorporation into the soil. Lightning must have cleaved the bonds that hold together diatomic molecules of nitrogen gas. Free radicals of nitrogen must have been incorporated into the soil. Because of nitrogen's abundance in the atmosphere and given the frequency of lightning, the opportunities for incorporating nitrogen into the soil must have been numerous.

The proliferation of greenhouse gases during the Cambrian and Ordovician Periods must have maintained temperatures at roughly 40°C, perhaps still too high for the transition of plants from an aquatic to a terrestrial environment.[14] Moreover, rain appears to have been scant between 45° north and south of the paleoequator. At the time, climate appears to have been dynamic. The transition to cooler, wetter environments depended on two factors. First, the concentration of carbon dioxide may have lessened due to a diminution in volcanic activity. Phytoplankton likely also absorbed carbon dioxide. Their burial at the bottom of the ocean prevented their decay and release of accumulated carbon dioxide. About 445 million years ago, glaciers began to form at the South Pole, suggesting a moderation in the climate.

RAPID EVOLUTION OF TERRESTRIAL PLANTS

The evolution of land plants appears to have occurred quickly, between 470 and 430 million years ago.[15] This was a time of important innovations for life on land: the evolution of structures that slowed desiccation, the evolution of cells to transport water and nutrients from soil to stems and branches, the evolution of support structures to keep stems erect, and the evolution of the ability to reproduce in a terrestrial environment. Early land plants reproduced by producing spores. The production of spores should not be confused with sexual reproduction, which was a later development. Meiosis divides a single diploid cell into four spores, each of which is capable of germinating into a new plant. Ferns, for example, continue to reproduce through spores. An important feature of spores was their resistance to decay, a fact that must have given early land plants a selective advantage. Wind could carry spores some distance, giving plants a greater geographical range with the passage of each generation. The first land plant likely arose near water but as spores spread inland, these plants filled new niches. Spores that did not desiccate may have arisen in semi-aquatic plants.

Another innovation was a layer of wax, known as a cuticle, that insulated plant structures and slowed the evaporation of water. The widespread appearance of the cuticle on all parts of the shoot suggests that terrestrial plants could not have evolved without it. The cuticle had evolved by about 430 million years ago.[16] The cuticle would have also protected the shoot for pathogens and abrasion, but it must have impaired the absorption of carbon dioxide and the release of oxygen. A thin cuticle, however, might have permitted gaseous exchange by diffusion, an ability that some mosses exhibit. About 408 million years ago, plants evolved the stoma (Chapter 1), an efficient structure that allowed the release of oxygen and water vapor and the absorption of carbon dioxide. As a rule, the number of stomata varied inversely with the concentration of carbon dioxide in the atmosphere. Intuition suggests as much. When the concentration of carbon dioxide increases, a plant needs fewer stomata to absorb it. When carbon dioxide is dear, plants with additional stomata have a selective advantage.

The need for specialized cells to transport water and nutrients throughout the shoot must have kept plants quite small in their earliest incarnation. The first land plants were not more than 2 cm long. They were horizontal like a vine rather than vertical like a corn plant. About 430 million years ago, plants evolved their first simple vessels through which water and nutrients could travel. One might think of these vessels as tube. The first were very narrow, possibly taking advantage of capillary action. In these tubes lay the origin of vascular plants. About 400 million years ago, specialized cells evolved to conduct water and nutrients. By the beginning of the Devonian Period, plants had made the transition from an aquatic to a terrestrial environment complete with a vascular system.

Water supported aquatic plants, although terrestrial plants needed to evolve new structures. Lignin might have provided some support, but probably did not suffice to cause a land plant to become erect. As we have seen, the first land plants grew horizontally rather than vertically. In this sense, the geometry of land plants depended on their requirements. That is, a plant needed maximum surface area to volume to maximize the amount of plant surface that intercepted sunlight and could exchange gases. Leaves would satisfy these requirements but were not part of the earliest land plants. The habit of indeterminate growth (Chapter 1) put a premium on the development of a long stem that, being green, photosynthesized. Early land plants, however, had short stems, probably to lessen the distance through which water and nutrients traveled. These early stems took the form of a cylinder. Over time, these stems became stouter and longer. Without leaves, we have seen that the stem had to do the work of photosynthesis. The base of the stem had the most photosynthetically active tissue. Much of a plant's potential for an erect habit came from the evolution of a stout, fibrous cell wall that was much more rigid than the cells of animals, insects included. The first solution, however, was the evolution of vessels that when flush with water held the stem stiff and rigid. The cells that conducted water were often in the middle of the stem.

EVOLUTION OF ROOTS

Another problem that evolution needed to solve was to develop some structure to anchor a plant to the soil. Intuitively, one fastens on roots as the solution, but the earliest plants lacked roots. In this sense, the simple definition of a plant as root, the underground portion, and shoot, the aboveground portion, did not apply to the earliest land plants. Nonetheless, plants needed a structure that would absorb minerals and water from the soil. The earliest fossil evidence for roots dates to about 408 million years ago.[17] This may be a late date because these roots were impressive at between 1.5 and 2 cm in diameter and as long as 90 cm. These roots could thus penetrate the soil to nearly 1 m in depth. Doubtless, the first roots must have been shorter. In any case, the evolution of roots appears to have been rapid, as the theory of punctuated equilibrium suggests. One hypothesis holds that as the soil clung to plant stems, these new underground structures evolved into roots. All these new structures arose during perhaps 60 million years ago, a burst of evolutionary activity.

VARIETY OF LAND PLANTS

By about 408 million years ago, vascular plants had colonized vast areas of the continents then known to exist. The process of diversification was well under way, with the radiation of plants into an array of environments. For example, members of the genus *Cooksonia* (Figure 2.4) were prominent in what are today North America and Europe. The earliest evidence for this genus comes from what is today Ireland about 428 million years ago.[18] The plant was erect and lacked leaves. The sporangia produced spores at the end of each branch. The plant was not quite 7 cm tall, although this is impressive early in the evolution of land plants. These plants may

Cooksonia

FIGURE 2.4 *Cooksonia*: a genus of plant, *Cooksonia*, was among the first land plants. (Adapted from K. J. Willis and J. C. McElwain. 2002. *Evolution of Plants* (Oxford: Oxford University Press).)

have inhabited swamps. They appear to have lacked a cuticle and may have absorbed carbon dioxide through underground tissue, an interesting if peculiar innovation.

Another European plant, *Aglaophytum major*, occupied what is now Scotland and is not plentiful in the fossil record.[19] It may have arisen about 400 million years ago and could elongate to 20 cm, about thrice the height of *Cooksonia*. Branches arose from an underground stem, a rhizome (Chapter 1). The rhizome persists in modern plants such as the potato. *A. major* had a cuticle. The rhizome and the structures that radiated from them formed a root system of sorts.

In what is today Australia, then part of the supercontinent Gondwanaland arose *Barogwanathia longifolio*. Its dating is controversial with some specimens thought to have originated about 420 million years ago. Another hypothesis proposes a later date of some 410 million years ago. These dates appear close enough not to warrant despair. The plant had a stem as thick as 2 cm and, at last, leaves, the ultimate architecture for photosynthesis. The stem reached about 1 m in height. Each leaf was thin and about 4 cm long. If one accepts the earlier date, it seems remarkable that a plant of such complexity arose not much later than the spare *Cooksonia*.

PATHWAYS OF EVOLUTION

By 400 million years ago, the transition from green algae to land plants had been remarkable. Algae had depended on an aquatic environment, but land plants inhabited a world in which the water content in the soil was not constant. Rainfall too was variable. The pathways by which green algae evolved into land plants are not clear. Scientists cannot be certain, given current evidence that a single precursor was the foundation of all land plants. One should be mindful that the photosynthetic pigments and the metabolism of land plants share much with algae. All green algae have chlorophyll (Chapter 1), starch, and cellulose in the cell walls. Green algae and land plants also share certain enzymes, notably, glycolate oxidase, which aids a plant in absorbing carbon. Some algae and all plants are multicellular. In fact, *Fritschiella*, a genus of green algae, was at least partly adapted to a terrestrial environment and may have been a precursor to land plants. Freshwater green algae must have been important transitional forms, given that large groups of land plants fare poorly in saline soils.

The evolutionary history of early land plants is complicated by evidence for nonvascular plants, vascular plants, and plants that were akin to hybrids of the two, although these early plants did not propagate sexually and so could not have produced hybrids. Perhaps, each of these three groups had its own precursor, so that evolution took multiple pathways. Current thought supports the notion that nonvascular plants arose before vascular plants. These nonvascular plants were the mosses, liverworts, and hornworts. Liverworts appear to have been the earliest land plants, with mosses and hornworts following later. Liverworts, however, do not appear to have evolved into all other land plants. The hornworts likewise appear to have split from the line or lines that gave rise to all other terrestrial plants. This line of reasoning may leave mosses as perhaps the best candidates as the precursors of vascular plants.

FIRST FORESTS

The period between about 395 and 286 million years ago witnessed the rise of vast forests, with trees more than 35 m tall.[20] This was an impressive change over the diminutive *Cooksonia*. This time witnessed movement in the continental plants and changes in the climate. About 400 million years ago, the climate was warm and humid, with all land free from ice. About 300 million years ago, the landmasses coalesced into the supercontinent Pangea. Yet, climatic change was well under way. By about 290 million years ago, during the late Carboniferous Period, the climate turned cooler and drier. Glaciers appeared at the South Pole.

Taking a closer look, one notes that between about 400 and 360 million years ago, the climate was warm enough that January temperatures did not dip below 20°C.[21] Temperatures at the paleoequator were about 32°C. After about 360 million years ago and before the formation of Pangea, the climate began to cool. Two periods of glaciation at the South Pole appear to have occurred sometime after 360 million years ago and again about 290 million years ago. Although the climate began to dry, lands near the paleoequator retained substantial rainfall. The current tropics likewise continue to receive abundant rain. The swamps near the paleoequator appear to have had the greatest biodiversity among the flora. This remains true of the tropics today. These immense paleoforests continue to play an enormous role in the present economy. During the Carboniferous Period, the huge biomass of dead plants, covered by earth and subjected to great heat and pressure, became the coal deposits on which humans depend, particularly for the generation of electricity. During the Carboniferous, coal deposits formed in what are today eastern North America, western Europe, and parts of Russia.

The reduction in carbon dioxide, no doubt drawn down by plants, and an increase in oxygen must also have influenced the climate, contributing to cooling. This transition is striking, given that about 400 million years ago, the atmosphere had about 10 times more carbon dioxide than it does today.[22] Between about 360 and 286 million years ago, carbon dioxide levels fell 10-fold, so that by roughly 286 million years ago, atmospheric carbon dioxide was near present levels. Plants caused this transition, as we have seen. Moreover, plants would have released acids that would have weathered rocks, causing carbon dioxide to react with calcium silicate ($CaSiO_3$) to yield calcium carbonate ($CaCO_3$) and silicon dioxide (SiO_2). In essence, atmospheric carbon dioxide became a carbonate mineral, namely, calcium carbonate. Water carried this mineral to the ocean, making it a giant carbon sink. Ironically, the carbon dioxide that is stored in coal is now returning to the atmosphere as humans burn it.

Between about 390 and 365 million years ago, the number of plant species rose markedly as plants radiated through the vast expanses of land. In just 20 million years, the number of plant species may have tripled, another evolutionary burst that appears to confirm the theory of punctuated equilibrium. By about 360 million years ago, however, extinction was claiming a number of these species. The process of radiation had given rise to reduction. This period of floral extinction coincided with the extinction of many species of fauna, suggesting a time when life was under duress. Of the plants that survived, the trend toward size accelerated. Vascular systems evolved the capacity to conduct water and nutrients over long distances, allowing plants to grow taller. Roots became larger and denser, able to anchor large trees to the soil. Leaves also lengthened and became more numerous.

Between about 390 and 354 million years ago, plants came to rely on leaves as the principal photosynthesizers. Leaves were then of two types. Microphyllis leaves tended to be small and attached directly to the stem. These leaves may have evolved from tiny spines on the stems of plants. Megaphyllis leaves were not directly attached to the stem, but rather to a petiole that was attached to the stem. The petiole remains today an important conduit between leaf and stem. These larger leaves may have evolved from the webbing between branches. Ferns, extinct horsetails, and all flowering plants have megaphyllis leaves. The decrease in carbon dioxide levels during the late Devonian and early Carboniferous Periods may have caused the evolution of long leaves, with presumably many stomata to absorb the declining amount of the gas. The leaves also appear to have benefited from a decrease in temperatures and carbon dioxide levels diminished.

EVOLUTION OF SEED PLANTS

Between about 395 and 286 million years ago, the first seed plants evolved.[23] The seed was an important innovation because the seed coat protected the embryo and harbored nutrients upon which the nascent plant could draw. Because of these advantages, seed plants spread over vast tracts of land and were particularly successful in drier, upland environments. During this time arose plants with large spores. By one conjecture, these large spores may have given rise to the female ovule, whereas small spores may have evolved into the male gamete (pollen). The large spore may have been a sport of the small spore, perhaps by a doubling of the genetic material in what had been the small spore. In similar fashion, the chemical colchicine has been applied to grapes to increase their size. The large spore was at least twice the size of the small spore, judging from fossil evidence.

Between 370 and 354 million years ago, seeds developed a tough coat better to protect the embryo.[24] This period witnessed the evolution of the ovule and pollen. Pollen appears to have evolved about 364 million years ago. The earliest pollen grains appear to have some traits from spores, not surprisingly. By about 310 million years ago, pollen came in two types. One had one or two bladder sacs with germination at the distal point. This type of male gamete resembles the pollen of extant gymnosperms. The other pollen, simpler, lacked bladder sacs instead having what appears to have been a long suture along the distal point. This must have been the site of germination. The seed coat appears to have encased the ovule. Early seed plants had structures to conduct pollen to the ovule where fertilization occurred. At this time, wind carried pollen to the structures that conducted it to the ovule. Apparently, insects were not instrumental in the early transfer of pollen to ovule. Some ovules had hairs to trap pollen. Wind remains important in the pollination of some plants, among them the great American grass, corn.

We witnessed early the origin and expansion of forests. Trees evolved about 380 million years ago. From this point until the late Carboniferous Period, forests contained one group of spore yielding trees and two groups of seed-bearing trees. Spore-bearing trees appear to have evolved between about 380 and 290 million years ago.[25] One may divide these trees into four groups: helycopsids, sphenopsids, filiaapsids, and ptogymnosperms. The last succumbed to extinction, but the other three all have extant species. Lycopsids appear to have evolved about 370 million years ago, some trees being as tall as 35 m. Trunks had a girth of 1 m. Already, these trees had xyla to permit the transit of water and nutrients over long distances. Roots penetrated to a depth of 12 m. Lycopsids appear to have been a large component of the boreal forests, comprising as much as two-thirds of these early forests. Not all lycopsids grew to great height. Some were shrubs and must have formed the understory of the early forests.

The sphenopsids have about 20 extant species, some adapted to the tropics and others to the temperate zones. Curiously, the sphenopsids had microphyllous leaves. Fossil sphenopsids from the Carboniferous and Permian Periods (about 354 to 248 million years ago) reveal a large number of extinct species. Many must have succumbed during the mass extinction that ended the Permian Period. The large horsetail tree was among the vanquished. This tree had a characteristic single vein that bisected each leaf. Some leaves resemble needles, as was common at this time, whereas others were lanceolates of less than 1 cm. Ferns arose about 360 million years ago.[26] Many in the fossil record resemble extant species, suggesting that fern architecture has not changed much over time. Tree ferns were numerous in the prehistoric fossil record. Tree ferns grew to 10 m. Leaves were numerous, suggesting that the tree ferns were proficient photosynthesizers. The trunk had no branches with the only foliage atop the tree. The foliage, being atop the tree, must have helped the tree fern compete for sunlight in the forest. The ptogymnosperms had about 15–20 genera, all extinct. Despite coming to a bad end, the ptogymnosperms may have been the ancestors of all modern seed plants. These trees grew to about 8 m and so too must have been part of the understory. They became extinct about 340 million years ago. Given their deep roots and megaphyllous leaves, the ptogymnosperms seem to have been well adapted to terrestrial conditions.

Between 354 and 248 million years ago, about six families of seed ferns evolved. The tree stood about 10 m, and the adventitious roots were well adapted to absorbing water and nutrients. The fossil record of the early Carboniferous gives one a glimpse at the diversity of the primeval forests. For clarity, one might divide the forests into five biomes. The ever-wet tropics included forests near the paleoequator, which spread along a landmass that includes what are today China, Scandinavia, and parts of Greenland and North America. Lycopsids were the dominant flora, with sphenopsids and pteridopsids also important. The summer-wet tropics confined much of their rainfall to summers. The land was swampy during the summer, with lycopsids and pteridosperms dominant. Seed ferns were also important colonizers. The subtropical deserts were barren. The warm temperate zones between 30° and 70° North and South latitude included what is today Siberia, which included small lycopsids. The cool and cold temperate zones were sparsely populated.

During the Permian Period, seed plants became the dominant flora. By about 260 million years ago, seed plants comprised more than 60% of the flora.[27] Lycopsids and sphenopsids had passed their zenith and were dwindling. The Permian Period witnessed the coalescence of the continents into Pangea. The climate appears to have warmed during the Permian Period. The interior of Pangea was comparatively arid.

Pangea had seasons with corresponding fluctuations in temperatures. Unlike today, lands along the paleoequator were even dry. Because much of Pangea was within the tropics, warm temperatures prevailed. During the Permian Period the seed plants cycads, benignities, and ginkgoes evolved. In the Southern Hemisphere, the seed ferns known as glossopterids expanded their range.

The Permian Period had seven biomes.[28] The cold temperate biome between 60° and 90° North and South latitude included what is today part of Siberia in the north and what are today Antarctica and southern Australia in the south. Cordiates and sphenopsids dominated in the north and glossopterids in the south. The cool temperate zone included what are today southern Africa, India, Madagascar, and much of Australia in the south and part of Siberia in the north. The flora was more diverse than that in the cold temperate biome. The trees were deciduous, underscoring the seasonal nature of the climate. The warm temperate zone included in the north what are today northernmost North America, Greenland, and Scandinavia and in the south what are today northern India, western Austria, and Chile. In the south, seed ferns were dominant. The north had a diversity of plants. The desert biome was barren of plants. The subtropical desert includes what are today northern Africa, northern South America, and parts of Europe and North America. Ginkgoes and conifers were numerous. The tropical ever-wet biome had perhaps the greatest diversity of plants during the Permian Period. Conifers dominated the summer-wet biome.

RISE OF FLOWERING PLANTS

Known as angiosperms, flowering plants may be divided into monocotyledons (monocots) and dicotyledens (dicots). See Chapter 1 for a discussion of the differences between these categories. Despite the persistence of the monocot–dicot divide, other divisions are possible. The fossil record bears traces of flowers about 127 million years ago in what are today Portugal and Australia.[29] These paleoflowers appear to have come from monocots. Early angiosperms may not have included perfect flowers, that is, a flower with both male and female gametes. The alternative is the unisex flower, which had only a male gamete or only a female gamete. Such a plant will produce both male and female flowers, or one plant will produce all male flowers and another plant all female flowers. The perfect flower may have been a later evolutionary development, although it is possible that the perfect flower may have been the earlier innovation. Another alternative, supported by the fossil record, suggests that perfect and unisex flowers arose about the same time, making the awarding of priority difficult if not impossible. About 120 million years ago, fruits evolved to encase the seeds. This evidence appears first in what are today Asia and North America.

By the era of angiosperms, pollen had shed the bladder sacs. This new type of pollen arose no later than 130 million years ago.[30] Fossil evidence places this pollen first in what is today Israel. What are today Libya and China also contain fossilized pollen, but the dating appears to be controversial. It is possible, at least in Libya, that the pollen grains are older than those in Israel. These grains are exceptionally small. Flowering plant leaves were megaphylls from the outset and as a rule had a number of veins. These leaves arose about 120 million years ago, leaving an extensive fossil trail in what are today Central Asia, Russia, Portugal, and the eastern United States.

These leaves were only about 2–4 cm in diameter. Their shape and structure suggest that early flowering plants grew in moist and even marshy soils. In fact, flowering plants would have formed part of the understory of forests. About 100 million years ago, flowering plants radiated into a number of species, whose biodiversity was impressive. Curiously, there is a paucity of evidence to classify these early flowering plants. Were they trees, shrubs, or herbaceous plants? Hypotheses for each viewpoint compete for partisans. The dearth of evidence for wood associated with flowering plants may point to the supposition that early angiosperms were herbs, although this paucity is not proof of an absence of wood. One must remain cautious.

About 70 million years ago, wood in association with flowers appears more frequently in the fossil record, suggesting that trees and shrubs must have had flowers. This is no surprise considering the large number of flowering trees today, including the apple, fig, date, cherry, peach, nectarine, plum, apricot, orange, lemon, lime, grapefruit, and many other agronomically important flowering trees. Dicots may have been the first plants with flowers. Again, one may point to an abundance of flowering dicots in modernity: the soybean, pea, chickpea, lentil, peanut, potato, sweet potato, and tomato merely scratch the surface. In contrast, the staple, grasses and sugarcane, represents modern examples of flowering monocots. For a very long time, dicot species have outnumbered monocot species by a factor of at least 6. The monocots were certainly present by the early Cretaceous and almost surely did not evolve from the dicots.

Flowering plants likely evolved within 30° North and South of the paleoequator.[31] This region would have been tropical or warm temperate as one moved toward 30°. Movement outside of these bounds took 20 or 30 million years, again a brief span that suggests the rapidity with which flowering plants radiated. Again one glimpses support for the theory of punctuated equilibrium. By about 127 million years ago, flowering plants had colonized what are today central Africa, Australia, Europe, and China. Yet there were limits. At high latitudes, conifers were more numerous than flowering plants. Interestingly, the rise of flowering plants coincides with global warming during the Cretaceous Period, the final era of dinosaurs. Angiosperms appear to have been suited to dry conditions, which may have prevailed away from the paleotropics. After about 144 million years ago, dinosaur herbivores may have denuded vast tracts of forests, opening the land to opportunistic flowering plants. It is well known that insect pollinators and flowering plants coevolved. The flowering plants of the early and mid-Cretaceous must have been the focus of insect pollinators.[32] These flowering plants had small anthers and surprisingly little pollen. Would wind pollination have sufficed? One glimpses in corn the abundant pollen that a wind-pollinated plant produces. Early flowering plants, however, must have depended on insects to transfer pollen from anther to stigma. Insects must have abetted cross-pollination, contributing to genetic diversity among flowering plants. It is also interesting to note the success of flowering plants in the tropics during the Cretaceous, a trend that continues today.

FLOWERING PLANTS AT THE END OF THE CRETACEOUS PERIOD

The end of the Cretaceous about 65 million years ago and with it the extinction of the dinosaurs ushered in a period recognizable to many of us. During the last 65 million

years, the continents assumed the present configuration. The key mountain ranges—
the Himalayas, Alps, Carpathians, Caucasus, Zagros, and Rockies—all formed. The
climate cooled to the point that both poles had ice. Initially, however, earth remained
warm in the years after the end of the Cretaceous. The cooling trend was not apparent until after about 45 million years ago. The two chief climate trends of the last
45 million years have been decreases in temperatures and rainfall. With Australia's
movement away from Antarctica, equatorial waters could no longer penetrate south
to the pole so that Australia became cooler. In the Northern Hemisphere about 30
million years ago, cold water from the Arctic Ocean spread south to North America.
The rise of mountains prevented moisture-laden clouds from penetrating the interior
of continents, creating more arid conditions. The rise of these mountains exposed
rock to weathering, so that carbon dioxide was again bound up in calcium carbonate
minerals. By the Miocene Period, atmospheric carbon dioxide approached its current
levels. The South Pole acquired ice by 30 million years ago, although the Arctic ice
cap had to await formation during the last 3 million years. During the last 3 million years, earth has endured repeated glaciations, with humans inhabiting the most
recent glacial era and the current interglacial interlude.

RISE OF THE GRASSES

The warm interlude between about 60 and 50 million years ago witnessed the radiation of plants from pole to pole, evidence that the poles must then have been warm.
Roughly during this period arose the grasses, a category of flowering plant.[33] This
moment marked another milestone in prehistory because humans would come to
depend on the grasses for sustenance. Rice, corn, wheat, barley, millet, sorghum,
oats, rye, and sugarcane are all grasses of great importance. They have made it possible for humans to bloat the planet with seven billion people. All grasses are members
of the Poaceae or Gramineae family. Today, the world has more than 10,000 species
of grasses. Grasses challenged the dominance of forests. Today, about 30% of earth's
surface is grassland rather than forest. This development would be important in the
evolution of humans.

The grasses probably arose between roughly 65 and 55 million years ago.[34] By
about 10 million years ago, earth witnessed the spread of grasslands. Grasses radiated at a time of increasing aridity and decreasing temperatures. One should not
assume, however, that all grasses are temperate plants. Far from it, one of humankind's great grasses, sugarcane, is a tropical grass unable to tolerate frost. The thick
cuticles of grasses suggest an adaptation to aridity. There is evidence that as the
climate dried, fires became more frequent. Grasses would have survived because
they regrow from a point just below the soil's surface. This phenomenon is evident to
anyone who has had to cut turf grass. It does not die but regrows, sometimes to the
exasperation of humans who wish to maintain a well-manicured lawn. These fires
also created the space into which grasses could colonize. Grazing also selected for
grasses, which regrow after the exposed leaf has been eaten. Other plants were not
so fortunate. In the temperate zones, the principal grasslands are the Great Plains of
the United States and Canada, the steppe of Eurasia, the Pampas of Argentina, and
the savannas of temperate Africa. Several of these regions have enormous agronomic

value; the wheat lands of the Great Plains nourish humans worldwide. The durum wheat grown in Minnesota and North Dakota may be the finest anywhere.

TROPICAL RAINFORESTS

By 10 million years ago, tropical rainforests girded the equator, holding sway in central Africa, northern South America, southern and Southeast Asia, and northern regions of Australia.[35] These forests were particularly large in the tropics, that is between 20° North and 20° South. The tropics in Brazil alone produced cacao and rubber trees and the pineapple plant, all important to humans worldwide. Human activity appears to be squandering the genetic diversity of these forests. These forests contain evergreen trees, flowering trees, herbs, palms, vines, and some conifers. Above 25° North and South, forests become less diverse, with evergreens and conifers the norm. Western Asia, including Arabia, and Central Asia, farther north from the equator, were wetter than today and had grasslands. The Sahara Desert was also wetter and had grasses.

EVOLUTION OF PLANTS IN PERSPECTIVE

Broad trends have shaped the evolution of land plants since their origin about 430 million years ago.[36] Plants have radiated into a large number of species over time. The origin of new species appears to be a rapid occurrence, perhaps in keeping with the theory of punctuated equilibrium. The evolution of new plant species may be linked to changes in earth's relation to the sun. Over time, the tilt of earth's axis has changed. The size and shape of earth's elliptical orbit and the procession of earth's orbit have all changed. These changes have affected how much sunlight earth has received over time and how sharp temperature fluctuations between seasons and at different latitudes have been. These factors have produced significant changes in the climate about 400,000, 100,000, 41,000, and 20,000 years ago. Evidence of these changes is at least 200 million years old and has probably occurred throughout plants' tenure on earth. These changes have altered the climate, causing rapid evolution in some plant lineages and retrenchment in others. Changes in flora, as the fossil record makes clear in what is today Hungary, correspond to orbital and other changes at 100,000, 41,000, and 20,000 years ago. At these markers, subtropical forests ceded ground to boreal forests, with subtropical forest returning during warm interludes. Another way to envision these changes is to note the retreat of plants toward the equator during cool and cold intervals and their return to temperate locales and even high latitude during warm interludes.

Plate tectonics also produce evolutionary bursts. That is, the movement of continents provides another opportunity for rapid changes in the environment and so in species. The movement of continents appears to confirm the theory of punctuated equilibrium, with long periods of stasis interrupted by brief periods of mantle expansion and subduction elsewhere. Interestingly, the first period of continental activity dates between roughly 460 and 430 million years ago, the later date coinciding with the evolution of land plants. In fact, five stages of continental movement are apparent in earth's prehistory. Each period coincides with an increase in the concentration of carbon dioxide because of volcanism. Between 375 and 350 million years ago, for example, volcanism coincided with the radiation of lycopsids, sphenopsids, and ptogymnosperms. Between 300 and 260 million years ago, the continents coalesced into Pangea. During this time, the gymnosperms arose and radiated. About 260 million years ago, the conifers evolved. Between about 170 and 160 million years ago, Pangea began to break apart. There appears to have been little change in the flora, although it is possible that the ancestors of flowering plants arose then. Within the last 160 million years and particularly since the last 65 million years, the continents began to assume their present configuration a time during which flowering plants and then grasses evolved and radiated. As one might expect, during periods when the land surface area was at a maximum, the diversity of plants was high. Diversity was also high when large landmasses were near the equator. In fact, most flowering plants arose in the paleotropics.

As a rule, temperatures and humidity have increased during periods of tectonic activity. Volcanism increased the amount of carbon dioxide in the atmosphere and the upwelling of hot mantle heated earth. Between 460 and 430 million years ago, the amount of carbon dioxide was about 15 times larger than that at present. One wonders whether the current episode of global warming will again benefit plants. In this context, it is interesting to note that the flowering plants, although evolving in warm climes, radiated only when the climate began to cool. Moreover, grasses seem to be adapted to an environment of comparatively low carbon dioxide concentrations. Flowering plants, including grasses, appear to photosynthesize more efficiently at relatively low concentrations of carbon dioxide. The current episode of global warming may be ominous.

HUMAN EVOLUTION

Humans have long been curious about their origin. For millennia, myth and religions provided speculative answers. In the West, the most well-known story comes from the first chapters of Genesis, part of the Hebrew scriptures.[37] In recent centuries, scientists have begun to provide their own answers to the question of human origin. Humans are no different from plants and all other members of the biota in having evolved to their present state. The place to begin our inquiry is with the primates.

Humans are primates, which are generally classified as mammals that have grasping hands and feet, the presence of nails rather than claws, and the presence of a relatively large head. Primates derive most of their power from their hind limbs and have opposable thumbs ideal for grasping objects. The eyes are close together, allowing for binocular vision. The brain is comparatively large. Babies tend to be helpless and so depend on parental care. Primates tend to live comparatively long lives. The jaws are small and the dentition compact. Being mammals, many primates have abundant hair, what one calls fir. Humans are a notable exception. The first primates differed from the apes and humans in being nocturnal. These early primates ate primarily insects. Primates are social animals, a trend that continues among humans. Today, there are nearly 200 species of primates.[38] Primates arose after the demise of the dinosaurs in the late Cretaceous Period. The founder species dates between 55 and 35 million years ago, arising during a period of global warming. Tropical conditions prevailed even in northern Europe. Even during this warm interlude, primates were absent from Australia and Antarctica. North America and Europe had distinct species of primates, although they are no longer extant. China and Southeast Asia appear to have had large numbers of primate species. During this time arose monkeys, although the place of origin is not clear. Early primates were fast, adept leapers, and well adapted to arboreal life.

As the Eocene Period came to a close about 40 million years ago, temperatures began to decline and forests contracted. Grasslands expanded, forcing primates to adapt to new conditions. No longer could they range over an unbroken forest of trees with the intrusion of grasslands. Some manner of terrestrial locomotion would become necessary. In North America and Europe, primates became extinct, although they clung to life in equatorial Africa and Asia, where the forests reigned supreme. Egypt was not then a desert. Remains of primates have been found in what is today the renowned Fayum Desert. These finds support the notion that something akin to monkeys and apes lived about 36 million years ago. Two species, collectively known as the Propliopithecids, resembled apes. In this context, it is worth noting that the chimpanzee and gorilla are native to Africa. Whereas monkeys have tails, apes do not. It is difficult to be certain of a fossil ape because the record is so fragmentary. Most often incomplete skeletons and perhaps parts of the skull are found. An animal's tailbones may not be present. It is possible that scavengers took the tail for sustenance, so that the identification of apes in the fossil record is not always certain. The absence of the tail in apes evolved as a novelty, separating apes and, later, humans from all other primates. Apes are also distinguished by the mobility of their arms. Humans share this innovation. Consider the range of motion in the shoulder, elbow, and wrist of a baseball pitcher.

Between 26 and 24 million years ago, in Kenya arose the first member of the Hominoidea, of which humans are a member. Despite similarities, humans differ from apes in important ways. Humans have a comparatively flat face, although the nose protrudes, an enlarged brain, bipedalism, very little sexual dimorphism, and thick enamel on the teeth. In seeking human ancestors, one looks for animals that approximate some of these traits. These traits did not evolve as a package. Only during the last 2 million years has the brain expanded and sexual dimorphism diminished.[39] These traits help us identify human ancestors but only comparatively recently. They do not help scientists identify earlier ancestors. In fact, the only characteristic that at first distinguished hominids from apes was the bipedal habit. This habit must have evolved when our ancestors came down from the trees to cross the grasslands. In this instance, bipedalism would have been an advantage. Human-like animals would have stood erect above the grasses to survey the land for predators and the possibility of finding food.

A partial skeleton of an ape-like creature that might have walked upright dates between 7 and 6 million years ago.[40] It had inhabited the renowned Lake Chad in what is today Chad, Africa. The date must be near the time when the lineages leading to humans and chimpanzees diverged. If this animal walked upright, bipedalism must have been a very early development in hominid evolution. The animal had a small skull and teeth, was short, and had an ape-like face and prominent brow ridges as one sees in several human ancestors. A similar find in Kenya suggests that bipedalism must have been a widespread adaptation.

AUSTRALOPITHECINES

About 4.2 million years ago arose the Australopithecines in Kenya. They were indisputably bipeds and played a central role in human evolution. In 1925, Australian anatomist Raymond Dart discovered the original fossil remains of an infant's skull in South Africa, assigning it to the genus *Australopithecus*, meaning "southern ape."[41] To punctuate his belief that this being might be a typical African specimen, Dart named it *Australopithecus africanus*. It is not always easy to infer adult characteristics from an infant because ape-like and human-like animals change over time. The infant's relatively flat face, however, seemed human-like to Dart.

In 1974 came one of the crucial finds in the prehistory of the Australopithecines. That year Donald Johanson, then a member of the Cleveland Museum of Natural History in Ohio, discovered a 40% complete skeleton, skull, and teeth of another *Australopithecus* in Ethiopia.[42] Johanson dated it to about 3.2 million years ago, making it a little more recent than Dart's find. Johanson christened a new species, *Australopithecus afarensis*. The hips and leg bones made clear that this specimen was an efficient walker who perhaps covered long distances over the course of a day. Because the hips were wide, Johanson judged the animal a woman, naming her Lucy in part because he had been listening to the Beatles' song "Lucy in the Sky with Diamonds" when he drove into camp with her fossilized remains. Below the neck, Lucy looks strikingly human, but her skull was not large. Perhaps, more than any other fossil, Lucy made clear that bipedalism predated a large brain by at least one million years. Other *A. afarensis* fossils populate Tanzania. Another striking find came from Laetoli, Africa, where British anthropologist Mary Leakey discovered two sets of footprints dating to about 3.2 million years ago.[43] These were *Australopithecus* tracks. One may have belonged to an adult and the other to a child. The two had walked close together. It is not difficult to imagine that the tracks came from a parent and a child. If this is true, Australopithecines formed close familial and societal bonds. Such behavior is altogether human.

By the time of Lucy, the line leading to humans had already split from that leading to the chimpanzee. Nonetheless, humans and chimps share about 99% of their DNA. What separates humans from chimps? Tool making is not the answer because chimps make crude tools. The answer may be language, although the origin of language is a mystery. Even though chimps may learn sign language, their vocabulary is limited to about 200 symbols. Chimps have no way of conceptualizing the past or envisioning the future. In this sense, language, certainly abstract thought, seems special to humans. One thesis holds that, judging from social cohesion and behavior, the first humans may have lacked language. Yet behavior is difficult to infer from the fossil record. Paleoanthropologists must be content to document changes in anatomy to trace our origin.

The Australopithecines make clear that a growth in brain size was not a feature of early hominid evolution. The Australopithecines had brains no larger than that of a modern chimpanzee. Because the brain began to enlarge only during the last 2 million years, it is fashionable to assert that this period represents the evolution of true humans. The Australopithecines, at least the adults, appear to have lacked a flat face, a chin, and probably a prominent nose. Yet their babies may have been helpless, depending on parental care as is true of humans. These traits did not all arise at once but in piecemeal fashion.

HOMO HABILIS

Some population of Australopithecines must have given rise to our own genus, *Homo*. Eighteenth century Swedish naturalist Carl Linnaeus classified humans within the species *Homo sapiens*, variously translated as the "man who knows" or the "wise man." Several members of our genus predated the putative wise man. The first member of our genus came to the fore in 1960, when Mary Leakey and her renowned husband Louis Leakey were working in Africa's Olduvai Gorge.[44] They had come to Africa about 30 years earlier, convinced that Darwin had been right to suggest that human ancestors had arisen in Africa, not in Asia (Chapter 1). Years of painstaking work had yielded little, but in 1960 came a significant find when the Leakeys believed that they had discovered the remains of the first toolmaker in the hominid line. Supposing it to have been the first member of our genus, Louis named it *Homo habilis* (handy man).

One might note that paleoanthropologists tend to divide our genus into a number of species. This nomenclature is easy to track and compartmentalize, but it may create a false sense of distinctiveness. When members of our genus came upon one another, there may have been a number of possible reactions. Perhaps, modern human behavior holds the key to these interactions. One needs to look only at the early modern era to examine the reactions among Europeans, Africans, and Native Americans. Yes, violence was a common element to the way Europeans treated Africans and Native Americans, but all three groups also interbred in large numbers. It is difficult to believe that one can find anything other than mixed lineages in places such as Brazil and the Caribbean. The point is that early people who encountered one another probably interbred. If the offspring were fertile, as they must have been in Neanderthal-anatomically modern human crosses, then it is difficult to maintain the walls of speciation among all these early people.[45] Species names may be more convenience than reality.

Examining the teeth and anatomy of their *habilis* find, the Leakeys concluded that the hominid could not have been an Australopithecine. The teeth in particular looked much more human than those of *Australopithecus*. *Habilis* also had a larger brain than the Australopithecines. The Leakeys believed that *habilis* was the first truly human-like ancestor. Many of their peers, however, doubted the Leakeys, believing that more evidence was necessary before one took the step of naming a new species within a new genus. Louis and Mary's son Richard, renowned in his own right, found important *habilis* remains near Lake Turkana in northern Kenya.[46] Scientific opinion swung in favor of the Leakeys' supposition that *habilis* was truly a member of our own genus.

Another find of a robust *Australopithecus*-like skull seemed to have traits intermediary between the Australopithecines and *H. habilis*. One thesis holds that the new find must have been another member of our genus, this one *Homo rudolfensis*. Its inclusion in our genus is controversial because *rudolfensis* had a brain on par with that of the Australopithecines. In addition, an Australopithecine dating to about 2.5

million years ago had human-like arm and leg bones and might have been a precursor to *habilis*. There may have been several species of *Australopithecus* and *Homo* that coexisted. Adaptive radiation must have caused the founder populations to have speciated.

HOMO ERECTUS

It is well known that Darwin had favored an African origin of humans. His contemporary, German naturalist Ernst Haeckel disagreed. Rather than seek the human lineage among the gorilla and chimpanzee of Africa, Haeckel proposed the gibbon and orangutan, the apes of tropical Asia, as the progenitor of *Homo*. Asia and not Africa was therefore the ancestral homeland of humans. In retrospect, Haeckel's ideas influenced many more scientists than they should have.

Taking Haeckel's ideas as gospel, Dutch physician Eugene DuBois secured a commission in the Dutch army for the sole purpose of gaining an assignment to Indonesia, where he believed the earliest human ancestors would be found (Chapter 1). Settling on Java, the largest Indonesian island, in 1899 DuBois was determined, energetic, and lucky. By 1901, after only 2 years' work, DuBois found a stout femur and part of a skullcap. Examining the femur, DuBois knew at once that this new species walked just as we do. To emphasize the specimen's bipedal gait, DuBois' followers named it *Homo erectus*, positing that it was the first true human ancestor.[47]

It is difficult to be certain what *H. erectus* (Figure 2.5) looked like. Artistic representations depict muscular men and beautiful women with graceful forms, prominent breasts, and dark skin.[48] It is possible that such reproductions tell us more about our own ideals of conventional beauty than about *H. erectus*. Because the bones were large, however, it makes sense to portray *erectus* as muscular, a trait that must have been true of women as well as men.

FIGURE 2.5 *H. erectus*: an early member of our genus, *H. erectus* appears to have been the first hominid to leave Africa.

Other scientists found additional *erectus* remains in other parts of Indonesia and China. A cave near Peking (now Beijing), China yielded more *erectus* fossils and evidence that the species controlled fire. This innovation made possible cooking meat to derive more nutrients from it. Because of the association with Peking, these finds were known as Peking Man, although they have more recently been identified as *H. erectus*.

With so much fossil evidence, it is possible to draw some conclusions about *H. erectus*. It had a long rather than bulbous cranium. Like Neanderthal (sometimes rendered Neandertal) after it, *erectus* lacked the chin that is distinctive in modern humans. The teeth were larger than those of modern humans. The robust skeleton suggests that *erectus* had large muscles and so must have led a strenuous life. The brain was about 75% as large as the brain of modern humans.[49] DuBois was never certain how old *H. erectus* was, although modern estimates put *erectus* on Java about 1.5 million years ago. It persisted until perhaps 100,000 years ago on the island. Given the appearance of anatomically modern humans about 200,000 years ago, it seems possible that the two species met. Violence and sexual attraction may have characterized these encounters. The two may have produced children.

Holding fast to Darwin's ideas, *H. erectus* must have arisen in Africa and was the first member of our genus to set foot in Asia. In this context, *H. erectus* appears to have displayed the mobility and restlessness that mark *H. sapiens*. *H. erectus* may have arisen in Africa about 2 million years ago and so must have been a contemporary of *H. habilis*. The Africa first model gained support from the Leakeys, who discovered in Oldulvai Gorge a partial *erectus* skull about 1.2 million years old. In perhaps his most dazzling discovery, Richard Leakey found a nearly complete skeleton of a *H. erectus* boy on the western side of Lake Turkana.[50] The place, Namiokotoma, Kenya, gave the boy his name. The Namiokotoma boy lived about 1.5 million years ago. Judging from the stage of maturation of his skeleton, the boy died between ages 9 and 12. An early death is not surprising, given the low life expectancy of prehistoric people. Starvation and diseases must have truncated many lives. Even in modernity, one finds appalling examples of early death. In the eighteenth century, only 10 of German composer Johann Sebastian Bach's 20 children reached adulthood.[51] The Namiokotoma boy was well over 5 ft tall at such an early age. Had he lived to adulthood, he likely would have surpassed 6 ft. The supposition that modern humans are taller and more muscular than their predecessors must be false. The boy may have been comparatively slender with the body proportions of a person adapted to a warm climate. In fact, humans, with their abundance of sweat glands, relative hairlessness, and pigmentation that darkens in sunlight, are well adapted to the tropics. One cannot pinpoint the Namiokotoma boy's skin color, but blackness would have been an advantage. In fact, when one compares skin color with geography, it is apparent that humans are darker near the equator and lighter as they inhabit higher latitudes.

Some paleoanthropologists believe that Asian and African specimens merit classification into different species. If this is so, the African variants of *H. erectus*, including the Namiokotoma boy, are members of *Homo ergaster* (working man). Outside of Africa, the first Asians developed the lineage that led to *H. erectus*. In this scheme, *H. ergaster* is an African phenomenon whereas *H. erectus* is an Asian variant.

TOWARD A VARIETY OF HUMANS

Much debate surrounds the comparatively recent events in human evolution. The movement toward anatomically modern humans involved the evolution of new adaptations: a chin, a bulbous braincase, an enlargement of the brain, a less robust skeleton and accordingly smaller muscles, small brow ridges, and a prominent nose. Yet, in this assemblage of traits lies the danger of racial typing. Scientists are in two camps. One holds that all humans contain far too many genes in common to warrant the classification into races, literally subspecies. The other camp holds that important physical differences warrant the division of humans into races. People in this camp point to the prominent nose in Europeans and to more diminutive noses of East Asians. The classification of humans into races poses problems when one applies labels of superiority and inferiority. One need to look no further than American history to supply innumerable instances whereby the people of European ancestry devalued the people of African ancestry. People of European ancestry reacted with similar contempt to the Chinese and Japanese immigrants who came to the United States in the nineteenth century.

Those who adhere to racial typing hold that *H. erectus* evolved into geographical variants (races) in Africa, Asia including Indonesia, and perhaps Europe. Variations made each group distinct. With its differences, each group in each geographical region evolved into *H. sapiens* in what is known as the "multiregional model of human evolution."[52] Products of this model, modern humans retain their racial differences. Popular in the 1980s and 1990s, the multiregional model appears to have fewer proponents today. According to the model, *H. erectus* evolved into Neanderthals in Europe, with Neanderthals in turn playing some role in the evolution of modern Europeans. The role of Neanderthals in this transition is difficult to pinpoint. The fact that modern Europeans do retain some Neanderthal genes proves that Neanderthals and anatomically modern humans had children together, but this does not tell us much about the evolution of anatomically modern humans. Neanderthals lacked a chin and were very robust and muscular. Why did these traits disappear in modern humans?

In contrast to the multiregional model, the Out of Africa Model supposes that *H. sapiens* arose in Africa and then colonized the world, displacing *H. erectus* in Asia and Neanderthals in Europe.[53] Displacement seems too simple a phenomenon. Again, *H. sapiens* must have interbred with the other people with whom they came in contact on their global trek. The transition to *H. sapiens* was neither sharp nor sudden. About 600,000 years ago, *H. erectus* or *ergaster* populations in Africa may have evolved into a new species, *Homo heidelbergensis. Heidelbergensis* may have occupied Europe about 500,000 years ago. It had a larger nose and brain than *H. erectus.* After about 400,000 years ago, *H. heidelbergensis* may have evolved into *H. sapiens* in Africa and Neanderthal in Europe. Anatomically modern humans arose in Africa perhaps 200,000 years ago and began to venture beyond the continent's confines about 100,000 years ago. They entered Australia about 50,000 years ago, Europe about 45,000 years ago, and the Americas perhaps 15,000 years ago. These are all round figures.

Because Neanderthals did not become extinct, and extinction is not the only possibility, until about 28,000 years ago, they must have been in contact with anatomically modern humans for some 20,000 years. The two clearly shared tool technologies and had children. Some fossil remains appear to be intermediate between Neanderthals and anatomically modern humans. The fact that Europeans carry Neanderthal genes proves interbreeding without further debate. If there is consensus, it appears to support interbreeding among anatomically modern humans, Neanderthals, and *H. erectus* in Europe, North Africa, and western Asia including Turkey.[54] Although it cannot be proven, modern humans almost certainly have genes from *H. erectus* in addition to Neanderthals. Modern humans became the species they are because they had sex with a lot of different people. In this view, anatomically modern humans did not replace Neanderthals so much as they swamped them with their genes, although this view is probably in the minority. We may have incorporated Neanderthals into our own lineage. Another hypothesis holds that Neanderthals went extinct but the early people of China and Indonesia reached modernity through interbreeding with anatomically modern humans.

The use of the phrase "anatomically modern humans" leads to the question of how they differed from modern humans. The obvious answer must be the development of a gracile skeleton and skull among moderns, but one wonders whether cultural factors might also mark the transition. The development of art appears to have had its origin among robust humans and so probably is not the dividing line. The development of agriculture was an important cultural innovation that might signal the rise of modern humans. If one wishes a later date, perhaps the development of writing cements the divide between anatomically modern and modern humans. If there is merit to these possibilities, then the transition must have been as much cultural as biological.

In tropical Asia, the story is a bit different. We have seen that *H. erectus* populated this region about 1.8 million years ago. The move into the temperate zone must have been difficult for a tropical animal. Substantial clothing was necessary to endure winter. People must have acquired skills as hunters or at least scavengers because plants would likely have died during winter. The same must have been true for Neanderthals, whose diets were about 85% meat.[55]

There is evidence of human-like animals in southern Europe as early as 1.5 million years ago, possibly *H. ergaster*. The early inhabitants of what is today Israel had mastered fire by 800,000 years ago. How did these early humans come to inhabit the Levant and Europe. The transition to the Levant must have occurred in a movement north and east from Egypt. From the Levant, humans may have crossed the Hellespont into southeastern Europe. The movement to Europe may also occurred from what is today Tunisia to Sicily and then to southern Europe. Another possibility may have brought humans from Morocco across Gibraltar into southern Spain. The sea routes must have been easier to cross during periods of glaciation when the drop in sea level must have exposed more land and less of the Mediterranean Sea to traverse. At least, one Roman historian has posited a long series of prehistoric migrations from North Africa to Sicily and into southern Europe.[56] Of course, the flow of people must not have been one way. People must have traveled south from southern Italy back to North Africa.

Human-like animals and humans have inhabited the Levant at least for the last 1 million years. Indeed, the first occupants may date to 1.5 million years ago.

The occupation of Spain may have occurred as early as 1.2 million years ago. Fossil evidence includes several individuals, most of them children. They show large brains and modern faces. Although it is difficult to infer the anatomy of an adult from that of a child, the principle of neotony applies, by which I mean the retention of juvenile features into adulthood. For example, a baby is relatively hairless, a trait that humans have retained into adulthood, distinguishing us in an important way from all other primates.

H. HEIDELBERGENSIS

In 1907, workers discovered a lower jaw near Heidelberg, Germany.[57] The teeth resembled our own but the jaw was more robust. The jaw displayed no chin and so, to use the terminology of paleoanthropologists, was "archaic." Scientists named it a new species in our genus *H. heidelbergensis*. This initial suggestion focussed on the possibility that *H. heidelbergensis* was a variant of *H. erectus*, one that had settled Europe before the Neanderthals and that must have made warm clothing to have inhabited such a northern latitude. Heidelberg man may have inhabited Europe roughly 500,000 years ago. In 1921, in what is today Zambia, southern Africa yielded a skull and partial skeleton, although it is not certain that all remains are from the same individual. British scientists named the remains *Homo rhodesiensis* to signify that it had arisen in what was then Northern Rhodesia, a British colony. The skull was long and the forehead very low, but the brain would have been as large as that of modern humans. The skeleton was robust so that the person must have been muscular. Other finds followed in France, Germany again, Hungary, Britain, Greece, Ethiopia, Tanzania, Kenya, and South Africa. Here must have been a case of African origin and subsequent radiation into Europe. Scientists now recognize all these fossils, even *H. rhodesiensis*, as *H. heidelbergensis*. The viewing of these remains may lead one to suppose that Neanderthals evolved from *H. heidelbergensis*. In fact, *H. heidelbergensis* may have been the last common ancestor of Neanderthals and *H. sapiens*. This language implies that Neanderthals and *H. sapiens* were different species. One might dispute this designation on the grounds that the two interbred to yield fertile offspring. It seems entirely possible that Neanderthals were a subspecies of *H. sapiens*. *H. heidelbergensis* may have originated in Africa about 600,000 years ago. The splitters believe that *heidelbergensis* evolved into Neanderthals in Europe and into *H. sapiens* in Africa.

NEANDERTHALS

The first evidence of the Neanderthals may locate to a cave in northern Spain. This cave has yielded a spectacular find in the search for human origin: more than 600 bones from 28 individuals including men, women, and children.[58] The secret to their preservation may lie in the probability that the Neanderthals were the first humans to bury their dead, the first to ritualize burial, and perhaps the first to anticipate an afterlife. These actions and beliefs are clearly modern and should dispel any temptation to marginalize the Neanderthals as somehow subhuman. Because the interior of the cave is extremely difficult to access, it cannot have been living quarters but must have been a subterranean mausoleum. These skeletons, at about 400,000 years old, may mark the transition from *H. heidelbergensis* to Neanderthals. Artistic renderings of Neanderthals are not easy to interpret. One depicts a nearly nude woman.[59] Her body is nearly hairless except for her head, a sign that neotony was at work. Her hair is curly and appears to have a reddish tint. She is muscular as all Neanderthals must have been. Her face is careworn from the rigors of life and her breasts sag, apparently with the advance of age. Yet she is more beautiful than the artist may have intended to convey. Women like her surely attracted the attention of the anatomically modern humans who ventured into Europe. The American Museum of Natural History in New York City too displays a robust, nearly nude replication of a Neanderthal woman. Were such women alive today one suspects that many men would seek their company.

Neanderthals derive their name from their accidental discovery in the Neander Valley in Germany. Despite artistic renderings, Neanderthals must have clothed themselves for the winter. There appears to be no way to decide whether they shed all or most of their clothing in summer. In contrast, not everyone is convinced that Neanderthals wore clothes even during winter. To date, paleoanthropologists have discovered more than 500 Neanderthal skeletons, many of them well preserved. Neanderthals were shorter than anatomically modern humans, more muscular, had a larger nose, more prominent brow ridges, and long braincases with brains in some cases larger than those of modern humans. One cannot doubt their intellect. The frontal region of the brain appears to have been smaller than that of the brain of modern humans, but the back of the brain was larger. Neanderthals had large, long teeth. Some teeth show wear, possibly from holding animal skins to free the hands to work the skins. Neanderthals lacked a chin. Their short stocky bodies suggest an adaptation to ice age Europe. This body shape would have retained heat much better than, say, could the Namiokotoma boy. The large nose must have warmed and moistened air with the inhalation of each breath. Neanderthals were not just a western European phenomenon. They also inhabited what are now Iraq, Syria, Israel, Georgia, Russia, and even Siberia. Neanderthal remains show the presence of cripples. Someone must have cared for them, a trait much needed in our troubled world. Fractured bones are common in the fossil record, suggesting that Neanderthals lived dangerous lives.

H. SAPIENS

Africa is the homeland of *H. sapiens*. About 400,000 years ago arose humans who resembled the Zambia find (*H. rhodesiensis*) throughout Africa. Chris Stringer marks the transition to anatomically modern humans between 400,000 and 130,000 years ago.[60] The second date is quite recent. Fossils dating to 160,000 years ago in Ethiopia appear to be early *H. sapiens*. South Africa also has anatomically modern remains, but these date to only 75,000 years ago. North Africa has produced several skulls and partial skeletons. One Moroccan find dates to 200,000, a date that several scientists accept as the origin of *H. sapiens*. Equatorial Africa has yielded nothing between 200,000 and 130,000 years ago, perhaps because the environment is not ideal for the preservation of human remains.

In Asia, *H. erectus* went as far east and south as Java and perhaps the Indonesian island of Flores but appears not to have settled Australia. Compared with Africa and Europe, Asia has a poor fossil record. A skull from India dates to 300,000 years ago. Where are its compatriots? The skull appears to be more modern than *H. erectus*, but it is difficult to draw conclusions from limited data. It is possible that *H. heidelbergensis* ventured into Asia and that it, rather than *H. erectus*, may be the ancestor of modern East Asians, but proof awaits. Western Asia provides a contrast because it was near Africa and Europe. Surely, its people interbred with the archaic people of Africa and Europe. It is possible that *H. erectus* ventured as far west as Syria. Opportunities for inbreeding must have abounded, in contrast to the relatively isolated populations in China and Java.

Early *H. sapiens* retained a long comparatively slender form ideal for an active life in warm climes. In this respect, they differed from Neanderthals. Yet anatomically modern humans did not have what we would consider fully modern faces. Neanderthals made jewelry but anatomically modern humans made art including figurines of voluptuous nude women. The libido must have been active throughout human evolution. Other groups of anatomically modern humans colonized Indonesia and Australia. The colonization of Australia is interesting because there has never been a time since the continents took their present configuration that a land bridge connected Indonesia to Australia. The first humans must have taken watercraft to reach Australia. Artistic depictions of the first Australians portray them to resemble *H. erectus*. Perhaps, the intent is to suggest that prehistoric men and women had dark skin and lacked clothing. Again, the women are beautiful with graceful bodies and well-endowed breasts. Perhaps, the art means to convey the continuity between *H. erectus* and anatomically modern humans in this part of the world. There appears to be every reason to suppose that *H. erectus* and anatomically modern humans interbred in Southeast Asia.[61]

Humans colonized Australia within the last 70,000 years. Colonization might have been accidental with wind and ocean currents driving humans off course. Humans must have reached Australia from New Guinea or Java. Human remains from New Guinea date to only 35,000 years ago, which may be too late for the colonization of Australia. Accordingly, Java must have been the more likely route. Indisputable evidence of humans in Australia dates to about 50,000 years ago. Two skeletons from southeastern Australia, both likely women, suggest a light build in concert with the gracile skeletons that modern humans exhibit. These skeletons date to about 40,000 years ago. Other individuals were more robust. Could two types of humans have coinhabited Australia? Perhaps, a founder population diverged into two types. Alternatively, it may be possible that *H. erectus* colonized Africa first, followed by anatomically modern humans, but these humans must have differed from Cro-Magnons in their build, with the moderns in Australia being gracile rather than robust. The later anatomically modern humans might have been descendents of *H. erectus* populations in China. One suspects that the latter group must have interbred with anatomically modern humans. None of these ideas quite account for the slight build of the moderns in Australia. Altogether, the early inhabitants of Australia pose many questions. The modification of stone tools in Australia appears to have lagged behind developments in Africa and Europe. In contrast to Neanderthals, the people of Australia practiced cremation rather than inhumation.

HOMO FLORESIENSIS

In recent years, scientists have challenged the notion that *H. erectus* alone inhabited Southeast Asia before the arrival of anatomically modern humans.[62] In the twenty-first century, a team of scientists identified sophisticated stone tools on Flores dating to 800,000 years ago. These tools raised the possibility that another type of human populated Indonesia. Spectacular finds followed, including compete skulls and nearly complete skeletons of a three-foot tall human with brain only as large as that of a chimpanzee. This human controlled fire, likely cooked meat, and hunted pygmy elephants and small game. The small stature may not be a mystery but rather part of the phenomenon of island dwarfing, in which all species on an island are small because they do not need to evolve size as protection against predation. Yet this explanation may not satisfy. Java is an island and *H. erectus* did not evolve small stature there. Scientists named this human *H. floresiensis* (Man from Flores). Perhaps, *H. erectus* arrived on Flores and then shrunk, but as we have seen this did not occur on Java. *H. floresiensis* appears to have inhabited Flores until about 17,000 years ago and so must have coexisted with anatomically modern humans by this late date. Again, one wonders whether the two interbred. Would two humans of vastly different sizes have found one another attractive? Given the recentness of *H. floresiensis'* bones, one wonders whether DNA might be extracted from them with the aim of comparing them with modern human DNA.

MITOCHONDRIAL EVE

DNA is found not only in the nucleus of a cell, but also in the mitochondria, known as the "powerhouse of the cell." Whereas humans contribute equal parts of the nuclear DNA to a child, only the mother passes down her mitochondrial DNA (mDNA) to her children. It is possible, at least in theory, to use mDNA to trace the lineage of humans through the maternal line to the point of origin. Using this technique, geneticists have traced the maternal line of modern humans to a group of women who lived in Africa about 200,000 years ago. The date is significant because many scientists take it as the origin of humans. The place is significant because it confirms the African origin of all humans.[63]

OUT OF AFRICA

If humans arose first in Africa, their descendents must have conquered the world. Some anatomically modern humans left Africa for the Levant about 120,000 years ago or perhaps sometime earlier. They moved east and south into Australia about 50,000 years ago, into Europe about 45,000 years ago, to China about 40,000 years ago, to Sri Lanka about 30,000 years ago, and to Madagascar, New Zealand, and Polynesia about 12,000 years ago. These dates admittedly are a bit confusing. How could humans have settled Australia before China, and why was Madagascar settled so late? Within these later years came the move to the Americas. The most recent ice age revealed a land bridge between northeastern Asia and northwestern Alaska. The forbidding climate that far north must have discouraged human habitation until the manufacture of warm clothing was commonplace. The first people who crossed this land bridge, perhaps in pursuit of game, happened upon two virgin continents in the sense that no other human had predated them in the Americas. An early group, the Clovis people, manufactured a distinctive arrowhead and may have reached the Americas between 13,000 and 12,000 years ago. The advent of humans in the Americas coincided with the extinction of elephants, horses, camels, and ground sloths. Had humans hunted them to extinction? It seems unlikely given that even with modern herbicides humans cannot eradicate the dandelion.

The presence of people in South America about the time of the Clovis migration suggests that a population, probably small, had preceded the Clovis people. In fact, multiple groups of humans may have entered the Americas over the period of a few thousand years. Those who advocate an early date suppose that the first Americans entered the New World as long ago as 30,000 years. There appear to be no fossil remains to justify so early a date. Curiously, the early people of the Americas differed physically in a number of respects and differed from the Native American populations that survived into modernity. The ancestors of the Amerindians may have arrived comparatively late, perhaps interbreeding with the original inhabitants.[64]

NOTES

1. Ernst Mayr, *Growth of Biological Thought*, (Cambridge, MA: Belknap Press) 360–362, 417.
2. Luther Burbank with Wilbur Hall, *The Harvest of the Years* (Boston and New York: Houghton Mifflin, 1931), 96–97.
3. Charles Darwin, *The Origin of Species by Means of Natural Selection or the Preservation of Favored Races in the Struggle for Life* (New York: Modern Library, 1993), 29.
4. Mayr, *Growth of Biological Thought*, 423–425.
5. Malthus, *Essay*, 13.
6. Darwin, *Origin of Species*, 86–90.
7. Ibid, 108.
8. L. M. Cook and I. J. Saccheri, "The Peppered Moth and Industrial Melanism: Evolution of a Natural Selection Case Strategy," *Heredity* 110 (2013): 207–212.
9. Niles Eldridge, *Life Pulse: Episodes from the Story of the Fossil Record* (New York: Facts on File, 1987), 81–82; Stephen Jay Gould, *Punctuated Equilibrium* (Cambridge, MA: The Belknap Press of Harvard University Press, 2007), 39–61; Niles Eldridge and Stephen Jay Gould, "Punctuated Equilibrium: An Alternative to Phyletic Gradualism," in *Models of Paleobiology*, ed. T. I. M. Schopf (San Francisco: Freeman, 1972), 82–115.
10. William J. Cook, "How Old Is the Universe?" *U.S. New and World Report* 123 (1997), 34; Raven, *Biology of Plants*, 2; Thomas N. Taylor, Edith L. Taylor, and Michael Krings, *Paleobotany: The Biology and Evolution of Fossil Plants*, 2nd edition (Amsterdam: Elsevier, 2009), 44–47.
11. Corrine E. Blank, "Origin and Early Evolution of Photosynthetic Eukaryotes in Freshwater Environments: Reinterpreting Proterozoic Paleobiology and Biogeochemical Processes in Light of Trait Evolution," *Journal of Phycology* 49 (2013): 1040; Yiannis Manetas, *Alice in the Land of Plants: Biology of Plants and Their Importance for Planet Earth* (Berlin and New York: Springer, 2012) 126.
12. K. J. Willis and J. C. McElwain, *The Evolution of Plants* (Oxford: Oxford University Press, 2002), 30; Taylor, *Paleobotany*, 123.

13. Ibid, 43.
14. Ibid, 46.
15. Ibid; Manetas, *Alice in the Land of Plants*, 132.
16. Willis and McElwain, *Evolution of Plants*, 51; Manetas, *Alice in the Land of Plants*.
17. Manetas, *Alice in the Land of Plants*, 134; Willis, *Evolution of Plants*, 56.
18. Willis and McElwain, *Evolution of Plants*, 56.
19. Ibid.
20. Lorentz C. Pearson, *The Diversity and Evolution of Plants* (Boca Raton, FL: CRC Press, 1995), 556.
21. Willis and McElwain, *Evolution of Plants*, 81.
22. Ibid.
23. Pearson, *Diversity and Evolution of Plants*, 487–490.
24. Willis and McElwain, *Evolution of Plants*, 93.
25. Pearson, *Diversity and Evolution of Plants*, 447–448; Willis and McElwain, *Evolution of Plants*, 99.
26. Pearson, *Diversity and Evolution of Plants*, 470.
27. Willis and McElwain, *Evolution of Plants*, 112.
28. Ibid.
29. Pearson, *Diversity and Evolution of Plants*, 526; Willis, *Evolution of Plants*, 112.
30. Willis and McElwain, *Evolution of Plants*, 170.
31. Ibid, 171.
32. Pearson, *Diversity and Evolution of Plants*, 544–546.
33. Willis and McElwain, *Evolution of Plants*, 201.
34. Ibid.
35. Ibid, 209.
36. Ibid, 286–289.
37. Genesis 1:1-2:25 (New Revised Edition).
38. Chris Stringer and Peter Andrews, *The Complete World of Human Evolution* (London: Thames & Hudson, 2012), 82.
39. S. H. Montgomery and N. I. Mundy, "Microcephaly Genes and the Evolution of Sexual Dimorphism in Primate Brain Size," *Journal of Evolutionary Biology* 26 (2013): 906–911.
40. Stringer and Andrews, *The Complete World of Human Evolution*, 107.
41. Erin Wayman, "Discovering Human Ancestors," *Smithsonian* 42 (2012): 38–39.
42. Michael D. Lemonick and Andrea Dorfman, "A Long Lost Relative," *Time* 174 (2009): 52–45; Donald C. Johanson and Tim D. White, "A Systematic Assessment of Early African Hominids," in *Evolution*, ed. Mark Ridley (Oxford and New York: Oxford University Press, 1997), 355.
43. Russell H. Tuttle, "The Pitted Patter of Laetoli Feet," *Natural History* 99 (1990): 60–65; Johanson and White, "A Systematic Assessment of Early African Hominids," 355.
44. Ann Gibbons, "Who Was *Homo habilis*—and Was It Really *Homo*?" Science 332 (2011): 1370–1371.
45. Carl Zimmer, "Interbreeding with Neanderthals," *Discover* 34 (2013): 38–44.
46. Ryan Shaffer, "Evolution, Humanism and Conservation," *Humanist* 72 (2012): 19.
47. George W. Stocking, "Eugene DuBois and the Ape Man from Java," *Isis: Journal of the History of Science in Society* 81 (1990): 785–786.
48. Stringer and Andrews, *The Complete World of Human Evolution* 135.
49. G. Philip Rightmire, "*Homo erectus* and Middle Pleistocene Hominins: Brain Size, Skull Form and Species Recognition," *Journal of Human Evolution* 65 (2013): 223.
50. Stringer and Andrews, *The Complete World of Human Evolution* 135.
51. Joseph Machlis, *The Enjoyment of Music: An Introduction to Perceptive Listening* (New York: W. W. Norton, 1955), 423.

52. Chris Stringer, "Modern Human Origins—Distinguishing the Models," *African Archeological Review* 18 (2001): 68.
53. Stringer and Andrews, *The Complete World of Human Evolution* 151.
54. John Hawkes, Gregory Cochran, Henry C. Harpending, and Bruce T. Lahn, "A Genetic Legacy from Archaic *Homo*," *Trends in Genetics* 24 (2008): 19–23.
55. Michael P. Richards and Ralf W. Schmitz, "Isotope Evidence for the Diet of the Neanderthal Type Specimen," *Antiquity* 82 (2008): 553.
56. Arthur E. R. Boak, *A History of Rome to 565 AD*, 4th edition (New York: Macmillan, 1955), 9.
57. Stringer and Andrews, *The Complete World of Human Evolution* 148.
58. Ibid, 152.
59. Ibid, 155.
60. Ibid, 158. For a contrary view see, Timothy D. Weaver, "Did a Discrete Event 200,000–100,000 Years Ago Produce Modern Humans?" *Journal of Human Evolution* 63 (2012): 121–126.
61. Stringer and Andrews, *The Complete World of Human Evolution* 171.
62. Kate Wong, "Human or Hobbit?" *Scientific American* 311 (2014): 28–29.
63. Alan R. Templeton, "Genetics and Recent Human Evolution," *Evolution* 61 (2007): 1507–1519.
64. Stringer and Andrews, *The Complete World of Human Evolution,* 194–197.

3 Early Uses of Plants

OVERVIEW

We noted in the last chapter that the first primates ate insects and so did not depend on plants for sustenance. But it is also true that these primates lived in trees and so depended on plants for shelter and protection against predators. In a very real sense, the long lineage of primates owes its success to plants in various ways. The line leading to *Homo sapiens* interests us most, and here we find strong evidence of plant use. The Australopithecines (Figure 3.1) (Chapter 2) were plant eaters, although they may have supplemented their diets with the scavenging of carcasses that carnivores had not eaten, at least not completely. When we arrive at our genus about 2–2.5 million years ago, controversy erupts. Many generations of male paleoanthropologists have asserted the importance of hunting in the belief that meat provided the calories and nutrients necessary for the human brain to enlarge.[1] Not only was hunting important, but also men drove our evolution because they were the hunters. This hypothesis relegated women to the margins. They cared for the children and awaited the return of men from the hunt. Women were content to supplement the meat diet with edible plants. They dug roots and tubers and collected edible leaves, berries, nuts, seeds, and other plant foods. This sexist view is difficult to sustain. For one thing, where women were the collectors in a region of abundant plant resources, they were in a sense the breadwinners and their status was high. There is also evidence from modern hunter-gatherers that men joined women in collecting edible plants. The same also appears to be true of hunting. The current emphasis on Neanderthal women (Figure 3.2) is that they participated in the hunt.[2] In some examinations of modern hunter-gatherers, women too join the hunt. Men were thus no more the drivers of evolution than women. Given these suppositions, this chapter assumes that women played a vital economic role. As important as meat may have been, humans were primarily plant consumers and to the extent that women participated in the procurement of plants, they deserve special attention for the role that they played in our evolution.

ROLE OF PLANTS IN HUMAN EVOLUTION

The first primates evolved in forests during the radiation of flowering plants. From an early date, flowering plants would provide sustenance to primates. Even plants whose role has been to provide humans with roots or tubers are flowering plants. The sweet potato and potato, for example, supply important nutrients and calories, even though humans do not use the flowers, seeds, leaves, or stems in any notable way. The early ancestors of humans ate roots, tubers, and the forerunners of pistachios, walnuts, and mango. Surely, wild acorns must have been prevalent in deciduous forests. The prominence of acorns in Native American mythology supports the notion

FIGURE 3.1 Australopithecine "Lucy": the Australopithecines ate primarily edible plants.

that the nut was an important source of calories, including essential fatty acids and protein.

Birds may have been the first to consume these foods. Perhaps by observing birds, the progenitors of humans must have quickly recognized the value of edible plants.[3] That the first primates ate insects is not surprising in biological terms because they needed ample calories as they were small. Small animals must burn more calories to generate heat because of the rapidity of heat loss. Larger animals—humans, gorillas, and elephants are good examples—can subsist on lower caloric foods such as plants. With the passage of time, and as primates grew larger, they focussed more on plant foods such as fruits and nectar. Seeds and nuts provided the protein that might have been missed in the transition from a diet of insects to one grounded in plants. One

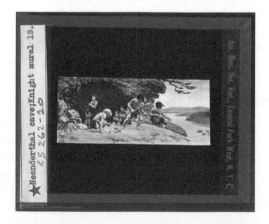

FIGURE 3.2 Neanderthal: Neanderthals appear to have derived only a minority of calories from edible plants.

peculiarity of some primates is their inability to manufacture vitamin C the way that humans manufacture vitamin D in the presence of sunlight. Although scientists would discover these and other vitamins only in the early twentieth century, these vitamins were no less important in prehistory despite being unknown. This uniqueness is remarkable, given the ability of other mammals to manufacture their own vitamin C. It is possible that primates had this ability early in their evolution but lost it when they began to consume a variety of fruits. The consumption of fruits must therefore have driven one aspect in our evolution. It is obvious today that citrus fruits, cabbage, pineapple, and potatoes are good sources of vitamin C, but these foods appear not to have been part of the early primate diet. Indeed, it may not have been until the evolution of humans and perhaps later that these foods were sought. The most obvious example is the grapefruit, which appears to have arisen only in the eighteenth century Jamaica or perhaps another Caribbean island as a sport. Its consumption, therefore, is part of only the most recent history of humanity. Nonetheless, fruits of many types must have been common to the diets of several primates.

When one compares primates, say the gorilla and human, the size differential is striking. One also observes the importance of plants in the gorilla's diet. It is thus possible for primates to attain large size on a diet of plants. The robust Australopithecines were probably similar to the gorilla in being large plant consumers, with or without any supplementation from scavenging.[4] The ridge of bone atop the skull, the large facial muscles that anchored to this ridge, and large molars suggest that these animals spent much time grinding tough, fibrous plant roots and leaves. These Australopithecines probably ate plants in bulk, as the gorilla does, to make up for the fact that the diet was not especially nourishing. The phenomenon by which the nineteenth century Irish laborer ate as much as 14 pounds of potatoes suggests that even among modern humans, the reliance upon plants may not in some cases be far removed from the situation among the robust Australopithecines and gorillas.[5] The robust Australopithecines likely went extinct without contributing genes to the current human population. Rather, humans likely trace their lineage to the gracile Australopithecines (Chapter 2). They, too, likely ate primarily plants.

A plant diet, important as it has been to our evolution, is risky because no single edible plant contains all nine essential amino acids that humans need to synthesize proteins. An alternative that arose in prehistory, among both hunter-gatherers and the first farmers, was the combination of a whole grain and a legume, which together supply all essential amino acids, the example of rice and beans being well known today. White rice will not do, however, because it lacks the nutrients in the bran and germ. Brown rice and beans supply all essential amino acids. But, this combination is recent in evolutionary terms because the beans usually chosen to accompany this dish are the *Phaseolus* beans of the Americas (Figure 3.3). Only when the Columbian Exchange (Chapter 6) brought these two foods from different hemispheres together was it possible to eat what we now know as rice and beans. Primates must have compensated for the dearth of amino acids by expanding their diet to include a large number of plants. The consumption of nuts and seeds must have been a starting point in the quest for protein.

A milestone in human evolution was mastery over fire. *Homo erectus* was probably the first human to control fire and to use it to cook plants and meat. Cooking

FIGURE 3.3 *Phaseolus* beans: an American genus, *Phaseolus*, contains among the world's most nutritious and popular beans.

made plants taste better, and the stomach and intestines more easily digested the nutrients from plants and of course meat. From rice and beans to pasta and much more, humans routinely cook food plants. A raw salad or a piece of fruit is an exception. Cooking must have aided early humans in preserving foods. Cooking also made plants easier to eat by requiring less grinding of the molars to process these plants. This advantage must have been important for the young and old, whose ability to chew was either in its early stages or well past its prime. Early hominids gravitated to plants with carbohydrates and fats. Nuts must have been a good source of fat, and fruits of many types contain carbohydrates. The latter is also true of roots and tubers, whose importance would grow with the advent and spread of agriculture. Fatty acids are important because the body cannot absorb sufficient protein without them. The brain and muscles require vast quantities of carbohydrates, although the consumption of any food can be overdone of course leading to obesity in the developed world.

Evidence from modern hunter-gatherers in remote parts of South America suggests a reliance on nuts, roots, and fruits, especially wild figs.[6] The true fig was unknown in the Americas before the Columbian Exchange. The fig would rise to global importance with the assent of Judaism and Christianity, both of which have emphasized the creation account in Genesis.[7] Hunter-gatherers keenly observed nature, finding place in a forest in which new seedlings had established themselves and were tender and reasonably succulent. At times, humans established simple and temporary dwellings from which they gathered the surrounding edible plants. The establishment of sedentism, however brief, would mark an important stage in the long transition to agriculture (Chapter 4). Human activity in constructing these dwellings may have denuded a small area, which new plants might colonize, providing a possible meal. Human excrement must have enriched the soil, hastening the growth of new plants. This excrement might have contained the seeds that humans had recently eaten. These seeds were doubtless the source of new plants. Fruits with a single seed, known as a stone or pit, when discarded might in turn germinate, giving humans insight into how seed plants propagate. Modern examples of such stone fruits include the peach, nectarine, apricot, cherry, date, and olive. One might note the importance of the olive tree and olives in the ancient Mediterranean Basin, where its oil would be the chief dietary fat (Chapter 4).

The earliest inhabitants of Australia (Chapter 2) processed seeds and probably nuts to derive more nutrients from them. This practice appears to have blossomed on the margin between desert and grassland. Seeds might have been crushed together to create something akin to flour. One might have consumed this flour right away or have added water and allowed the food to ferment to yield alcohol. These gatherers probably numbered more than 50 people per tribe. Population density appears to have increased only with the transition to agriculture. The early inhabitants of Australia preferred roots and fruits to other parts of plants. Roots and fruits were desirable because they could be eaten without processing. Nuts and seeds, in contrast, required effort to process. It is interesting that throughout much of Southeast Asia, particularly the Philippines and Indonesia, hunter-gatherers had only late contact with grains, which one often considers the food of modern societies. Rather, roots and tubers were more important in much of the tropics. Of course, both the Philippines and Indonesia would become leaders in rice production in the transition to an agrarian order.

At high latitudes, near the Arctic Circle, for example, hunter-gatherers tended to favor meat and fish. Nevertheless, plants played a role, with hunter-gatherers harvesting wild crowberries, bilberries, and cowberries, all repositories of vitamin C. The creeping willow supplied vitamin C in its leaves. The Inuit and other hunter-gatherers in Alaska left these berries and leaves outdoors to freeze. In this form, the foods stored well and were available during the lean months of winter. As in Australia, the Inuit used a stick to dig edible roots. One suspects that this was an activity possible only when the soil had thawed. The Nabesna of Alaska stole the roots that muskrats had amassed for winter.[8] These people dried berries in the sun to desiccate them, a condition that favored preservation. In other cases, they covered berries in oil to preserve them. The source of this oil is unclear. Perhaps, it came from fish.

The Amerindians who settled near the Great Lakes continued to rely on hunting and gathering. They ate the seeds of a plant that resembles rice but is not. Rice cannot be grown at high latitudes and was unknown to the Amerindians during the pre-Columbian era. Only during the seventeenth century would rice come to the Americas and then from Africa (Chapter 6). Asian varieties of rice came even later. The point is that the Amerindians of the pre-Columbian period did not eat "wild rice." Hunter-gatherers also used plants for clothing, the prototypical example perhaps coming from the creation account in Genesis. Other uses included as weapons, dwellings, medicine, dyes, poison to tip their spears, and hallucinogens. There is evidence, for example, that sugarcane was first used not as a sweetener, but to thatch a hut. Only later did humans, probably first in New Guinea, begin to chew the pith of a stem to derive sweetness. Sugarcane has a tumultuous history that will form part of Chapter 6.

About 1200 CE, the Amerindians of Arizona had not adopted agriculture. In fact, hunter-gatherers remained prominent in North America until European contact. The Arizona Native Americans subsisted on a diet of roots, nuts, and possibly a large amount of insects and small game. If this diet was primarily meat, then these Native Americans may have resembled Europe's extinct Neanderthals. Even so plants were important. These Amerindians, and hunter-gatherers worldwide, depended on about 25 species of wild plants for food. The Hopi of the American Southwest used about 30 wild plants for food and another 40 as medicines. Some of these hunter-gatherers relied on wild plants to enhance the flavor of meat. Among edible plants, hunter-gatherers have, depending on geography, eaten acorns, amaranth (a cultivated plant in some areas), burdock, cattail, pine nuts, prickly pear, purslane, yucca, dandelion, goosefoot, stinging nettle, curly dock, tea made from pine needles, and wild fruits of several descriptions.

WOMAN, THE GATHERER

Among early hunter-gatherers (Figure 3.4), women may have played a large role in gathering edible plants.[9] They collected yams with a stick to penetrate the soil, preparing the yam for hand extraction. This was possible only in warm climates because yams do not tolerate the cold. These actions were still those of hunter-gatherers. Root and tuber agriculture had yet to develop. Curiously, the step toward agriculture would be tardy in Australia. The aborigines never made the transition. Aboriginal women were also instrumental in harvesting water lilies during the rainy season. When rainwater was scarce, women exploited water chestnuts and cycad nuts. In prehistoric Australia, the Philippines, and Indonesia, humans detoxified and ate the nuts of the cycad palm. As in so many regions of the world, the women in the Americas did the work of acquiring plants, in this case, the pseudo rice. With women accomplishing so much, one is tempted to wonder what men did.

Among hunter-gatherers in the tropics, women continue to play an important role as gatherers and sometimes as hunters.[10] Men in turn help women gather edible plants. In fact, plants, not meat, fostered cohesion as early humans shared nuts and other edible parts of plants. Women played an important role in this context, sharing food with their children and other relatives. The search for and the sharing of edible plants were thus communal activities. The gathering of edible plants predated hunting by millennia, if not millions of years. Even scavenging, we have seen, predated hunting. The instinct to gather edible plants is sound because it requires the expenditure of fewer calories than does hunting. Of course, the payoff from hunting must have been substantial on occasion. A base of plant foods gave humans the energy they would need to hunt.

Today, hunter-gatherers in the tropics collect abundant plant resources, an important source of food should the hunt fail. Gathering edible plants must have been

FIGURE 3.4 Female field workers: whether as gatherers or as farm workers, women have always shouldered a heavy burden.

easier in the remote past because hunter-gatherers lived everywhere. Today, they are confined to marginal and inhospitable lands where the search for plants is necessarily difficult. Yet even today, hunter-gatherers do well in collecting plants, which continue to constitute the bulk of calories. The sharing of these plant foods is nearly ubiquitous among mothers and children, whether fully grown or not.[11] Women also likely used plants to make the first basket in which to carry other foods. The collection of edible plants must have shaped human societies in other important ways. If women foraged for plants, other people, including men, had to care for the children while women were away. The acceptance of childcare responsibilities among men must have intensified social cohesion and gender relations. It seems possible that women who gathered plants entrusted their children to young women, who thereby gained experience raising children before the onset of menses and would make them candidates for motherhood. This model of plant collection as a method that solidified a society makes difficult the explanation of why aggression and hierarchy arose, but both are clear from the first civilizations to America's current bombardment of Iraq and Syria, an ironic situation given that the people of the Levant and Mesopotamia were among the first to leave the Stone Age.

Some hunter-gatherers (Figure 3.5) display gender divisions, in that women are more likely to share edible plants with other women than with men.[12] Such societies may be matrilineal. Other societies lack this division. The Agta of the Philippines is such a group. Its members, women and men, gather edible plants. Both also hunt. In fact, the choice of labor stems from personal preference. Women and men who prefer to gather plants pursue this activity. Those drawn to the hunt act accordingly. Neither gathering nor hunting is a gender-coded activity. Among the early inhabitants of Australia, women had the responsibility of gathering plants, but they hunted as well. Curiously, only in the temperate zones has gathering to some extent ceded ground to hunting, probably because plants cannot be gathered well during winters.

FIGURE 3.5 Hunter-gatherers: throughout much of human prehistory, hunting and gathering have been the principal economic activities.

This may have been the context in which Neanderthals emerged as primarily hunter, if in fact they were primarily hunters. In many hunter-gatherer societies, women adept at collecting plants gain status. Because women contributed to the production of food by gathering plants and sometimes by hunting, they, by definition, had status in a plant-based proto economy. Only with a division of labor and the concept of private property have hierarchies arisen among hunter-gatherers. As a result of their process as plant collectors, Agta women can make choices about their future, deciding whom to marry, how many children to have, and whether to divorce a husband on any grounds. Plant gathering societies therefore tend to respect women. This insight challenges old assumptions that hunting made us human and that men enjoyed higher status than women. Among hunter-gatherers who have the concept of private property, it is rare for a woman not to own property to which her husband has no right. Women may disperse this property as they wish.

PLANTS AND HUMAN EVOLUTION

The foraging for the plants with the highest food value contributed to the uniqueness of humans in having a large brain and cognitive capacities, a long childhood and adolescence, and a lifespan greater than any other primate.[13] In these ways, plants shaped human evolution. This diet, according to one expert, may have been 30%–50% meat and 50%–70% plant with neither dairy nor grains, at least not until the end of the Paleolithic Era.[14] One thesis holds that men had the status of primary food producers, an idea at odds with the previous section, underscoring the degree to which paleoanthropologists disagree. In Tanzania, among the Hadza hunter-gatherers, however, women play the principal role as gatherers and as specialists in gathering and processing roots and fruits.[15] One sees the emphasis on root and fruit crops in modern agriculture whereby both have sustained humans. Consider only the sweet potato, yam, and cassava as examples of the importance of root crops to modern humans. In cases in which women gather edible plants, it is common for children to assist their mothers. By late adolescence, these children will be adept gatherers. The act of gathering plants thus strengthens bonds between one generation and the next.

The diet of the Australopithecines (Chapter 2) has been a matter of contention. Australian anatomist Raymond Dart discovered the first specimen in 1925 in what is today South Africa. He supposed that *Australopithecus* was an aggressive hunter.[16] With additional discoveries, paleoanthropologists conclude that the robust Australopithecines were plant eaters. The difference between robust and gracile Australopithecines may have come down to diet. The Australopithecines had large molars and strong facial muscles to grind tough plants. Large fruits were also important, although their demands on the teeth should have been minimal. It is possible that Australopithecines consumed some grit, perhaps dirt that had not been removed from roots or tubers. Robust Australopithecines may have been better crushers and grinders and may have eaten more seeds, whereas gracile Australopithecines may have consumed more fruits. The robust species may have consumed seeds from wild date palms, palm nuts, and even bark. The gracile species may have obtained more than half of their calories from fruits and 18%–28% of their calories from young, tender leaves and stems.[17]

Worldwide, gathering plants was the primary means of subsistence until the spread of agriculture. Even then, the transition to agriculture was not rapid. Bipedalism and tool making increased the efficiency of gathering and processing plants and bringing them home. In contrast to the long prehistory of foraging for plants, hunting may be unique to our species. By about 4 million years ago, hominids shared plant foods with women provisioning their children.[18] By about 3 million years ago, the Australopithecines shared food with the extended family. Early kin networks must have been larger than the nuclear family, a more recent development. By 2 million years ago, our genus scavenged and perhaps hunted to supplement a plant diet. The consumption of plants was basic with or without scavenging and hunting. Because the Australopithecines inhabited the savanna, they must have eaten grass seeds, which one might parallel to the emphasis on grain agriculture, certainly in southwestern Asia and other parts of the Mediterranean Basin. The prime evolutionary adaptation was bipedalism, which allowed the Australopithecines to range some distance in gathering plants. Indeed, modern hunter-gatherers may cover 12 miles per day in search of berries, seeds, and other edible plant materials.[19] In contrast, chimpanzees in the forest probably had greater opportunity to catch and kill small animals than did *Australopithecus*, who may have been more herbivores than chimpanzees. Bipedalism allowed for carrying sticks with which to extract roots. *Australopithecus* may also have used stones to attempt to dig roots. Stones allowed one to crack the shells of nuts to make available the edible portion.

The earliest members of our lineage may have been vegetarians. As a rule, the lineage leading to humans went through three stages. First, these people gathered edible plants. Second, they scavenged to supplement gathering. Third, they scavenged and hunted to supplement gathering. It is possible in this context that Neanderthals ate substantial amounts of plants, although this topic is open to interpretation. According to American paleoanthropologist Pat Shipman, men tended to hunt and women to gather plants, a view that has been strongly challenged.[20] Wherever plant resources were abundant, women gained status as the provider. In this case, women as gatherers may have driven human evolution.

As a rule, hunter-gatherers subsist on four principal types of food: nuts, roots, seeds, and meat. Three of the four categories derive from plants, underscoring their importance in the diets of pre-agricultural people.[21] In some hunting and gathering societies, plants provide up to 70% of a person's daily calories. The Gwi San of the Kalahari Desert in southern Africa consume large quantities of fruits. Fruits remain important to agricultural people worldwide. Melons, particularly as a source of water and vitamins, are important to the Gwi San. Indeed, watermelon probably evolved in the Kalahari Desert. The Nukak of Colombia eat the fruits from numerous palm trees. So important are these fruits in the diet that the Nukak do not hesitate to climb the palm trees to acquire the food. The human specialization on roots and nuts appears to provide substantial calories, in both carbohydrates and fatty acids, and nutrients. The difficulty of acquiring some of these foods appears to have prolonged adolescence, the period when the young gain skills and strength to find their own food. For example, extracting the heart of a palm tree requires expertise and strength as a person must know where to remove the wood of a tree with an axe and then exert his/her strength in getting to the heart.[22]

The collection of fruit from palm trees is also arduous work because the forager must not only acquire the fruit but must also carry it home over a distance of 10–20 km. Between ages 20 and 40, women are at the peak in gathering edible plants. The availability of edible plants during the year affects pre-agricultural people in important ways. For example, when fruits are abundant, Ache children of Paraguay eat five times more calories than when fruit is in shortage. Yet even these numbers are not lavish. When fruits are abundant, Hadza girls consume about 1650 calories per day. The dry season may reduce their consumption of food to just 610 calories per day. If obesity is a problem in the developed world, one cannot say the same of modern hunter-gatherers. When averaged over the year, Hadza girls derive more than half of their calories from fruits. Ranging over a large portion of equatorial Africa, the Kongos, specializing in gathering berries, are another example.

Curiously, the evolution of primates is notable for the movement away from eating insects and toward the consumption of plants. The gorilla and chimpanzee demonstrate this progression. The lineage leading to humans appears to have demonstrated this trend, although during the last 2 million years, the growth of the human brain may stem partly from the importance of meat in the diet. The shift toward the consumption of fruits and leaves may have selected for primates, including humans, who see in color, at least the colors of the visible spectrum. Consider the tomato plant. Humans know that the fruits are ripe when they turn red or yellow, depending on the variety. Agribusiness has, however, misused this principle to trick the sense of sight. Growers pick tomatoes green and then expose them to ethylene gas to redden them. But such tomatoes are not truly ripe and lack the nutrients and flavor of a vine-ripened tomato. Humans and other primates are adept at selecting fruits and other parts of edible plants for ripeness and thereby peak nutrient content and at avoiding or modifying toxic plants. Cassava is a comparatively modern example. This selectivity must have driven the enlargement of the parts of the brain that process images. This selectivity must also have demanded an increase in the brain's ability to store and retrieve memories. Once a cluster of fruit plants had been identified, it could be revisited many times with the aid of memory.[23]

Among apes and the earliest members of our lineage, wear patterns on teeth suggest that these animals ate primarily fruits. The chimpanzee is a particularly eager fruit eater, a trait that humans retain. That is, the common ancestor of chimpanzees and humans must have eaten fruits. The habit of eating fruits must have enlarged the brain because the fruit would be ripe for only a short duration, during which birds and other animals would compete for it. The hominoids needed a brain that enabled them to monitor the environment closely to determine when the fruit was ripe before other animals grasped the situation. In the same way, I monitor my cherry trees to determine the ripeness of fruit, so I may gather it before the birds stake their claim. The ability to process nuts also demanded intelligence. Thus the ability to predict and locate edible plants likely increased brain size among primates. The fact that fruits may be seasonally abundant or scarce given climatic and other factors must have prodded the brain to process information so that primates could find other plant resources during times of scarcity. The primates that subsist primarily on leaves, the gorilla, for example, do not need such abilities because leaves are more or less omnipresent, at least in the African tropics, where the gorilla resides. The consumption of leaves may be perilous because some contain toxins. Modern examples include tomato and potato leaves. The leaf eater, again the gorilla is an example, must have large stomachs and intestines to extract maximum nutrients from a source that is often sparse in nutrients. Modern exceptions are collard, mustard, and turnip greens. In fact, cabbage and Brussels sprouts are bundles of nutritious leaves.

Dental patterns also signify the importance of plants in the primate diet. Only the eruption of the first molar and later its followers prepares a youth for an adult diet. Because molars grind plants, the importance of edible plants must have been great in the primate diet and in primate evolution.[24] The human is a good example of this pattern. The importance of a high-quality diet of plants, meat, fish, and other organisms appears to have lengthened lifespan and with it the longevity of the woman's fertility with menopause a late development. Key in this development was the provision of ample edible plants and some meat to lactating women, who were temporarily free from the demands of plant collection. The scarcity of these foods must have led to the practice of infanticide. From an early period, humans must have been sensitive to the dietary needs of pregnant and lactating women. The shift of plant and other food resources to them surely ensured the reproductive fitness of our species. Remember that the goal of evolution, if one may use such language, is to pass one's genes to the next generation. In this regard, the diet of women is particularly important. Among Neanderthals, evidence suggests a wide sharing of plants and meat. Even people too badly injured to collect their own food received aid from their community. This behavior must have enabled their long tenure in Europe and the Levant. They were not at all evolutionary failures.

A long lifespan allows many humans to live to see not only their children but also grandchildren. The grandparents assisted their children and grandchildren in providing edible plants to these younger generations. Even though no longer fertile, the grandmothers could still contribute important plant resources to the community. The importance of the grandmother in this context may help explain why women tend to outlive men. Of those who become centenarians, almost all are women. In another sense, this pattern appears strange because a man remains potent well after a woman enters menopause. In this context, why do men not live as long as women? In hunter-gatherer societies, older women may remain quite skillful and productive plant gatherers late in life. Among the Hadza, older women are generally more productive in gathering plants than younger women.[25]

As a rule, a diet of plants produces greater wear on teeth than does a meat diet, particularly when humans use stones to grind seeds into flour, a process that introduces grit, small flakes of stone, into the flour. Tooth abrasion was severe among ordinary Egyptians before and after the development of agriculture because of the probability of incorporating sand into the process of grinding seeds. The use of sandstone tools to grind seeds produced the worst wear. Wooden implements, however, reduced the presence of grit. The diets of hunter-gatherers are also notable for producing few dental caries. The advent of agriculture, at least in southwestern Asia, led humans to consume barley on a regular basis. The barley kernels stuck to the teeth, inviting bacteria to degrade teeth in the process of consuming the barley on the enamel. Among the hunter-gatherers in what is today Nevada, the diet consisted largely of plants. Cattail formed a larger part of the diet than pine nuts. This diet produced few caries. Interestingly, among some hunter-gatherers, women display more caries than men. Diet must play a role that is yet to be articulated.[26]

EDIBLE PLANTS IN THE AMERICAS

Agricultural production may have been low in pre-Columbian North America, leading even those who farmed to continue to collect wild plants. Part of the problem was that the cultigens of tropical America did not easily transfer to higher latitudes.[27] It may be possible to describe the relationship between hunting and gathering and farming in North America as large-scale foraging for wild plants and hunting, supplemented by small-scale gardening. In this context, hunting and gathering may have been akin to the notion of "the state of nature." It was the activity that united many groups of Amerindians in North America. Gathering roots was an important activity in large areas of North America. In what is today the northwestern quadrant of the United States and parts of Canada, hunter-gatherers relied on more than 25 species of edible plants for sustenance.[28] Cooking time was long, particularly for fibrous roots, providing a source of starch. The wild "prairie turnip" is an example of a starchy root, although it is not truly a turnip. The turnip is an Old World food that was not introduced into the Americas until the Columbian Exchange. According to one thesis, the digging for and eating of roots may have been a precursor to the development of agriculture in the American Southwest, the Great Plains, and eastern North America. Other sources of food included small seeds, fruits including berries, nuts, and leaves. Here again, foraging was part of a general food acquisition strategy that would come to include agriculture. Agriculture, however, did not supplant gathering in pre-Columbian America. The Amerindians pursued both methods.

Given the breadth of plants that the Native Americans used, it is not always easy for archeologists to identify the most important plants in the diet by region. In some places, the consumption of berries, roots, and corn demonstrates the connection between foraging and farming. Between about 400 and 1300 CE, the hunter-gatherers of California and the Pacific Northwest coexisted alongside the farmers of what are today Utah, parts of Nevada, Colorado, Idaho, and Wyoming.[29] Different seasonal patterns and elevations dictated what plants foragers pursued and when they pursued them in lands near the Pacific Ocean. The Amerindians were exceptionally mobile in their search for edible plants. As a basic model, hunter-gatherers in North America ate leaves in spring. As the year progressed, a surplus was stored in expectation of a lean winter. In early summer, the focus was on seeds, and these too were stored for winter. In late summer, the Amerindians of the Pacific Northwest concentrated on roots and berries. Pine nuts entered the diet in autumn. In what is today British Columbia, the Amerindians favored the seeds of several grasses and the collection of a number of berries, among them gooseberries, currants, elderberries, serviceberries, chokecherries, elder, raspberries, and strawberries. In the Upper Sonoran, Native Americans ate pinon nuts and sagebrush seeds. In the Lower Sonoran, people ate parts of the Joshua tree, mesquite, screw bean, cactus, and yucca. At higher elevations, the Amerindians exploited camass root, bitterroot, yampa, sego lily, wild onion, spring beauty, and wheat grass. The last is problematic. Wheat is a grass, but wheat grass cannot be wheat because the grain was unknown in the Americas before the Columbian Exchange. In what is today southern California agave was often stored for winter consumption. In spring, these people ate the buds and fruits of yucca, onion, cactus, goosefoot, and cats claw. In summer, the forager turned to ocotillo, mesquite, and screw bean. In autumn, berries, yucca, and cactus tended to be plentiful. The diet expanded in late autumn to include acorns, the seeds of many grasses, chia, saltbush, pinon nuts, palms, thimbleberry, raspberry, blackberry, juniper berry, and chokecherry. In the Pacific Northwest, fruits, many of them berries, vegetables including leaves, seaweed and sprout, roots, bulbs, tubers, rhizomes, and the interior of certain trees formed the basis of the diet. As elsewhere, the Amerindians of the Pacific Northwest took their cues from the environment.

In the spring, these Native Americans ate roots, namely, lupines, sea milkwort, wild carrot, and springbank clover. The terminology is confusing. The lupine and clover are Old World legumes. The carrot is an Old World root crop unknown in the Americas before the Columbian Exchange. Other spring plant foods include eulachon, salmonberry, thimbleberry sprouts, nettle, firewood shoots, seaweed, and cambisicun. In early summer, the focus was on berries and other fruits, prominent among them are salmonberries, strawberries, buckle berries, blueberries, Saskatoon, soapberries, gooseberries, currants, elderberries, raspberries, blackcaps, thimbleberries, and salal. In late summer, attention turned to bulbs, including blue camas, onions, and Indian rice, although strictly speaking rice was unknown in the Americas before the Columbian Exchange. In autumn, the Amerindians of the Pacific Northwest ate roots, hazelnuts, crabapples, cranberries, kinnikinnick berries, rose hips, and buckle berries. Where salmon was available, it complemented a diet of roots. Mushrooms and other fungi were also important, although they are not plants. The Amerindians extracted roots with a stick and boiled, roasted, or steamed them. In other cases, Native Americans coupled roots with meat from small game. The people of northern California roasted lily bulbs. Their preservation for winter made available an important source of vitamin C. Again, where possible, the Amerindians ate berries, fresh or dried. Throughout California, nuts of several kinds were important. The people roasted pine nuts. Cooking extended the shelf life of pine nuts to 1 or 2 years. Acorns were also important. Seeds were ground into flour or boiled. People engaged in proto agriculture by watering food plants, sowing wild seeds, fertilizing the soil, and eliminating weeds.

Between roughly 500 and 1000 CE, Amerindians became more sedentary as they relied increasingly on stored foods: acorns and other nuts, wild cherries, and seeds.[30] Fishing appears to have been more intensive to complement a plant diet. The Great Basin between the Sierra Nevada and Cascade Mountains in the west and the Rockies in the east was another region of gathering. Part of the region is desert. People either mobile in foraging for edible plants or comparatively sedentary put their efforts into collecting plants in bulk to store against lean times. The gathering of seeds was intensive about 9500 years ago. Hunting accompanied the gathering of seeds. Between roughly 4500 and 1000 years ago, the emphasis was on hunting. Between 2100 and 500 years ago, however, the gathering of wild plants intensified to supplement corn agriculture.[31] As a rule, people spent more time foraging than farming. In this instance, the gathering of plants appears to have hindered the transition to agriculture. About 1000 years ago, people from southern California migrated into the Great Basin. They gathered seeds and pinon nuts and hunted deer and bighorn sheep. An intriguing notion suggests that nowhere in pre-Columbian North America did people depend solely on agriculture for food. Plant agriculture, the collecting of wild plants, and hunting all combined. North Americans subsisted on indigenous amaranth, chenopodium, Kentwood, and the tropical imports corn, beans, and squash. Where farming was not competitive, it was not practiced.

PARTIAL AND TENTATIVE MOVEMENT TOWARD AGRICULTURE WITHIN THE CONTEXT OF HUNTING AND FORAGING

No region has attracted so much attention as southwestern Asia, probably because the Levant is holy to the monotheistic religions.[32] In the Levant, the presence of animal bones prejudices the archeological record to favor hunting, or at least scavenging. Throughout the Mediterranean Basin, though, of which the Levant is part, and in the tropics and subtropics, plants were the primary source of calories. Archeological evidence from near the Sea of Galilee (a lake rather than a sea) demonstrates that humans ate wild emmer wheat, barley, oats, peas, lentils, vetch, almonds, figs, grapes, and olives. This is a nutritious package. Early agriculture would not change the consumption of plants but rather make them more available. These people used stones to grind grains, again imparting grit to the diet. Other sites show the consumption of seeds from grains, leaves, fruits, legumes, acorns, and pistachios.

The gathering of plants persisted in parts of Russia, the Baltic States, and Scandinavia, well after farming had taken hold in southwestern Asia.[33] In large parts of Europe, farmers and gatherers coexisted without an immediate transition to agriculture. Belgium, the Netherlands, and Germany witnessed the coexistence of hunter-gatherers and farmers. Again the transition to agriculture was not abrupt, though interactions among people helped hunter-gatherers to understand the methods of agriculture. By 5000 BCE, farming was established in large parts of Europe, although hunting and gathering persisted. Some people assimilated farming while remaining hunter-gatherers. Some hunter-gatherers both cultivated crops and gathered wild plants. Wild berries and nuts prevailed, though by the Neolithic Revolution they were no longer staples as grains came to the fore. Hunter-gatherers maintained small fields of wheat or barley. Oats and rye came later. Throughout the Neolithic, hunter-gatherers probably derived fewer calories from emmer and einkorn wheat and barley. Emmer and einkorn wheat are no longer grown. It is possible that people grew grains for fermentation into beer while they gathered plants to feed themselves. When agriculture arose in Britain and Ireland, it was perhaps a secondary activity to supplement hunting and gathering.

The first occupants of what is today Malaysia were gatherers more than hunters.[34] Foragers must have collected wild rice as the Natufians in southwestern Asia gathered wild barley and wheat. Trade may have been the basis of interactions between gatherers and farmers. There were also areas where the two groups did not overlap. The rainforests of Southeast Asia were inhospitable to rice culture and so were the domain of gatherers.

NOTES

1. Frances Dahlberg, "Introduction," *Woman the Gatherer*, ed. Frances Dahlberg (New Haven and London: Yale University Press, 1981), 1.
2. Nicholas Wade, "Neanderthal Women Joined Men in the Hunt," *New York Times*, 2006 http:///www.nytimes.com/2006/12/05/science/05nean.html?_r=0.
3. David R. Harris, "Origins and Spread of Agriculture," in *The Cultural History of Plants*, eds. Ghillean Prance and Mark Nesbitt (New York and London: Routledge, 2005), 3.
4. F. E. Grine, "The Diet of South African Australopithecines Based on a Study of Dental Microwear," in *The Human Evolution Source Book*, eds. Russell L. Ciochon and John G. Fleagle (Englewood Cliffs, NJ: Prentice Hall, 1993), 145–152.
5. "Glossary," in *The Prendergast Letters: Correspondence from Famine Era Ireland, 1840–1850*, ed. Shelley Barber (Amherst and Boston: University of Massachusetts Press, 2006), 193.
6. Harris, "Origins and Spread of Agriculture," 7.
7. Genesis 3:7 (New Revised Version).
8. Harris, "Origins and Spread of Agriculture," 10.
9. Ibid, 9.
10. Dahlberg, "Introduction," 2.
11. Ibid, 8.
12. Ibid, 11–12.
13. H. Kaplan, K. Hill, J. Lancaster, and A. M. Hurtado, "A Theory of Human Life History Evolution: Diet, Intelligence and Longevity," in *Evolutionary Ecology and Archaeology: Applications to Problems in Human Evolution and Prehistory*, eds. Jack M. Broughton and Michael D. Cannon (Salt Lake City: University of Utah Press, 2010), 48.
14. Robert L. Anemone, *Race and Human Diversity: A Biocultural Approach* (Upper Saddle River, NJ: Prentice Hall/Pearson, 2011), 127.
15. Kaplan, "Theory of Human Life History Evolution," 51.
16. Grine, "The Diet of South African Australopithecines," 145.
17. Ibid, 146.
18. A. Zihlman and N. Tanner, "Gathering and the Hominid Adaptation," in *The Human Evolution Source Book*, eds. Russell L. Ciochon and John G. Fleagle (Englewood Cliffs, NJ: Prentice Hall, 1993), 220.
19. Ibid, 224.
20. Pat Shipman, "Early Hominid Lifestyle: Hunting and Gathering or Foraging and Scavenging," in *The Human Evolution Source Book*, eds. Russell L. Ciochon and John G. Fleagle (Englewood Cliffs, NJ: Prentice Hall, 1993), 279.
21. Kaplan, "Theory of Human Life History Evolution," 58.
22. Ibid, 60.
23. Ibid, 70.
24. Ibid, 72.
25. Ibid, 73.
26. Tim D. White, *Human Osteology*, 2nd edition (San Diego: Academic Press, 2000), 434.
27. Kenneth E. Sassaman and Asa R. Randel. "Hunter-Gatherer Theory in North American Archaeology," in *The Oxford Handbook of North American Archaeology*, ed. Timothy R. Pauketat (Oxford: Oxford University Press, 2012), 18–19.
28. Deborah M. Pearsall, "People, Plants, and Culinary Traditions," in *The Oxford Handbook of North American Archaeology*, ed. Timothy R. Pauketat (Oxford: Oxford University Press, 2012), 73–74.
29. Ibid, 74.
30. Ibid, 78.

31. Christopher Morgan and Robert L. Bettinger, "Great Basin Foraging Strategies, in *The Oxford Handbook of North American Archaeology*, ed. Timothy R. Pauketat (Oxford: Oxford University Press, 2012), 194.

32. Ofer Bar-Yosef, "Upper Paleolithic Hunter-Gatherers in Western Asia," in *The Oxford Handbook of the Archaeology and Anthropology of Hunter-Gatherers*, eds. Vicki Cummings, Peter Jordan, and Marek Zevelebil (Oxford: Oxford University Press, 2014), 260–261.

33. D. C. M. Raemaelers, "The Persistence of Hunting and Gathering amongst Farmers in Prehistory in Neolithic North-West Europe," in *The Oxford Handbook of the Archaeology and Anthropology of Hunter-Gatherers*, eds. Vicki Cummings, Peter Jordan, and Marek Zevelebil (Oxford: Oxford University Press, 2014), 805.

34. Huw Barton, "The Persistence of Hunting and Gathering amongst Farmers in South-East Asia in Prehistory and Beyond," in *The Oxford Handbook of the Archaeology and Anthropology of Hunter-Gatherers*, eds. Vicki Cummings, Peter Jordan, and Marek Zevelebil (Oxford: Oxford University Press, 2014), 857, 859.

4 The Neolithic Revolution

OVERVIEW

The origin and spread of agriculture during prehistory is often termed the Neolithic Revolution. One should approach this phrase with caution. Anyone who expects the word "revolution" to mean an event or a series of events that occur with great rapidity will be disappointed with the Neolithic Revolution. In fact, one should reflect on the prospect that many putative revolutions occurred only gradually. The Scientific Revolution unfolded over centuries. Scientists and scholars did not overnight embrace the heliocentrism of Polish astronomer and mathematician Nicolas Copernicus. Only the work of German astronomer and mathematician Johannes Kepler brought a modification of Copernicus' ideas to the fore. Even then, these ideas were difficult to explain until British polymath Isaac Newton devised the notion of gravity to explain the behavior of the planets. The Industrial Revolution is another example of a series of events that unfolded over centuries. Indeed, one senses that humanity is entering a new phase of the Industrial Revolution in which manufacturers flee nations with putatively high wages, such as the United States, to settle in the low-wage environments of India, China, Vietnam, and other nations.

The Neolithic Revolution spanned not merely centuries but millennia. One must keep in mind two ideas. First, agriculture did not originate in one place and then spread to the rest of the world. Even today, agriculture has not conquered the globe. Chapter 3 reminded us that even today the world contains hunter-gatherers who never made the transition to agriculture. Rather, agriculture appears to have originated in several regions of the world. In this sense, the focus on the rise of agriculture in southwestern Asia is illusory. Papua New Guinea appears to have an agricultural tradition every bit as ancient as that of southwestern Asia. Secondly, even where agriculture sunk early roots, the commitment to it was equivocal. Many people, the Amerindians are good examples, did not embrace agriculture as the be-all and end-all of food acquisition. Rather, they practiced it along with the older strategies of hunting, fishing, and gathering wild plants. At first, then, farming was part of a mixed economy. It largely replaced the older means of subsistence only over millennia. Within these parameters, it seems fair to continue referring the origin and spread of agriculture in prehistory as the Neolithic Revolution.

Humans, as we have seen, have always depended on plants. Even as hunter-gatherers, they must have obtained a large portion of their calories and nutrients from edible roots, tubers, grasses, nuts, and berries. Between 10,000 and 4500 years ago, people in several regions of the world began to cultivate plants. They planted seeds and portions of a root or tuber, weeded the soil to keep other plants from competing with theirs, irrigated their plants, and harvested them. The deliberate cultivation of plants marked a watershed in human existence. As a rule, agricultural societies produced more food than did hunter-gatherers, making possible large populations. Migration was no longer necessary to follow game. Some farmers became sedentary and founded civilization. All of the subsequent inventions—writing, the wheel, the plow, the monotheistic religions, the automobile, the airplane, the computer, and so much else—were the product of plant cultivators.

If humans arose about 200,000 years ago, the invention of agriculture was a late development and ultimately a reorganization of the way people interacted with the local environment. As we have seen, agriculture was not a single, uniform development. Implicit in the ideas of French botanist Alphonse de Candolle and explicit in the work of Russian agronomist Nikolai Vavilov (Figure 4.1) was the notion that the world contained several centers of agricultural innovation, places where different peoples domesticated different suites of edible plants.[1] This notion is strikingly true of the Americas and Papua New Guinea, where people had no knowledge of agricultural developments in southwestern Asia. The New World and Papua New Guinea, to take extreme examples, invented agriculture *de novo*. Yet, there were instances of borrowing. The crop assemblage in southwestern Asia differed little from the suite of crops that Egypt and other parts of the Mediterranean Basin adopted. As a rule, grains were a bit more important at higher latitudes, whereas roots and tubers tended to be more important in the tropics. Although potatoes are a crop of the tropics, they were domesticated in the Andes Mountains of South America and so are important in cool climates. Moreover, the grain rice was and is a crop of warm climates.

The pairing of a grain and a legume made nutritional sense in an era before the rise of nutrition as a science. Neither a whole grain nor a legume provides all nine essential amino acids that the body needs to synthesize proteins. But together, these

FIGURE 4.1   Nikolai Vavilov: a prominent Russian agronomist, Nikolai Vavilov, identified several centers in which he believed that agriculture had independently developed.

foods provide all essential amino acids. On their own, legumes are a fine source of protein. In many other regions of the world, the combination of a whole grain with a legume—corn and *Phaseolus* beans, rice and soybeans, barley and faba beans, and rice and mung beans—was central to the development of agriculture. Where agriculture arose early, it may be connected to climate change, notably, the retreat of the glaciers at the end of the Pleistocene era about 11,000 years ago. As Chapter 3 made clear, the gathering of plants preceded and overlapped with the cultivation of crops. If not at first, in the long run, the trend toward agriculture led humans to rely on a comparatively small number of domesticates among the grains, legumes, roots, tubers, and fruits. Among the grains (Figure 4.2), wheat and barley were tied to the foodways of Asia, North Africa, Egypt, and Europe. Rice, sorghum, and millet were important to Africa and Asia. Corn, potatoes, beans, cassava, and sweet potatoes were important to the Americas.

The transition to agriculture required no fewer than three steps. First, humans had to cultivate edible plants. These plants were at first wild and humans simply cleared land to plant them. As cultivation became important, humans took the additional steps of removing weeds, added fertilizer (human and animal dung), and protected their plants from predators. If all went well, the harvest and processing followed. Where rainfall did not suffice and where it was possible, humans irrigated their plants. Secondly, over a long period, the continual selection of these plants, likely for yield, flavor, and other desiderata, led to domestication, the point at which the nascent crop could no longer reproduce without human assistance. Corn is a good example of this phenomenon. The third step is the rise of what we mean by the term "agriculture," the complex set of social, economic, and political arrangements that ensure the perpetuation of the cultivation of domesticates. In many respects, this transition led to the rise of the first civilizations. Iraq and Egypt are good examples.

It is important to note that as long as people have access to abundant food, the incentive is weak to try cultivation or to adopt crops from other areas. It is only as wild plants and game became scarce that humans became willing to experiment. It seems possible in this light to suppose that humans turned to agriculture, when they finally did, to maximize the consumption of calories. The transition to agriculture, in this case, is a Darwinian process. The mechanisms of natural and human selection, Darwin understood, led to the human creation of and transition to agriculture. He understood that the domestication of plants and animals involved human selection and was at its heart a biological process. Indeed, Darwin emphasized the importance of human selection in producing a vast diversity of plants and animals.[2]

FIGURE 4.2   Wheat: an Old World grain, wheat is important in making bread.

Interest in how humans have accumulated and used food is increasing in importance as scholars are coming to appreciate the role of food in shaping culture. Through domestication, humans have altered plants, but agriculture has shaped humans just as profoundly. The rise of sedentism and agriculture made possible large gatherings of people and called for the "domestication" of humans in an attempt to curb aggression and violence. This goal has yet to be achieved. One model of agricultural development holds that wherever the elites adopted agriculture the masses followed.[3] Another possibility is that women, who may or may not have held elite status, invented agriculture. In this model, women were the first to exploit and experiment with plants, and over a long gestation, they came to cultivate them. The Barasarna women of Colombia appear to fit this model.

SOUTHWESTERN ASIA

The first people to cultivate plants are unknown, although much research has focussed on southwestern Asia, in a region known as the Fertile Crescent, bound by the Mediterranean Sea in the west and the Zagros Mountains in the east. The Levant, and what are today Turkey, Iraq, and Iran, participated in this development. It is difficult to pinpoint one locale as the catalyst for the origin of agriculture in southwestern Asia. Tell Abu Hureyra in Syria, along a tributary of the Euphrates River, is surely among the early sites.[4] It seems possible that this village may have cultivated wheat and barley as early as 12,000 years ago. The first cultivated plants were barley (Figure 4.3), wheat, lentils, chickpeas (Figure 4.4), and the garden pea. (The modern varieties of sweet garden peas should not be confused with sweet pea, a different legume.) These plants are all valued for their edible seeds, leading one to suppose that it was no coincidence that the people of southwestern Asia practiced seed agriculture, a very different model than what would arise in Papua New Guinea, for example. It is also interesting to contrast the Andean potato with the rise of seed agriculture. The potato does in fact yield seeds, but they are inedible. Moreover, there are no records that the people of the Andes Mountains planted seeds. Rather, as is the practice today, they planted sections of the tuber itself, with each section containing at least one "eye."

In northern Syria, attention has focussed on the Balikh River, a tributary of the Euphrates River. Soils in northern Syria are alluvial near the river and high in calcium. Other soils tend toward clay as the subsoil with a layer of loam atop. As a generalization, these soils favored agriculture. Before the advent of agriculture,

FIGURE 4.3   Barley: an Old World grain, barley is important in fermenting into beer.

FIGURE 4.4   Chickpea: an Old World legume, the chickpea is an important source of protein. (From Shutterstock.)

northern Syria was grassland. With only about 25 cm of rain per year, northern Syria was on the margin of dry farming.[5] Because rainfall fluctuated year to year, crop failure must have led to local famines. One has this sense in reading the prediction in Genesis of 7 years of famine following 7 good years, at least in the case of Egypt. About 10,000–8000 years ago, rainfall may have been more generous in northern Syria, providing the conditions for agriculture. From the earliest period, barley and wheat were in cultivation. Farmers would not have planted wheat without the expectation of rain. Two-rowed barley was the staple along with emmer and einkorn wheats. Northern Syria had sufficient success to produce surpluses for storage. Later came bread wheat and durum, the latter having more protein than any other type of wheat. Archeological evidence confirms the importance of durum in northern Syria. The region also grew flax. Other Syrian sites omit the presence of durum and bread wheat. Emmer and einkorn must have held their own. During the sixth millennium BCE, the climate appears to have dried, lessening the reliance on wheat, although in some places, again, emmer and einkorn held on. Wheat may have contracted to lands along the Balikh. In other areas, barley retreated as wheat asserted itself as the food of Syrians. By the Bronze Age, barley was again important. It seems possible that population pressure led farmers to plant barley in hopes of averting crop failure. By the time of the Assyrians, barley and wheat appear to have been equally important and lentils were the chief legume. Chickpea and pea must also have been important.

Between roughly 11,000 and 10,000 years ago, the climate cooled and became drier, a period known as the Younger Dryas.[6] Wild plants and perhaps even small game may have become comparatively scarce, pushing humans in the direction of adopting farming from Syria. About this time, at least three sites in what is today Israel witnessed the cultivation of grains and legumes. By about 9500 years ago, the climate warmed again, and it was during this period that farming began to spread to several other sites in southwestern Asia. The warmer, wetter climate must have favored agriculture, particularly in a region not blessed with abundant rainfall. As elsewhere, the initial adoption of agriculture was not an ironclad commitment. The people of southwestern Asia simply practiced agriculture as one of several subsistence strategies. Gathering wild plants, hunting, and fishing remained important.

About 9000 or 10,000 years ago, the inhabitants of Jericho in the Jordan Valley, and Jarmo, Iraq, may have joined the ranks of farmers, cultivating plants with emmer wheat being possibly the earliest domesticate.[7] The pea, chickpea, and lentil may have been domesticated about this time. Einkorn wheat, native to Turkey, may have been domesticated near Damascus, Syria about 9700 years ago and along the Euphrates River about 9500 years ago. At the same time, barley became an important domesticate. Farmers grew barley with two rows of seeds near Damascus about 9700 years ago and barley with six rows of seeds approximately 9500 years ago. The domestication of wheat and barley over the 300 years between 10,000 and 9700 years ago appears to have been a comparatively rapid event. From these early beginnings, the cultivation of wheat and barley spread throughout the Fertile Crescent and Turkey by 6000 BCE. The role of these grains has been contested. They probably were food, but they must have played a large role in the brewing of beer. At any rate, one cannot make bread from emmer and einkorn wheats. The adoption of bread wheats must have been necessary in this context.

If southwestern Asia took the first strides toward agriculture, a point debated, it at first relied on groundwater and rainfall rather than irrigation, a risky strategy in a semiarid region of the globe. Change came in the seventh and sixth millennia BCE, when people along the Tigris and Euphrates Rivers in Mesopotamia began to irrigate crops.[8] They may have used cattle, probably the ox, to pull plows. After the sixth millennium, the people of Mesopotamia emphasized the cultivation of tree crops such as dates and figs. Both tolerate drought well. Experimentation led to the vegetative propagation of fruit trees. Because these trees and many other fruit trees like them cross-pollinate, the seed is an unknown quantity. It will likely be heterozygous for a number of traits and so deviate to some degree from the parents. This is clear in other cross breeders such as humans, whose children may differ in many ways from the parents. From the perspective of horticulture, it is better to identify elite fruit trees and then vegetatively propagate them so that every new tree is a clone of its elite parent, ensuring the uniformity and quality of the harvest. This Mesopotamian invention has been central to the propagation of nut and fruit trees into the present. Given the importance of tree fruits and nuts, it is possible that they were used first for ritual purposes and later for horticulture. This idea appears to be implicit in the Genesis creation account.[9] According to it, the first humans dwelled in a garden before the invention of agriculture. Fruit trees were so important in this account that the author or authors of Genesis forbad the first humans from eating the fruit of one tree. Disobedience led to their expulsion from the garden and marked the onset of agriculture. People who claimed ownership of fruit and nut trees tended them, a movement toward domestication. In this context, trees assumed the status of property.

The early use of barley to make beer gave the people of southwestern Asia something akin to liquid bread, and the availability of bread wheat allowed the making of bread, an early staple in southwestern Asia and Europe. In 1953, Robert Braidwood may have been the first to link the adoption of barley and wheat not to the making of bread for sustenance but for making beer for pleasure. This idea may hold for the domestication of grapes and their link to wine in the Mediterranean Basin. The first wine likely derived from dates, although grapes would assert their superiority throughout the Mediterranean Basin. Wine in Greece and Rome commingled with notions of revelry and sex.[10] It was the staple of many brothels in Pompeii alone. By 4000 BCE, the olive tree and date palm were common on lands near the Dead Sea. Olive oil would emerge as the chief dietary fat in the Mediterranean world. Turkey appears to have specialized in grapes and pomegranates. These tree crops were important as one moved north along the Tigris and Euphrates Rivers. The Levant specialized in olives, figs, and almonds. Near the Persian Gulf, date palms proliferated. The Phoenicians may have spread olive trees throughout the Mediterranean Basin.

The Mediterranean Basin includes Jordan, the subject of recent agricultural research. The region has a typical Mediterranean climate with hot dry summers and cool moist winters. Most of Jordan, similar to other areas in southwestern Asia, is semiarid, receiving just enough rainfall to sustain grain culture, with the land divided between crop agriculture and pasture. The best soils hold moisture and nutrients and so are suited to agriculture. The farther north one goes in Jordan the better the soils. As Egypt emerged as the granary of Rome, Jordan was early Syria's source of wheat and barley.[11] Damascus was the destination of Jordan's surplus. Ancient Israel also appears to have imported grain from Jordan. As should be obvious, Jordan inherited many of the crops of the initial agricultural revolution in southwestern Asia: wheat, barley, chickpea, vetch, and lentil. When the November rains came, farmers planted wheat and barley. Between December and February, vetch and lentils were planted and in April, chickpeas. Land tended to alternate between grain one year and fallow the second, although some lands were doubtless reserved to plant legumes. A 3-year rotation was possible: grain, legume, and fallow. As elsewhere farms were small.

EUROPE, EGYPT, AND NORTH AFRICA

From western Asia, agriculture spread, as we have seen, to Mediterranean Europe, Egypt, and North Africa between 6000 and 4500 BCE. Trade between the Levant and the rest of the Mediterranean brought hunter-gatherers in contact with the farmers of the Near East. This intercourse between farmers and hunter-gatherers brought the idea of agriculture and the seeds of cultigens to the latter. Italy and Greece were early adopters of plant cultivation, and agriculture in Mediterranean Europe may have predated its rise in Egypt about 4500 BCE. Sub-Saharan Africa depended on a different suite of crops: millet, sorghum, and rice. Millet and sorghum must have been prized for their drought tolerance. Farmers may have domesticated sorghum about 4000 BCE in the central Sahara Desert and millet about 3000 BCE in the southwestern Sahara. Africans domesticated rice, a subspecies different from Asian rice, about 200 CE along the Niger River.

The people of Egypt gathered wild sorghum before domesticating it. About 6600 BCE, Egyptians began eating wild sorghum.[12] When they began planting sorghum, it was still a wild plant, a commonality of the origins of agriculture in various regions of the world. Egyptians planted land with a variety of seeds, selecting the best for further propagation. About 4000 BCE, Egypt cultivated sorghum. The grain may have been the earliest grass that the Egyptians cultivated. These farmers may not have domesticated sorghum until the third century CE. In this context, southern Arabia and India appear to have domesticated sorghum before Africa did. Domestication in Yemen may date to 2000 BCE.[13] Oman may be even earlier at 3000 BCE. In these regions, including Egypt, barley and bread wheat were grown. The southwestern Asian crop assemblage spread to Europe, South Asia, and East Africa, including Egypt. From Egypt, agriculture spread west across North Africa. In the first century CE, Mesopotamia double-cropped land, likely irrigating grains and legumes. The Greeks interplanted wheat and legumes. Faba beans, peas, emmer wheat, and spelt were important in Spain. Northern Europe planted rye and oats together to create maslin. In some parts of Europe, wheat fed humans and clover went to livestock. Britain favored grains and legumes. Egypt practiced double cropping about 300 BCE. Southwestern Asian crops may have traveled from the Levant to Egypt or by sea from southwestern Yemen to East Africa. Perhaps, humans pursued both routes. By the second millennium BCE, East Africa and India were exchanging crops. In many parts of southwestern Asia, people preferred emmer to einkorn wheat. The growing of grains, legumes, and fruit trees was common among the Levant, Turkey, and Mesopotamia. At some sites, these people may have gathered wild oats and rye to supplement farming.

SUB-SAHARAN AFRICA

One school of thought supposes that agriculture spread from southwestern Asia to Egypt.[14] This supposition makes sense, but this hypothesis goes farther in asserting that from Egypt the southwestern Asian assemblage moved south along the Nile River into Sudan and Ethiopia, from where agriculture might have moved west. Yet, the southwestern Asian package did not transfer to sub-Saharan Africa. South of the Sahara Desert, people must have independently invented agriculture. If sub-Saharan Africa marked a new departure for agriculture, the same cannot be said of Egypt and North Africa, both of which benefited from the southwestern Asian packet of crops. One sees continuity among the agricultures of southwestern Asia, Egypt, and North Africa but discontinuity in sub-Saharan Africa. West Africa appears to have invented agriculture about 7000 years ago, from where it likely spread east to Sudan.[15] Sorghum and pearl millet were the dominant grains. The cowpea was the chief legume. West Africa had two legumes, both labeled "groundnuts," a confusing appellation given that the English-speaking world refers to the peanut as the groundnut. Yams and oil palm were also important in West Africa. Rice deserves special mention because African rice is not identical to Asian rice. The grain of African rice tends to be long and red or short and white, whereas the grain of Asian rice is long and white, once polished. There is no mistaking the two or the fact that West Africa independently domesticated rice. In parts of Africa, the package was sorghum and finger and pearl millets. Africa may have given India finger millet, sorghum, and pearl millet in that order. The American agronomist Jack Harlan believes that the people of Africa domesticated sorghum before 1000 BCE.[16] Yet, the Africa first thesis abuts against the possibility that Saudi Arabia domesticated sorghum as early as 3000 BCE.[17] Harlan places the domestication of sorghum in Chad and Sudan.[18] It is also possible that the people of easternmost Africa, along the coast of Red Sea as far south as the horn of Africa, may have domesticated sorghum first. From this region, the progression of sorghum to southwestern Arabia required the passage of just a short distance. It is also possible that Africa gave sorghum to Arabia, whose people thereafter domesticated it. The process may have worked both ways. Arabia may have given Africa teff, which farmers domesticated in Ethiopia. Chickpeas and sorghum were important in several regions of sub-Saharan Africa.

EAST ASIA

It is difficult to pinpoint the origin of farming in East Asia because archeological research has been less intensive than that in, say, southwestern Asia. As in southwestern Asia, Europe, Egypt, Africa, and the Americas, grains have been important in East Asia, with much attention focussing on rice, at least in southern China, Southeast Asia, and South Asia. The focus on rice appears to be appropriate, given that it provides more calories to more people than any other food. In East Asia, the valleys of the Yellow and Yangtze Rivers may have been the regions of the earliest plant cultivation. The earliest cultivation of rice in East Asia is not settled. One thesis holds that farmers domesticated rice along the Yangtze River about 6500 BCE and later spread it to Southeast Asia and India by 4000 BCE.[19] Other dates place the origin of rice culture along the Yangtze River about 6000 BCE, both dates being within the margin of error. At first, farmers grew rice in paddies and only later adopted cultivars suitable for dry land cultivation and others suitable for cultivation in deep ponds. Millet appears to have been first cultivated about 5500 BCE along the Yellow River. Given the many species of millet, one wonders whether the term "millets" might be more appropriate. About 5000 BCE, the people along the Yellow River grew three species of millet. By this date, farming began to spread into central China.

The soybean appears to have been a later addition to these grains, although it was important enough by the second millennium BCE that one Chinese text named it one of the five sacred grains. The labeling of the soybean as a grain was wrong. The soybean, similar to beans, peanuts, peas, chickpeas, and lentils, is a legume whose seeds are edible. Perhaps, the Chinese grouped soybeans with the grains because all were seed crops, although the roots of a soybean plant are different enough from those of grains to call attention to the fact that soybeans are not grains. Similar to rice, the soybean is now a world crop.

The Korean peninsula must have imported rice from China about 3200 years ago.[20] This date is comparatively late and probably stems from the fact that Korea has a harsher climate than that of southern China. Korea needed new varieties that matured quickly during the brief summer. In Korea, millet likely preceded rice. The cultivation of rice became important partly because the elites taxed it to increase their wealth, to widen the social chasm, and to erect large architectural spaces. This trend was well known in Egypt. Korea was similar to the Americas in absorbing crops slowly. Rice did not therefore sweep through Korea as early research had supposed. Indeed, Korea adopted rice no earlier than the second millennium BCE. It is possible that new people entered Korea with rice, so that an old cultigen came with a new culture. Rice predated the construction of monumental architecture by about 700 years.[21] It is well established that rice was not native to Korea as it was not native to the Americas. Africa, China, and Southeast Asia were the early centers of rice culture. When rice reached Korea, it had been a domesticate for millennia. Before rice, as we have seen, millet was well established in Korea. Millet cultivation in Korea probably dates to the sixth millennium BCE. Northern China grew millet by 7000 BCE, from where it migrated to Korea.[22] About this time, southern China domesticated rice. Northern China appears to have traded with Korea about 3000 years before southern China did, leading one to conclude that millet must have preceded rice. Northern China was of course closer to Korea than was southern China. Rice also had climatic barriers that slowed its advance into Korea. The grass could succeed in Korea only when farmers had selected varieties with short growing seasons and some adaptability to a cooler climate. All early rice in Korea were the *japonica* varieties that had such an adaptation. As early as 2000 BCE, Koreans used stone blades for the harvest. By the first century BCE, rice culture moved as far north as what is today Pyongyang, North Korea. This assumes an introduction from the north but early success only in the comparatively warmer south. Rice was also grown near the Han River and of course farther south, which must have been the best region for cultivation. Kumgokdong near Pusan is among the early regions of rice culture in Korea, dating to about 2000 BCE. By this time, Korea cultivated sorghum and broomcorn millet. At the time, Koreans clustered in small villages near their farms. Near the Han River were other early sites of rice culture. In South Korea, the emphases were on rice, barley, foxtail millet, and sorghum. It is possible that rice was first cultivated in the south about 1400 BCE. Fossilized pollen has established rice cultivation at several Korean sites between 2000 and 1500 BCE. In 1986, Kim believed that agriculture began in Korea with the introduction of millet, but was not widespread.[23] Perhaps, as few as 2000 Koreans grew millet at its inception. These were the Chulmum people. Immigrants, the Mormun people, brought rice later. With the introduction of rice, population density appears to have increased. Because the Bronze Age emerged only about 1000 BCE in Korea, stone tools must have been exclusive in cultivating and harvesting rice before this date. The tendency was to plant millet in lands near rivers and without irrigation. But these rivers tended to overflow during the rainy season, a circumstance that would imperil rice because of its need for still water. The new layer of silt deposited from these rivers suited millet. Farmers tended to plant rice in the uplands near small streams. Koreans terraced hills to prevent erosion. In this case, rice tended to cluster inland. The fact that dwellings tended to be larger inland suggests that the rice surplus must have fed more mouths.

In the preagricultural period, Japan appears not to have had a vast assemblage of wild edible plants and so must have been eager to adopt cultigens from elsewhere. As a rule, dryland millet culture preceded the adoption of paddy rice in southern Japan by 1000–1500 years. China was the source of millet in both Japan and Korea. Japanese agriculture may have emerged in a similar fashion to that of the American Southwest. Both imported crops from other regions of the world: the American Southwest from Mesoamerica and Japan from China and perhaps ultimately from India in the case of buckwheat. The beginnings of agriculture in both regions did not produce notable social, economic, or political changes, at least not at first. When agriculture intensified in both, populations increased. Nomadic patterns appear to have persisted in Japan and the American Southwest. The Apache, for example, planted corn, resumed the hunt, and returned in autumn to harvest corn. The same was true in Japan, with nomads planting millet, which needed little attention, returning later for the harvest. Foxtail, broomcorn, and barnyard millets were particularly hardy and drought-tolerant. The last two are among humankind's most drought-tolerant grains and may be grown at elevation and in semiarid lands. Foxtail millet is adaptable to a range of soils, and all millets do well in poor soils. Millet needs only about 6 weeks from planting to harvest. Asian rice probably originated in Southeast Asia and southern China, from where it spread to southwestern Japan, thereafter climbing north. Chinese immigrants likely brought paddy rice to Japan. The initial spread of rice must have been rapid, conquering the south in only a few hundred years. Thereafter, rice spread slowly. Not until the nineteenth century CE did parts of northern Japan adopt rice. This delay was notable of the coastline, whose people, content to fish, did not perceive agriculture as a benefit. Some researchers do not accept such a late date for rice. Northern Japan appears to have focussed on millet, barley, and buckwheat (which is not a type of wheat), crops that predated the introduction of rice. The introduction of rice into northern Japan must have met the approval of farmers already accustomed to nurturing new crops. Japan appears to have adopted crops in four stages. First, the islands appear to have used small-scale gardening to grow indigenous crops during the Jomon period. This period may have witnessed the first crop acquisitions from China and Southeast Asia. The gourd may have been among the first cultigens, along with great burdock and hemp. About 5800 BCE, peach trees and buckwheat were in cultivation. Tsukada believes the peach to have been an import from Yunnan, China. It is possible that buckwheat originated in northern India or northern China. About 5000 years ago, farmers began cultivating barnyard millet, great burdock, gourd, and the mung bean.[24] About 4500 years ago, Japan began cultivating nut trees, yams, taro, and possibly rice and oats. An increase in seed size over 4000 years suggests human selection of foxtail and barnyard millets. This period witnessed the clearing of forests to open more land to agriculture. About 3000 BCE, most archeologists are confident that rice and barley were established.[25] The earliest rice pollen may date to 3200 BCE. Secondly, Japan about 1300 BCE adopted paddy rice during the Yayoi period. By then, the surplus sufficed to make rice an item of trade. The third and fourth stages include separate migrations of crops from south to north. Japanese agriculture was defined more by the introduction of domesticates rather than by the domestication of native plants. The first rice introduced into Japan may have been dryland varieties. The rapid movement of rice to north must have stemmed from Japan's familiarity with cultivating an array of crops. When agriculture arrived in Japan, people continued to gather wild plants, catch fish, and hunt game. Agriculture must have first been a peripheral activity. Archeological evidence appears to confirm the minor contribution that agriculture made to the diet in the early centuries. In fact, only in the fourth century BCE was rice an important crop in Japan. Climate along with other economic activities may have played the role of delayer.

About 3000 BCE, the climate cooled in parts of Japan and Korea, perhaps leading to the extinction of deer and water buffalos. Hunting no longer sufficed, leading to a turn toward agriculture. This transition necessitated the acquisition of domesticates from China. Millet was an early domesticate in Korea, although gathering wild plants persisted. As agriculture intensified so did fishing. Millet was also important in western Japan. About 2800 BCE, Koreans grew rice. Four hundred years later, rice had settled in the south and millet, barley, and wheat in the north. About 1000 BCE, rice was introduced into Japan. About the third century BCE, rice cultivation intensified. Japan's population increased first on the coast, possibly due to fishing, and later inland, perhaps due to plant agriculture. By about 100 BCE, Japan may have invested more time and effort on rice than on fish. Between about 100 BCE and 300 CE, Japan's population rose, probably due to rice culture. Rice farms stationed along rivers, probably for irrigation or inundation in Japan and Korea.

SOUTHEAST ASIA

Southeast Asia arose as a separate site for plant cultivation. Farmers there may have domesticated taro, yam, arrowroot, coconut, sago palm, citrus, banana, and breadfruit.[26] To this suite of crops, most scholars believe rice to be a latecomer, although Southeast Asia is today a land of rice farmers. About 5000 years ago, rice spread from southern China to Southeast Asia.[27] The date appears to be late given that the climate and soils of Southeast Asia favored rice and given that the migration from southern China to Southeast Asia cannot have taken long. The spread of rice to Indonesia came later still. In Southeast Asia, taro and yams, both swollen roots, were important. It seems possible that the focus on root crops delayed the introduction of seed agriculture that rice represented.

An important debate concerns the relationship among Indonesia, Papua New Guinea, and Australia. The tendency to see agriculture in Papua New Guinea as an offshoot of Indonesian agriculture is probably wrong, given the antiquity of farming in Papua New Guinea and the fact that the suite of crops was not identical. In particular, Papua New Guinea had sugarcane long before the crop spread to Indonesia. In addition, molecular studies appear to confirm that Papua New Guinea independently domesticated yams and taro.[28] Here is a case in which root agriculture prevailed over grains. There is even more debate about the transfer of sweet potatoes to Papua New Guinea. The sweet potato originated in South America, far from Papua New Guinea. One can scarcely envision its transpacific voyage to Papua New Guinea in the absence of oceangoing ships. The idea of a pre-Columbian transfer and thus an early date of cultivation must be wrong. Papua New Guinea may also have domesticated breadfruit and the banana, although the banana is in dispute. Australia remains a mystery. It might have acquired agriculture from Papua New Guinea or Indonesia, but the transfer never occurred. Only in the eighteenth century CE would the British bring agriculture to Australia.

Papua New Guinea boasts among the oldest agriculture in the world, dating at least 9000 years ago. In this context, the islands must have been an independent center of plant domestication. This idea remains startling to those who had assumed that agriculture spread from Southeast Asia, perhaps from Indonesia to Papua New Guinea only about 3500 years ago. The difficulty for archeologists is that the roots and tubers that were important to Papua New Guinean agriculture have not preserved well in the tropics, making it difficult to pinpoint the origin of agriculture in this region.

THE PACIFIC ISLANDS

About 3500 years ago, humans brought rice to several islands in the Pacific Ocean. Recent research has focussed on New Caledonia, an archipelago in the South Pacific on the southern border of the tropics.[29] The soil is not ideal for agriculture, and the main island was well forested in prehistory. Anatomically, modern humans settled the northern islands of Melanesia about 30,000 years ago, where they ate wild taro. It seems unclear whether these first inhabitants brought taro with them or found it when they arrived. The latter seems more probable. Yet, New Caledonia was settled only about 1000 BCE, a much later date, for example, than the settlement of the Americas. The radiation of humans into these Pacific Islands may have come from a founder population of Australians, who did not practice agriculture. The advent of agriculture witnessed the clearing and burning of forests to open land to cultivation. Probably coming from New Guinea, breadfruit, sugarcane, banana, and yams were the initial package. Here is an important example of an agriculture based not on grains but on fruits and roots. Taro was probably grown for medicine as well as for food. The hunting of game may have led to the extinction of at least one herbivore, and it seems possible that agriculture may have arisen in the context of diversifying the food supply as meat became less available. The collection of nuts resulted from foraging not farming. As elsewhere, farms were initially small enough not to tax unduly the labor of a family. In the first millennium CE, agricultural surpluses sufficed to enable populations to grow. The tendency to fallow land for long periods must have decreased and farmers brought more land into cultivation. They terraced hills to ease cultivation and to minimize erosion. Sweet potatoes were a late introduction and must have been a product of the Columbian Exchange.

Warfare may have stimulated the development of agriculture in Fiji, leading noncombatants to amass stocks of food to carry them through periods of conflict and to farm land that was easy to defend.[30] The warrior elites also taxed agricultural surpluses, making essential the maintenance of productive fields. In 1983, Hassan asserted that agriculture, by increasing populations, intensified warfare.[31] If this is true, at least in the case of Fiji, farming must have been the precondition for war. Perhaps from Australia, people settled Fiji about 3500 years ago. From the outset, they appear to have combined hunting, fishing, gathering, and farming. Yams, taro, and the giant swamp taro were the basis of early Fiji agriculture. Again, the story is one of roots, not grains. People terraced hills for planting, sometimes irrigating these crops, and farmed widely in the lowlands. Farmland belonged to the family or larger kin network. The Western concept of private property was absent from Fiji. As in the West, Fijian warriors plundered farms. That is, they must have stopped long enough to dig root crops. This practice was uncommon in the West until the eighteenth century, when northern European armies plundered farms by digging potatoes. Fiji warriors uprooted taro and yams before besieging a village, leading one to wonder why farmers cultivated land outside the village's fortifications. The coconut was a Fijian tree crop.

Banana and plantain were important in the tropical Pacific.[32] The fruit must have been an early domesticate because humans long ago selected it for seedlessness. Because the plant produces no seeds, it is underrepresented in the archeological record. Yet, the tropics, first in the Old World and during the Columbian Exchange in the New, are hard to imagine without the banana and plantain. In prehistory, these fruits were grown on several Pacific Islands and Papua New Guinea, where they could be grown up to 2000 m. The Solomon Islands may have been the site of banana domestication. From Papua New Guinea, the banana spread west to Burma (now Myanmar). Domestication may have taken more than 5000 years. Early in the process, humans hybridized banana species, meaning that sexual reproduction was then important. The best offspring of these crosses were used in asexual reproduction in the selection of suckers (clones) of the parent tree. In this case, the new plants had characteristics identical to the parent. At some point, the banana spread to tropical South Asia and Africa. The plantain may have originated in India, although the Philippines remains a possibility. The plantain reached Africa about 1000 BCE. Madagascar was apparently not a corridor to the continent. South Asia must have introduced the plantain to Africa through Indian Ocean trade. Papua New Guinea domesticated several species of banana before 2500 BCE. These bananas spread throughout Southeast Asia. About 1000 years later, the banana moved west to South Asia and East Africa. This date appears to mark also the introduction of yams and taro.

SOUTH ASIA

It is possible that India independently domesticated rice rather than borrowed it from China.[33] Established in the Ganges River valley about 4500 years ago, rice spread to the Indus River valley and south into the peninsula. India also grew legumes, probably the mung bean, and millet, activities that may have predated the arrival of rice. In India and what is today Pakistan, the emphasis was on crop dispersal. India appears to have borrowed wheat, barley, peas, lentils, flax, and probably chickpea from southwestern Asia. These crops were likely part of the northern assemblage. In the peninsula, India borrowed rice, probably from China. The millets came from Africa or East Asia. Independent of these developments, India and Pakistan domesticated local roots and tubers. The two regions appear to have domesticated about 80 such plants of perhaps 650 species that might have been amenable for domestication. These roots and tubers may have been eaten raw. Others were boiled or otherwise cooked to ease mastication and digestion and to remove potential toxins. India and Pakistan appear to have adopted taro and yam early. Once dried, yams had the advantage of long storage against lean times. Indians and Pakistanis tended to eat edible leaves raw. India adopted several species of millet. These were desirable crops because they did well in poor soils. The practice of cultivating more than one species of millet in successive plants led to year-round harvest. In fact, India may have domesticated what is known as Italian millet, suggesting that India not only imported millets, but also experimented with new ones. In the wild state, Italian millet may have originated in China but arrived in India perhaps before the introduction of wheat and other crops from southwestern Asia. This is not surprising given the proximity of India to China. Agriculture may have begun in India and Pakistan in lands near the Belan and Ganga Rivers. Rice may have been the first cultigen in the two regions about the mid-sixth millennium BCE. By the mid-second millennium BCE, rice was the principal crop in large areas of India and Pakistan. By then, if not before, the southwestern assemblage had a foothold in Pakistan. These crops were grown in winter in India and Pakistan. From Africa came sorghum, finger millet, and pearl millet about 2000 BCE. The African staples were suitable for summer cultivation, allowing India and Pakistan to crop the land year round. Legumes and grains were in rotation. The people of India and Pakistan recognized the value of legumes and in many cases tended to grow them year round where the climate permitted. As elsewhere, India made the transition from emmer, dwarf, and club wheats to bread wheats, which were favored for yield and suitable for making bread. India and Pakistan adopted oats for winter cultivation. Finger millet displaced Italian millet over time. The two regions recognized barley as more drought-tolerant than wheat.

Bear in mind that not all Pakistan was suitable for farming. Summer temperatures of 40°C desiccated plants and soils.[34] Wheat, barley, lentils, and chickpeas may have been the earliest cultigens, crops that must have moved east from southwestern Asia. Where agriculture was risky, land was converted to pasture. Because of its drought tolerance, barley was favored over wheat, although Pakistanis preferred wheat bread. Barley of course is not well suited to making bread. In prehistory, Afghans led their herds into Pakistan to graze, activity that must have disrupted agriculture. Genesis recounts part of this conflict in the murder of Abel.[35] Despite Pakistani preferences, barley was widespread in the region with archeological evidence suggesting that barley played a larger role in the diet than wheat. Lentils were also important, although they do not appear to have been grown on the scale of barley. Grains were planted in autumn with the expectation of a late spring harvest. Livestock raising furthered the advance of agriculture by providing dung for the enrichment of soils and the transporting of crops. India planted legumes and millet in rotation to maintain soil fertility. Nepal planted legumes with mustard and rice.

THE AMERICAS

In the Americas, the transition to agriculture occurred between 7000 and 3500 years ago in Mesoamerica, the American Southwest, the eastern woodlands, parts of Canada, the Andes Mountains, and Amazonia.[36] These regions developed important staples that would become world crops during the Columbian Exchange. If one wishes to assert that the Americas were relative latecomers to agriculture, one must acknowledge that North America and South America were the last continents that people settled and then only about 15,000 years ago (Chapter 2). The worldwide importance of corn, *Phaseolus* beans, potatoes, sweet potatoes, cassava, chocolate, tobacco, and other crops inspire awe at the achievements of the Amerindians.

Southern Mexico was probably the first region of the Americas to invent farming.[37] The people of this region domesticated corn, *Phaseolus* beans, squash, avocados, tomatoes, and peppers. The people of southern Mexico also ate papaya, guava, sapote blanco, sapote negro, and ciruela, all fruits. The Tehuacan valley southeast of Mexico City may have been the cradle of American agriculture. In the 1960s, American archeologist Robert MacNeish discovered fossilized corncobs, beans, squash, chili peppers, avocados, and bottle gourds in a sequence of caves.[38] Southern Mexico may have domesticated squash as early as 10,000 years ago, giving American agriculture great antiquity. The fossilized corncobs date to roughly 6500 years ago. Beans date to about 2300 years ago, a comparatively late date for such an important crop. Corn, beans, and squash may not have been farmed together until about 2000 years ago, accentuating the fact that the transition to agriculture often took millennia. All three crops, known as the Three Sisters, spread north but not rapidly. Corn and squash reached the American Southwest about 3300 years ago. From the American Southwest, corn and squash spread east to the woodlands and north into the Midsouth, the Midwest, and the dry areas of the plains. Corn reached the eastern woodlands about 200 CE, whereas beans only arrived about 1000 CE.

In the New World, corn, beans, and squash dominated agriculture from Argentina to southern Ontario. In South America, farmers grew cassava and sweet potato and in the Andean highlands potato and quinoa. Farmers in Mesoamerica were the first to cultivate corn, beans, and squash more than 5000 years ago.[39] Of the three, corn was probably the earliest domesticate, being grown in the Tehuacan Valley of Mexico as early as 5000 BCE. The people of Mexico and the Andes independently domesticated beans, including the lima bean. In Mexico, corn and beans may have been domesticated around the same time. The people of the Andes domesticated four tubers: oca, moshua, ullua, and potato, the last being a world staple. In the Central Andes, the potato may have been domesticated between 3000 and 2000 BCE. From its center in southern Mexico, corn migrated both south and north, reaching the American Southwest by 1200 BCE and the eastern woodlands by the time of Christ. In Mexico, South America, and eastern America, farmers domesticated goosefoot about 2000 BCE. To goosefoot the woodlands, Native Americans added march elder and sunflower about 2000 BCE. They also grew squash, an import from Mexico. About 1000 BCE, farmers grew corn and squash in the Southwest, though only by the time of Christ was corn important in this region. Only about 1000 CE was corn dominant in the eastern woodlands.

The people of Central America may have grown cassava (Figure 4.5) before 5000 BCE.[40] They also cultivated a single species of yam, arrowroot, and leten, all roots. They may also have grown a type of corn no longer in cultivation. When corn reached the eastern woodlands in what is today the United States, it may not have been embraced initially, given the probability that it was more labor-intensive than the gathering of edible plants. Humans must have selected more productive varieties of corn so that corn agriculture triumphed eventually, but again the transition was gradual. As a tropical plant, corn must have been difficult to adapt to the temperate zones. Millennia of selection must have been necessary to develop temperate

FIGURE 4.5   Cassava: an important American root crop, cassava is now important in tropical Africa and Asia.

varieties of corn. Note that one may speak of varieties of corn, not species because corn is only a single species, *Zea mays*. The people of eastern North America adopted corn sometime before European contact. In the Midwest and South, Native Americans grew corn on a small scale along with sumpweed, sunflower, and chenopod. An analysis of skeletal remains of Native Americans suggests that corn was far from being the dietary staple, although after about 800 CE, corn became increasingly important in North America.[41] Even in this context, one senses that corn's rise to prominence was gradual. In southeastern North America, corn was a latecomer but by about 900 CE it was becoming important. As corn became important, other Native American crops appear to have declined in the diet. Sunflower is an example.

Given the respectable yields of corn, one may evince surprise that the people of North America had not adopted it sooner. We know, of course, that the climate barrier was insuperable for millennia. Given this circumstance, it seems probable that the earliest varieties of corn did not yield well in North America. In the proper context, corn's plasticity is remarkable. Its relative, the grass sugarcane, has never left the tropics because it will not tolerate frost. Corn, in contrast, is much more malleable. Another possibility is the old practice of hunting. If the number of small game decreased, the people in North America must have put more effort into raising corn. It also seems possible that *Phaseolus* beans might have, when interplanted with corn, increasing yields because the beans, through the process nitrogen fixation, put ammonium and nitrate ions in the soil, enriching its nitrogen content. Keep in mind, though, that beans had even greater difficulty transitioning to the temperate zones than corn. The modern example of the efficacy of nitrogen fixation is the rotation of the grass corn with the legume soybeans.

The American Southwest adopted corn more quickly than did eastern North America.[42] The people of the American Southwest may have begun to cultivate corn as early as about 1600 BCE. This early agricultural period (1500 BCE–500 CE) witnessed a gradual acceptance of corn agriculture, although the dates are early enough favoring a relatively fast movement of corn from Mesoamerica to the American Southwest. By about 400 BCE, the people of the Southwest were committed to corn. During this period, surpluses were large enough to demand storage against lean times. Yet hunting, fishing, and gathering persisted. Farmers supplemented their diet with amaranth, rice grass, drop seed, wild rye, bent grass, hoe grass, purslane, mustard, bee weed, wild sunflower seeds, sumpweed, juniper, pinon nuts, mesquite, yucca, walnut, acorn, cactus, blackberry, wild onion, bulrush, panic grass, little barley, devil's claw, and knotweed. Rice grass was not a type of rice, because rice was native to Africa and Asia and unknown in the Americas before the Columbian Exchange. Wild rye was not a type of rye because rye was unknown in the Americas before the Columbian Exchange. The same may be said about the disconnect between little barley and barley and between wild onion and onion. The Anasazi of the Southwest continued to eat wild plants to supplement agriculture into the period of European contact.

In the eastern woodlands, corn yields must have been adequate by 800 CE to prompt Native Americans to create storage units, many of them pits in the ground, in which to store the surplus. The ease with which corn could be stored, once dried, must have hastened its success as an important crop throughout the Americas by the time of European contact. Much of the surplus nourished humans through the winter, although a portion was necessary for next year's planting. Around 800 CE, Native Americans specialized in the construction of hoes, which must have been used to weed the fields of several crops. These hoes were essential for people who lacked the plow and draught animals. The work of breaking ground must have been intensive. One has contemporary accounts of African slaves exhausted by the constancy and intensiveness of hoe agriculture. The people of the Northeast supplemented agriculture with the consumption of wild plants: acorn, walnut, hickory nut, black gum, hackberry, buckbean, chenopod, bunchberry, grape, gourd cherry, and spikenard. Throughout the eastern woodlands, humans continued to hunt small game and collect hickory nuts, pecans, hazelnuts, acorns, chestnut, beechnut, persimmon, plum, papaw, raspberry, strawberry, maypop, hawthorn, grape, elderberry, crabapple, blueberry, blackberry, chenopod, knotweed, amaranth, may grass, little barley, wild rice, wild gourd, sunflower, sumpweed, ragweed, wild beans, vetch, and pervine, all of these supplements to agriculture. Between about 3200 and 1800 BCE, the people of the eastern woodlands domesticated squash, chenopod, marshelder, and sunflower.[43] There was also an attempt to plant knotweed, little barley, and ragweed outside of their native lands, suggesting advancement in cultivation. As a rule, the more corn, beans, and squash one ate, the less reliant one was on acorns. The inverse was also true. Between about 200 BCE and 200 CE, the people of the middle woodlands moved toward corn culture.

Corn was also important to the Fremont culture that included the people of what are today Idaho and parts of Colorado and Nevada.[44] This culture appears to have flourished between about 700 and 1200 CE. These people cultivated their own varieties of dent corn, the basis of livestock feed in more recent times. The cob was apparently small by today's standards. The earliest evidence of corn culture in this region dates to about 100 BCE, although it is possible that the forerunners of the Fremont culture cultivated corn a little earlier. By the Fremont period, the surplus of corn required the digging of large pits for storage. Farmers may have irrigated corn during this period. At some point in the transition toward corn, natural selection must have favored corn culture because those who grew corn had more calories and gave birth to more children than those who had not adopted corn agriculture. The people of this region transitioned to corn only when it produced more food than traditional methods of hunting, gathering, and fishing. Again, one notes that the transition from hunting and gathering to corn agriculture was gradual in this region. One sees here evidence that the adoption of corn was often a halfway measure, one that initially complemented rather than replaced hunting, gathering, and fishing. In this sense, corn agriculture may have been a secondary activity.

The retention of dietary breadth suggests that corn must have been comparatively unimportant in the Midwest and Midsouth, both now important regions of corn culture. In contrast, in the Ohio River valley, where we witness a narrowing of the diet's breadth, corn became a staple. As people ate more corn, they ate fewer other foods. The differences between the Ohio River valley and the Midwest Proper are difficult to explain. Because corn often is a productive and reliable source of food, people who had experienced food shortages in the past must have been inclined to plant corn to ensure food security. By about 1200 CE, corn was important in both the Midwest and Northeast. The early farmers near the Mississippi River appear to have eaten a diet deficient in iron, probably because they ate plants to the exclusion of sufficient meat. Judging from these people, life expectancy must have been low. Subsistence on a small number of food plants must have contributed to nutritional deficiencies.[45]

The adoption of corn culture and of agriculture in general was not a simple, linear process. One needed to prepare the land without a plow or draught animal, plant the corn kernels (seeds), eliminate weeds, apply fertilizer (animal and human dung), irrigate where rainfall did not suffice, and harvest and process the crop. These steps took time and the expenditure of calories, exactly what humans needed to conserve in a food-scarce environment. In this context, one wishes to know corn yields in prehistory but these are difficult to gage. One is tempted to seek parallels among poor farmers in Central and South America. For example, most farmers in Guatemala lack draught animals and plows, as did the prehistoric people of the Americas. Yields in present-day Guatemala are about 10–50 bushels/ha.[46] Small gardens and putatively virgin land yield the best. In the case of gardens, Guatemalans are able to lavish more attention on a small number of plants. To the second point, no land colonized by plants is truly virgin, but land that has not been subjected to monoculture is generally more productive than land that has been planted to corn year after year. The American indigene tobacco is a good example of a crop that depletes the soil of nutrients relatively quickly. Corn yields best in tropical lowlands and on lands near rivers. Yields decline as elevation increases.

The comparison with Guatemala suggests that in prehistory the caloric investment in corn agriculture was within the range of return on investment among hunter-gatherers. If true, it is very difficult to decide why humans bothered to cultivate corn in the first place, unless it was a secondary strategy. That is, corn agriculture may make sense only as part of a broader subsistence strategy. The tendency in such situations must have been to minimize time and energy in farming corn. Slash and burn is one such minimalist strategy. Subsistence farmers in Guatemala and Peru exert about 300–500 h/ha/year on corn.[47] The average in Latin America appears to be about 400 h/ha/year. Women may have played a crucial role in the transition to corn agriculture. Latin American women often plant and harvest corn, although men may do some other farming chores. The making of corn kernels into a kind of flour has been entirely women's work. It is possible that women rather than men made the fundamental decision to plant corn, thus initiating the gradual transition to agriculture. It may be possible that women gained an economic incentive by planting corn.

Although the transition to corn was uneven in the Americas, cultural practices were relatively uniform. People planted corn in hills, interplanting it with beans in the species *Phaseolus vulgaris* in much of eastern North America and with tepary and lima beans in the American Southwest. These beans are all in the genus *Phaseolus*. That is, all American beans comprise one genus. Squash was another companion crop, making the Three Sisters. Native Americans used corn in various stages of development, sometimes eating parts of a plant without allowing it to produce tassels and silk. Such action must not have been widespread because it truncated the development of seeds. Native Americans processed corn for storage, mindful of winter's approach. Kernels were often dried in the sun. Squash and beans were also processed for storage, although beans required almost no processing if allowed to dry on the plant. As a culinary phenomenon, corn was eaten with wild plants, meat, and fish. The Iroquois ate corn, beans, and squash with a large number of other foods, including a variety of meat and fishes. Crab apples, wild cherries, chokecherries, grapes, and about 20 species of berries (grapes are a berry) rounded out the diet. Native Americans traditionally roasted, baked, or steamed corn picked early. They shelled corn or ate it on the cob. They ground dried kernels into a type of flour, although corn has never been as suitable in making bread as have been wheat and rye. Native Americans used lard from game or sunflower oil to fry corn.

The transition to *Phaseolus* bean culture in North America came late, only between 1000 and 1200 CE.[48] In several parts of North America, the domestication of sunflower and even squash may have been independent of these activities in Mesoamerica. In North America, corn, beans, squash, sunflower, and tobacco had become important before European contact. The Great Plains were the site of the bison hunt. Other game was also important, but so were plants, including prickly pear, goosefoot, sedge, sunflower, pine nuts, and chokeberry. In this region, hunting and gathering persisted past 1700 BCE. Agriculture may have spread to the Great Plains from the eastern woodlands. By the time of European contact, corn had spread through the Great Plains north into Canada.

In the eleventh century CE, the first agricultural villages formed in what is today Missouri and on the Great Plains.[49] This development does not preclude an earlier resort to agriculture, only that yields must have been large and stable to permit village life by the eleventh century. These communities still depended on bison and fishing, but they grew corn, beans, squash, sunflower, marshelder, and tobacco. This culture, known as the Middle Missourian, had a large enough farm surplus to trade with neighbors and with people over considerable distances. The surplus freed some people from the need to farm. Artisans made elaborate pottery and jewelry, which furthered the ambitions of traders. Farms tended to be small. The adoption of corn may have been relatively rapid in what are today Nebraska, Kansas, Colorado, New Mexico, Oklahoma, and Texas. By 1200 CE, agriculture had spread to parts of Texas and the Oklahoma panhandle. Bottomlands proved the best centers for agriculture and were occupied first. As one moves from west to east in the plains, the climate becomes moister, favoring the development of agriculture. As the Dust Bowl would make clear in the 1930s, farming in arid parts of the plains was risky. The natives of this region tended to establish farms near rivers. The agricultural surplus may not have been large because population appears to have been sparse. By about 1100 CE, what is today Kansas had adopted corn and other crops, although the time of the arrival of beans is still in dispute. By about 1200 CE, farmers grow tobacco in the southern and central plains. The climate dried between 1250 and 1450, although corn culture remained viable. This arid period appears to have led people to turn more intensively toward corn. The reliance on the cultivation of marshelder and sunflower declined. Beans and squash appear to have held their own during this dry interlude. The reliance on corn deepened in the first half of the fifteenth century as people migrated east in search of a moister climate. Lands became so dry in the West that even bison migrated east. Crop yields may have fluctuated during this period. Some people doubtless hunted bison when corn, beans, and squash yields were not high enough. Populations decreased after 1450 CE, suggesting a decline in the agricultural productivity, a surprise given that 1450 marked the end of the driest period. The difficulty of farming in the West seems to have depopulated the region. There is evidence of corn culture in the West, but it must have been minimally productive.

In the high plains, conditions were often too dry to permit agriculture. The people of the American Southwest appear to have exported corn surpluses north and west. On the eastern plains, farmers appear to have put more effort into growing corn for subsistence and where possible for trade. The fact that storage pits on the eastern plains became larger in the fifteenth century leads one to assume that the corn surplus must have been large, although this supposition is difficult to square with the dry climate and the population diminution in the West after 1450 BCE.[50]

In South America, the Andes Mountains cover parts of Ecuador, Peru, Bolivia, northern Chile, and northwestern Argentina. This region was another center of agricultural innovation. We have already considered the importance of potatoes, one of the five tubers that the Andeans domesticated. The grains quinoa and canihua were important. The lima bean was an Andean domesticate, and several of the beans in the species *P. vulgaris* appear to have been domesticated in both Mesoamerica and the Andes. By about 3000 BCE, the Andeans had domesticated quinoa, squash, and the bottle gourd; by 2800 BCE the lima bean; and by 2000 BCE other *Phaseolus* beans.[51] Curiously, Andean agriculture, like that of Papua New Guinea, did not spread widely or quickly. Perhaps, the cool climate of the Andes contrasted with the heat of the lowlands, although, as we have seen, beans were originally a tropical crop and so should have spread throughout South America in prehistory.

As important as the Andes Mountains were, the transition to agriculture and the long period of acquisition of crops, particularly on the arid Pacific coast of Peru, took millennia, stretching between 8000 and 1450 BCE.[52] One thesis supposes that population pressure led to the invention and adoption of agriculture in Peru and along its coast. The migration of people brought crops from one region to another. Population pressure may have led Peruvians to domesticate several tubers for their richness in carbohydrates. Along the Pacific coast, fishing was important and might have delayed the development of agriculture. One thesis holds that nomadic hunter-gatherers were more likely to adopt agriculture than sedentary foragers, a circumstance that was true on Peru's coastline. On the Peruvian coast, people knew of the agricultural developments in the highlands but were slow to adopt them. Climate must have played a role, given differences between the mountains and lowlands. Potatoes, for example, must not have been well suited to the South American lowlands. Flavorful, fattening, or spicy plants may have been the first to be domesticated because of human cravings for these qualities. This possibility may explain why the chili pepper was an early domesticate. Food plants such as chili peppers added a new dimension to the diet and were not necessarily selected because of their caloric contributions. The interesting point about the Peruvian coast is that it had virtually no cultigens, relying instead on outsiders to bring new crops. Between about 8000 and 6000 BCE, early dates, the people of coastal Peru began adopting plants, probably from neighbors, to cultivate. Only by 1400 BCE was the transition to agriculture complete, although people continued to rely on fish for protein. By 1400 BCE, these people had solved the problem of aridity by irrigating crops. The same phenomenon, of course, had applied to Mesopotamia at an earlier date. Over the millennia, the Peruvians adopted cotton, cassava, potatoes, chili peppers, *Phaseolus* beans including the lima bean, avocado, guava, gourd, and corn. The presence of the potato in the lowlands seems strange because it is not a crop of warm climes. Some of these crops descended from the mountains to the lowlands. Others came south from northern regions of South America. The earliest plant to have been cultivated along the Peruvian coast may have been the chili pepper, which may have been domesticated in southern Peru or eastern Bolivia. The lima bean, native to Peru, reached the coast before other American beans. By one account, all *Phaseolus* beans were native to Peru. In the mountains, cotton and gourd were probably domesticated. Both may have grown wild on the coast before cultivation there. The cotton familiar to the Peruvians may be the product of ancient hybridizations between cottons native to the Americas and those native to Africa. This occurrence came millennia before the introduction of Africans and their crops into the Americas. Cotton and gourd agriculture came late to the Peruvian coast. For millennia, humans had harvested wild cotton. Corn too was a late addition. Corn may have migrated south along either the western or eastern coast of South America. A southern migration along the Pacific coast would have been the direct route. A southeastern route terminating on the west coast would have necessitated the traversal of the Andes Mountains. Corn is an adaptable grass that does well in a variety of environments. Of the tubers and roots, cassava was the first to reach the Peruvian coast. Many points of origin in the Americas have been proposed. Certainly, a South American origin would have been convenient from the perspective of the Peruvians. As a perennial, cassava must have been attractive because it supported a year-round harvest. The root, however, was protein-deficient. Along the coast, it must have been paired with fish. The tuber achira followed cassava, but unlike cassava and the potato, it never became a world staple. The potato must have moved from highlands to coast. The avocado came to the Peruvian coast from northern South America. The fruit must have been desirable because of its protein and fat content. Remember that the body cannot absorb protein, except in the presence of fatty acids. The yield is year round, and the tree remains productive for decades. Similar to many fruit trees, the avocado does not yield fruit in its first years, requiring an investment of time and effort in expectation of a harvest down the road. Guava necessitated a similar investment. The tree came from middle elevations in the Andes to the coast. Beans and chili pepper populated northern Peru. To the south, in the Chilca valley, potatoes and cassava were important. Beans appear to have been more widely consumed than the potato. In fact, the potato and other tubers were sparsely consumed until about 3000 BCE.[53] Between roughly 6000 and 4200 BCE, beans became increasingly common in the valleys of southern Peru. Beans must have been heavily consumed and may have been cultivated to the exclusion of some other crops. Chili peppers were favored for their spiciness, remaining important to this day. Roasted beans were important throughout South America in the way that Americans eat roasted peanuts, another South American crop. As one moved along the coast and over time, beans appear to have become ever more important. About 2500 BCE, the chili pepper had emerged as a ceremonial plant in addition to a food. People burned chili peppers to create smoke at religious events. Because the smoke irritated the eyes, one wonders why Peruvians would have taken this step. Perhaps, they thought that the smoke would drive out demons and purify the body. Guava was largely confined to northern Peru. Religious rites and the quest for flavor may have driven the Peruvians to favor the potato, although some people find it insipid. Beans, peppers, and squash and gourd, cotton, archira, and guava were often grown as companion crops. Less frequent was the combination of corn, avocado, potato, and cassava. The potato and cassava preserved poorly and may have been more abundant than the archeological record suggests. Cassava may have been important because it tolerates poor soils. The root appears to have taken hold only after 2000 BCE. By then, avocado and corn were cultivated in small areas, suggesting perhaps only intermittent culture. Accordingly, the archeological record for corn is not uniform. It appears in parts of Peru but not in neighboring areas. By 1400 CE, however, corn was ascendant. Nonetheless, one senses a pattern of episodic, piecemeal cultivation of crops. There was not one agricultural revolution in Peru, but a sequence of waves of adoption. Along the Moche River, people ate avocado, beans, and chili peppers, but not corn. South to the Vira Valley people farmed corn but not avocado. Where agriculture intensified, one finds the remains of large stone buildings, suggesting dense populations. Political and economic elites must have dictated the choice of crops and the cycle of planting, weeding, watering, harvesting, and processing.

The lowlands of the Amazon and Orinoco River basins domesticated cassava, sweet potatoes, arrowroot, and possibly one species of yam and chili peppers. Amazonian agriculture may date no earlier than 400 CE, although dates are problematic in the tropics because preservation is always an issue.[54] The foragers and farmers of Amazonia must have exchanged cultigens through marriage into families outside of the local community. The husband and wife brought plants to their home for cultivation. Experimentation and curiosity may have motivated these new unions to pursue agriculture. Plants were regarded as communal assets and were shared with neighbors to build solidarity. This sense must have developed over the communal eating of corn, cassava, potatoes, sweet potatoes, beans, and other American crops. The status of food plants as communal property must have reinforced their cultivation. A community's familiarity with the local environment led it to select desirable plants and over time to cultivate them.

Even in South America, no crop seems to have garnered as much interest as corn. The fact that corn came to dominate the Western Hemisphere before the arrival of Europeans has fascinated archeologists. Corn succeeded even where tubers and roots were the mainstay of the economy. Corn was important even in tuber-rich Peru. Corn could not have conquered two continents in the pre-Columbian period until it was at least as productive as the local crops. Corn succeeded only by filling a niche among already established crops and by demonstrating its suitability for storage. In Ecuador, the transition to agriculture was a movement toward corn culture. From an early date, corn was more important than cassava. The people of Ecuador grew corn before 2600 BCE. Ecuador appears as well to have early adopted some varieties of beans, cotton, and fruit trees, although it is possible that pre-Columbian people did not domesticate cotton. Ecuador early grew three tubers, the potato not among them. In fact, Ecuador grew more tubers than beans. Important early in Ecuadorian agriculture, corn appears to have played a diminishing role over time. Ecuadorian agriculture was much too diverse to have settled on a single staple, a fact that may explain why corn, though important, never rose to the status of staple. In this way, Ecuador guarded against catastrophe. If one crop failed, Ecuador could count on others. About 350 CE, the situation appears to have changed, with corn becoming increasingly important, providing a larger part of the diet than beans, tree fruits, roots, and tubers. By the sixth century CE, corn trended downward again, only to rebound later. During these centuries, cassava increased in importance. Corn remained attractive because of its suitability for storage.

NOTES

1. Alphonse de Candolle, *Origin of Cultivated Plants* (New York and London: Hafner, 1964), 1–7; Nikolai I. Vavilov, World Resources of Cereals, Leguminous Seed Crops and Flax, and Their Utilization in Plant Breeding (Moscow: The Academy of Sciences of the USSR, 1957), 77–90.
2. Charles Darwin, 1993. *The Origin of Species by Means of Natural Selection or the Preservation of Favored Races in the Struggle for Life* (New York: Modern Library) 24–64.
3. Sarah Milledge Nelson, "Megalithic Monuments and the Introduction of Rice into Korea," in *The Prehistory of Food: Appetites for Change*, eds. Chris Gosden and Jon Hather (London and New York: Routledge, 1999), 147.
4. David R. Harris, "Origins and Spread of Agriculture," in *The Cultural History of Plants*, eds. Ghillean Prance and Mark Nesbitt (New York and London: Routledge, 2005), 14.
5. William van Zeist, "Evidence for Agricultural Change in the Balikh Basin, Northern Syria," in *The Prehistory of Food: Appetites for Change*, eds. Chris Gordon and Jan Hatler (London and New York: Routledge, 1999), 353.
6. Harris, "Origins and Spread of Agriculture," 15.
7. Ibid, 16.
8. Andrew Sherratt, "Cash-Crops before Cash: Organic Consumables and Trade," in *The Prehistory of Food: Appetites for Change*, eds. Chris Gosden and Jan Hatler (London and New York: Routledge, 1999), 23.
9. Genesis 3:1-3:24 (New Revised Edition).
10. Sherratt, "Cash-Crops before Cash," 26.
11. Carol Palmer, "Whose Land Is It Anyway? An Historical Examination of Land Tenure and Agriculture in Northern Jordan," in *The Prehistory of Food: Appetites for Change*, eds. Chris Gosden and Jan Hatler (London and New York: Routledge, 1999), 291, 295–296.
12. Randi Haaland, "The Puzzle of the Late Emergence of Domesticated Sorghum in the Nile Valley," in *The Prehistory of Food: Appetites for Change*, eds. Chris Gosden and Jan Hatler (London and New York: Routledge, 1999), 398.
13. Ibid, 409.

14. Ann Butler, "Traditional Seed Cropping Systems in the Temperate Old World: Models in Antiquity," in *The Prehistory of Food: Appetites for Change*, eds. Chris Gosden and Jan Hatler (London and New York: Routledge, 1999), 463; Cultural History of Plants, 19.
15. Harris, "Origins and Spread of Agriculture," 19.
16. Jack R. Harlan, *Crops and Man*, 2d ed. (Madison, WI: American Society of Agronomy, 1992), 140–141.
17. Haaland, "The Puzzle of the Late Emergence of Domesticated Sorghum," 411.
18. Harlan, *Crops and Man*, 140–141.
19. Harris, "Origins and Spread of Agriculture," 17.
20. Ibid, 18.
21. Nelson, "Megalithic Monuments and the Introduction of Rice into Korea," 148.
22. Ibid, 150.
23. Ibid, 160.
24. Catherine D'Andrea, "The Dispersal of Domesticated Plants in North-Eastern Japan," in *The Prehistory of Food: Appetites for Change*, eds. Chris Gosden and Jan Hatler (London and New York: Routledge, 170).
25. Ibid, 171.
26. Harris, "Origins and Spread of Agriculture," 19.
27. Nelson, "Megalithic Monuments," 170.
28. Harris, "Origins and Spread of Agriculture," 19.
29. Christopher Samuel, "From the Swamp to the Terrace: Intensification of Horticultural Practices in New Caledonia, from First Settlement to European Contact," in *The Prehistory of Food: Appetites for Change*, eds. Chris Gosden and Jan Hatler (London and New York: Routledge, 1999), 252–263.
30. Robert Kuhlken, "Warfare and Intensive Agriculture in Fiji," in *The Prehistory of Food: Appetites for Change*, eds. Chris Gosden and Jan Hatler (London and New York: Routledge, 1999), 270.
31. F. Hassan, "Earth Resources and Population: An Archaeological Perspective," in *How Humans Adapt: A Biocultural Odyssey*, ed. D. Ortner (Washington, D.C.: Smithsonian Institution Press, 1983), 204.
32. Edmond De Lanche and Pierre de Maret, "Tracking the Banana: Its Significance in Early Agriculture," in *The Prehistory of Food: Appetites for Change*, eds. Chris Gosden and Jan Hatler (London and New York: Routledge, 1999), 377–392.
33. K. L. Mehra, "Subsistence Changes in India and Pakistan: The Neolithic and Chalcolithe from the Point of View of Plant Use Today," in *The Prehistory of Food: Appetites for Change*, eds. Chris Gosden and Jan Hatler (London and New York: Routledge, 1999), 139.
34. Ken Thomas, "Getting a Life: Stability and Change in Social and Subsistence Systems on the North-Western Frontier, Pakistan, in Late Prehistory," in *The Prehistory of Food: Appetites for Change*, eds. Chris Gosden and Jan Hatler (London and New York: Routledge, 1999), 306.
35. Genesis 4:1-4:8 (New Revised Edition).
36. Harris, "Origins and Spread of Agriculture," 21–23.
37. Ibid, 21.
38. Robert S. MacNeish, "Ancient Mesoamerican Civilization," *Science* 143 (1964): 531–537.
39. Harris, "Origins and Spread of Agriculture," 21.
40. Ibid, 22.
41. Kristen J. Gremillion and D. Stein, "Adaptation of Crops in Evolutionary Perspective," in *Evolutionary Ecology and Archaeology: Applications to Problems in Human Evolution and Prehistory*, eds. Jack Broughton and Michael D. Cannon (Salt Lake City: University of Utah Press, 2010), 341.

42. Ibid, 343.
43. Deborah M. Pearsall, "People, Plants, and Culinary Traditions," in *The Oxford Handbook of North American Archaeology*, ed. Timothy R. Pauketat (New York: Oxford University Press, 2012), 82.
44. K. Renae Barlow, "Predicting Maize Agriculture among the Fremont: An Economic Comparison of Farming and Foraging in the American Southwest," in *Evolutionary Ecology and Archaeology: Applications to Problems in Human Evolution and Prehistory*, eds. Jack Broughton and Michael D. Cannon (Salt Lake City: University of Utah Press, 2010), 372.
45. Robert L. Anemone, *Race and Human Diversity: A Biocultural Approach* (Upper Saddle River, NJ: Prentice Hall/Pearson, 2011), 129.
46. Barlow, "Predicting Maize Agriculture," 376.
47. Ibid, 385.
48. Pearsall, "People, Plants, and Culinary Traditions," 82.
49. Mark D. Mitchell, "The Origins and Development of Farming Villages in the Northern Great Plains," in *The Oxford Handbook of North American Archaeology*, ed. Timothy R. Pauketat (New York: Oxford University Press, 2012), 360.
50. Ibid, 380.
51. Christine A. Hastorf, "Cultural Implications of Crop Introductions in Andean Prehistory," in *The Prehistory of Food: Appetites for Change*, eds. Chris Gosden and Jan Hatler (London and New York: Routledge, 1999), 36.
52. Ibid.
53. Ibid, 42.
54. Ibid, 43.

5 Rise of Empiricism in Antiquity
An Overview

Empiricism is the common sense position that the senses do not deceive us. That is, they are the direct and true path through which humans obtain knowledge about the world. So powerful and self-evident does this proposition seem that it is difficult to imagine objections to it. Greek philosopher Plato, however, made clear his objections in *The Republic*, a dialog in which he places a group of men in a cave, chaining them so that they can see only one wall of the cave.[1] Behind them is a fire that they cannot see and a person who passes a series of images behind them, but they can see only shadows of these images, which through their senses they perceive as reality. This is the human condition in which the senses cannot apprehend reality. All they can manage is a shadowy, impermanent, inconstant sense of what they erroneously perceive as reality. In fact, they are twice removed from reality because the fire that provides illumination is not the ultimate source of light. One must break free, pass the fire, and ascend to the opening of the cave and into the sunlight, which is the ultimate source of light. Even then, the person who manages to reach the opening of the cave will need time to adjust to the light. If he descends back down the cave to tell his comrades that he has apprehended reality as it is, they may be apt to kill him for attacking what they perceive to be knowledge derived from the senses. How then does one apprehend reality? Plato gives his answer in the *Phaedo*, in which the intellect seeks its own space, where apart from the body and the senses, it contemplates, by reason alone, reality as it is.[2] That is, the intellect alone and apart from the senses is the source of knowledge about reality. At a deep level, Plato struck a chord. All religions assume the existence of a reality beyond the senses, and Christianity borrowed heavily from Plato's ideas.

In contrast, Aristotle (Figure 5.1), Plato's most gifted pupil, turned against his mentor, returning to common sense empiricism. Aristotle's convictions were important because, more than Plato, Aristotle was a practicing scientist, whose work made empiricism the foundation of the sciences. Aristotle's science has had no lasting contribution to modern science, but his empiricism remains a bedrock of science.

Here our interest is the rise of empiricism as a guide to the formation of the plant sciences in antiquity. Two traditions are important. The Chinese took an intense interest in the plant sciences, particularly as they related to the practice of agriculture. The Chinese tradition may be the world's oldest application of empiricism to plant agriculture, appearing to predate the works of the Greeks and Romans, if only by a few centuries.[3] The agricultural treatise was the vehicle for imparting knowledge in both China and the Greco Roman world. The output was prodigious in China, where an average

FIGURE 5.1   Aristotle: Plato's brightest pupil, Aristotle was the leading empiricist of his day.

of one new treatise appeared every 5 years. The average masks fluctuations. Prosperity and peace resulted in the writing of many agricultural treatises. Economic and political decline and warfare nearly ended these literary efforts. In the West, the reader is likely to be familiar with the contributions of the Mediterranean world, specifically Greece and Rome. If this world did not produce as many treatises, those that have come down to the present influenced the practice of agriculture into the Enlightenment. Time and again, the Greek and especially the Roman writers emphasized that their recommendations derived from their own experiences as practical farmers.[4] They were empiricists at the core. Keen observation and practical insights set these writers partly apart from the literary exercises of other writers, though even poets like Virgil and statesmen like Cicero had something to say about agriculture.[5]

GRECO ROMAN AGRICULTURAL WRITERS

Greece may lay claim to producing the first agricultural treatises in the West. Theophrastus authored these works. Significantly, Theophrastus was Aristotle's most promising pupil. From Aristotle, Theophrastus gained his notion of a scientific method of sorts and of the primacy of empirical knowledge. Through his voluminous writings, Theophrastus may be accounted the founder of botany as a science, although his considerations were not purely academic. His third volume of *On the Causes of Plants* devoted much of its substance to agriculture.[6] The historian, however, finds little unequivocal material about Theophrastus. No contemporary commentary exists. Aristotle wrote nothing about his pupil, although this is not surprising for a man who wrote so little about his own mentor Plato. Although he wrote a series of biographies on eminent Greeks and Romans, Plutarch did not include Theophrastus. The first records about the latter date to about 400 years after his death. Diogenes records that Theophrastus was born in 379 BCE at Eresos on the island of Lesbos.[7] Taking root in Athens, Theophrastus first studied under Plato and then under Aristotle. His study of the science of classification, what one may call taxonomy, informed his work on plants, leading to the conclusion that Theophrastus was the first in the West to treat botany as a rigorous science. When Aristotle died, Theophrastus took charge of his teacher's manuscripts and appears to have headed the Lyceum, Aristotle's creation. Theophrastus' treatment of plants was so varied perhaps because Macedonian conqueror Alexander the Great's army brought him many specimens from as far away as India. Various ages are given for Theophrastus' death, Saint Jerome giving the improbable sum of 107 years.[8]

The first Roman to concern himself with a practical and empirical treatment of agriculture was Marcus Porcius Cato (Figure 5.2), known as Cato the Elder to distinguish him from his grandnephew Cato the Younger.[9] Roman biographer and historian Plutarch deemed Cato the Elder of such importance that the historian wrote a biography of the plebian turned large landowner and statesman. Cato observed food plants since childhood on his father's farm. A man of stern and traditional values, Cato praised the virtues of Rome's farmers. By his own account, Cato amassed so large an estate that he needed slave labor to cultivate the land. In this respect, Cato was ruthless, making clear that he valued slaves only for the labor they produced.[10] A sick or an injured slave, and probably old ones too, should receive less food than those fit for labor because they contributed less to the farm. The farmer should sell old or injured slaves. This sort of language invites comparison with agricultural slavery in the New World, both systems being unjust and exploitive. There remains the question of just how prevalent agricultural slavery was in the Greek and Roman world. The recognition that Greece and Rome were aggregates of large numbers of small farms suggests that there cannot have been many agricultural slaves in Greek and Roman antiquity. Cato disdained Greek culture as effeminate, preferring simple rustic values. Although he amassed vast wealth, Cato claimed to dislike the Roman plutocrats of which he was a member. Cato must have been a prolific author. According to one account, he was the most distinguished Latin author and the first Latin author to write about agriculture.

Perhaps, the best-known Roman agricultural writer and certainly the most politically active was Marcus Terentius Varro, who one might consider a bridge between Cato and Columella.[11] Born in the Italian town of Reate of Sabine ethnicity, Varro enjoyed an upbringing of great wealth and all the opportunities it afforded. A voracious reader, Varro entered politics as an equestrian, an order of nobility. He studied under a tutor in Rome and another in Athens, Greece, perhaps at the Academy

FIGURE 5.2   Roman Farming Scene: Idyllic farm scenes were a popular subject of Roman art.

that Plato had founded in the fourth century BCE. As an ambitious young man, Varro allied himself with military commander Pompey the Great, serving as tribune, aedile, and praetor, all important offices in the Republic. This alliance with Pompey might have proved dangerous as Julius Caesar worked to consolidate power in his hands. At the time of the First Triumvirate, Pompey allied with Caesar and Crassus, a ruthless senator, Varro followed the instructions of Pompey and Caesar that he assign plots of land to retired veterans so that they could earn a livelihood as farmers. In 76 BCE, Varro accompanied Pompey's army to Spain and in subsequent campaigns against pirates in the Mediterranean Sea. In 49 BCE, Varro was again in the field, this time commanding troops against Caesar. Caesar won the battle but spared Varro, a fortunate circumstance at a time when the victor usually killed or enslaved the vanquished. Varro rejoined Pompey in Greece, although he did not see action. When Varro returned to Rome, he met Caesar again. Events went well enough that Caesar assigned Varro the task of creating a library in the capital to rival what the Greeks had established in Alexandria, Egypt. In 43 BCE Varro found himself in trouble. Antony, Caesar's able general, turned against Varro, ordering his death. Antony's forces dismantled the library that Varro had worked so hard to build. Fortunately, for Varro, Octavian, Caesar's grandnephew and the man who would become the first emperor as Augustus, remembered Varro's service to Caesar and hid him from Antony. Varro thus spent his last years in solitude. A prolific writer, much of Varro's output dates to these years in seclusion. Some ancients believed that Varro was as prolific as Pliny the Elder. Varro claimed to have written 500 books, though most have not survived. German philosopher Friedrich Nietzsche's mentor Friedrich Ritschl estimated that Varro wrote about 620 books. The three extant books on agriculture represent one-third of the nine that Varro wrote. Aside from these books, only six others survive, none being complete. Varro wrote history, letters, technical books of which his agricultural output is a part, biography, philosophy, politics, military affairs, music, medicine, rhetoric, and grammar. Varro appears to have come late to agriculture, taking up his pen at age 80. His *On Agriculture* has a charming dedication to his wife Fundania, who had just bought a farm and turned to Varro for practical advice. Indeed, Roman law permitted women to own property. Farming was not simply a male activity. The treatise takes the form of a dialog, though Varro's refusal to give each speaker his own paragraph or paragraphs makes the dialog difficult to follow. One cannot imagine that *On Agriculture* compares well with any of Plato's dialogs. It is clear from the treatise that Varro had read Cato's *On Agriculture* and that subsequently Virgil, Columella, and Pliny would all read Varro and Cato.

Comparatively little is known about Lucius Junius Moderatus Columella. He lived in the first century CE, although his dates of birth and date are not firm.[12] Born in southern Spain, Columella spent part of his childhood on his uncle's farm, where he began to amass first-hand knowledge of agriculture. At one point in adulthood, Columella owned five farms in Italy. He may have authored 16 books. The survivors are his 12-volume treatise *On Agriculture* and a treatise on trees. Xii Modern additions of *On Agriculture* pack all 12 volumes into three, with each of the modern volumes holding four "books." *On Agriculture* is the most comprehensive treatise about agriculture in Latin. Among the works that no longer survive is a treatise on the religious rituals that should accompany the tasks of agriculture.

PLACE OF AGRICULTURE IN THE GRECO ROMAN WORLD

Climate and soils governed many aspects of agriculture in the Mediterranean Basin.[13] The climate was hot and dry during summer and cool and rainy during autumn and winter. The northern provinces had a cooler climate, heavier soils, and more rain. Olives and grapes could be grown at elevation in the basin and grains on the lowlands. Soils in the northern provinces tended to be fertile clays or loam. The northern provinces could not grow olives, and grapes were less widespread. North Africa, Egypt, and southwestern Asia provided a surplus, largely grain, and olive oil for distribution to the capital. The Romans believed agriculture to be the greatest profession, praising rural values of hard work, piety, and thrift. Early in the Republic, Rome was a collection of small family farms, the ideal that Thomas Jefferson would promote in eighteenth and nineteenth century America.

Southern Italy and Etruria, once the land of the Etruscans, cultivated grain. The conflict between Rome and Carthage in what is modern Tunisia during the Republic stemmed partly from competition for olive oil and wine markets. Rome converted Carthage to grain lands. Sicily, Sardinia, and parts of Spain also produced grain. Early in the Republic, Rome monopolized grape production by forbidding its culture outside of Italy. Some Romans, Columella among them, wondered aloud whether wheat would remain profitable in competition with wine and olive oil. In the early empire, Augustus permitted the provinces to grow grapes. Both grapes and olives prospered throughout the Mediterranean Basin. Gaul emerged as an important grape region. Spain and North Africa specialized in olives. The provinces may have caused a glut of wine and olives on the market. North African olive oil tended to be cheaper than Italian oil. Spain boasted higher quality than Italy could match.

The troubled third century CE led to economic and especially agricultural decline. To avoid taxes, some farmers abandoned their land. The emperors in this period encouraged private citizens to rehabilitate the land, and Diocletian required workers of all classes, including farmers and farm laborers, to remain at work and to pass down their profession to their children. By law, the children of farmers could only be farmers. Herein lay the roots of serfdom. Not everybody fared horribly. In southwestern Asia, farmers abandoned only about 1% of their land. Conditions were more serious in North Africa. In the fourth century, Britain righted the ship and improved agriculture.

The Mediterranean Basin concentrated, as climate dictated, on dry farming with its emphasis on water conservation. The two-field system was prevalent. One might plant grain one year and fallow the land the second year. In many cases, the Romans favored planting between October and December to take advantage of the winter rains. Because weeds competed for water and nutrients, the Romans appear to have spent an inordinate amount of time weeding their fields, the hoe ever at their side. The grain harvest occupied June and July, with the sickle the only reaping instrument. The land, then fallowed, received many plowings—Pliny the Elder recommending nine—and hoeing to eliminate weeds. Next October, the land would be ready for planting again.[14]

The use of dung as fertilizer was important, but it may have been in shortage. By the first century CE, the more enterprising farmers introduced legumes in rotation to improve soils. One might rotate faba beans with grains. Campania in Italy may have had soils of sufficient fertility to permit year-round cultivation. Olive trees were sometimes planted with grains. For this purpose, grapes were seldom planted. With time, the emphasis swung from grains to olives and grapes, a trend one perceives as early as the writings of Cato the Elder. To compensate for the lack of manure, the Romans used the residues of plants to make compost. In Spain, tree fruits, including the olive, and grapes tended not to need irrigation. As a rule, the Romans did not irrigate grains or legumes. In some areas, it was possible to tap water from aqueducts for irrigation. This circumstance required farmers to purchase water from the state. North Africa, Egypt, and Syria relied on irrigation.

Grains and legumes were broadcast. Grapes, olives, nuts, and other fruit trees—figs and dates, for example—were planted. Wheat, spelt, and six-row barley were sown in autumn. The Romans do not appear to have favored barley, perhaps because they consumed wine rather than beer. Only when the grain crop failed did Rome attempt to grow wheat and spelt in spring and summer. Egypt, southwestern Asia, and Gaul, but not Italy, appear to have planted millets. Rye was easy to grow but did not satisfy the Roman palate. Stockmen fed alfalfa and vetch, both legumes, to cattle. Alternatively, cattle might graze on the stubble of fields or on fields of oats, barley, and emmer wheat. The Romans ate turnips, although the bulk of the crop went to livestock.

The fig was an important fruit, especially in winter. Turkey, Greece, and Italy grew figs. The Roman diet tended toward vegetarianism, with a reliance on faba beans, turnips, rape, radish, carrot, pea, lettuce, almonds, and hazelnuts. The last two were renowned in southern Italy. Central Italy specialized in apples, pears, cherries, and peaches. The nursery was important to fruit culture. Olive seeds or cuttings, planted in the nursery, were transplanted to their permanent location only at age 5. Because olive trees took several years to mature, they required an investment of time and effort in expectation of a deferred reward. Grapes too were started in the nursery. Grapes transplanted facing the south received the maximum sunlight. Italy tended to trellis vines, although in the provinces support was atypical. Olive trees and grapevines were pruned in October or spring. August witnessed the harvest in southern Spain and September in Italy and Gaul. Itinerant labor took in the harvest. The Romans grew grapes for table and for making wine. A grape vineyard might cover 0.3 ha. Olive groves may have been a little larger. Gaul pursued a mixed agriculture of grain, olives, and other tree fruits and grapes. Spain specialized in grain, the olive, and the grape. Sicily and Sardinia provided grain for Rome, although the islands had grapes and fruit trees. In Sicily and Greece, grain production declined as ranchers converted grain land to pasture. In both places, olives and grapes remained important.

In the late empire, Pannonia was a grape and wine exporter. Dalmatia specialized in grain. Early, Pliny commented on the growth of massive estates. He mentioned that six men owned half of North Africa. At least in Italy, archeological evidence points to the existence of small farmers throughout the imperial period. Egypt became so important as Rome's granary that the emperors, beginning with Augustus, declared it their personal property. No private Roman citizen, especially not a Senator, could enter Egypt without an emperor's permission. In North Africa to the west, farmers grew grain near the coast where rainfall was more plentiful. They irrigated olives, which they grew almost as far south as the Sahara Desert. Tenants farmed much of this land. Rome used agriculture partly as a political and administrative tool, urging nomads to settle on farms, which were easy to tax. Olives were also important in semiarid Syria. Other areas of the Levant produced dates, grapes, and olives. Irrigation must have been important in these regions. Irrigation was crucial in Egypt, whose farmers tapped the Nile River. The narrow strip of fertile land along the Nile supported year-round agriculture. By the first century CE Rome imported one-third of its wheat from Egypt. The Levant had wheat and possibly rice under irrigation. If the empire really grew rice, it might have acquired the species native to sub-Saharan Africa. The Nile Delta grew papyrus, flax, hemp, and cotton. With irrigation, Syria produced peas, beans, cabbage, radish, plums, dates, pistachios, and figs. Grapes and olives crowded the coast. Romans prized Syrian wine. Olive trees could not grow as far north as Britain or in northern Gaul. Britain was too cold for grapes, at least in Roman antiquity. Gaul produced both grapes and grains. Even in Roman times, the Gallic region of Bordeaux was prized for its grapes and wines. According to Pliny, the people of Gaul were the first to attach wheels to the plow.[15] The Roman plow lacked a moldboard for turning the soil and did not penetrate the soil deeply, making it well suited to the thin soils of the Mediterranean Basin. The plowshare might be wood or iron. Two oxen pulled each plow. The farmer must have pressed down on the handles so that the share penetrated the soil.

EMPIRICISM AND THE FIRST STIRRINGS OF THE PLANT SCIENCES IN THE WEST

Theophrastus, the earliest agricultural writer in the West, divided plants into two categories. Nature had created wild plants, but cultivated plants depended on human agency to attain the fullness of perfection "to achieve [their] goal."[16] Theophrastus observed that some trees and shrubs "reject cultivation." Careful in his use of terminology, Theophrastus (Figure 5.3) defined agriculture as an art that supplements the work of nature. The differences between plants are not surprising because various plants have different formal, material, final, and efficient causes, to borrow the lexicon of Aristotle.[17] The aim of the farmer is to bring forth food in quantity and

FIGURE 5.3 Theophrastus: Aristotle's most consequential student, Theophrastus was arguably the founder of botany.

quality. Plants that reject cultivation cannot be domesticated because the attempt deviates from the nature of such plants.[18] Nature places plants in their proper location so that they are suited to their soils, climate, and rainfall. Theophrastus admitted that humans do not know why all plants are not subject to cultivation. Agriculture is a matter of shaping plants to satisfy human needs. Farmers must use reason rather than simply ape what others did in the past. It is possible, however, for a farmer to succeed without knowing why. The complete farmer uses both reason and habit in forming his art. Because lands and the climate vary, one cannot rely on generalizations. The expert must tailor specific advice to a particular region. Theophrastus advised farmers not to rely on astrology to determine when and what to plant, a foible that Columella retained into the first century CE. Facts must undergird precepts. The use of irrigation and manure and the preparation of the soil were the universals for plant growth, although different plants might need the application of other methods.[19]

In his *On Agriculture*, a single volume that reads as a set of instructions that details the day-to-day operations on a farm, Cato asserted that the finest Romans were farmers.[20] Varro believed that the farmer was superior to the urbanite because the city dweller was lazier than the farmer.[21] Conscious of this divide, aristocrats spent 7 of every 9 days on their country estates. Yet, Varro might have reflected on the fact that the aristocrats delegated all the work to their subordinates and so were really not engaged in farming. The countryside, Varro believed, was healthier than the city, a statement that was probably true given the nature of contagion. Varro lamented that he lived in an era when so many Romans abandoned the plow and sickle, taking more interest in the theater than the farm. Roman agriculture had come to such straits that Rome had to import grain from North Africa and Sardinia. The reason for this change, however, stemmed from Rome's transition to a market economy. The farms near Rome sold fruits and ornamentals because these fetched high prices. Varro reminded the reader that shepherds had founded Rome and had taught their children to farm. Yet in this age, farmers planted pasture where they once grew grain. This is not the optimistic vein with which Varro treated agriculture throughout most of his treatise. Rome might prosper without theater and dance, but not without agriculture, noted Columella.[22] Neither plundering from warfare nor trade were respectable livelihoods. Money lending, oratory, and law were little better. Agriculture alone was the noble pursuit, one that Columella for the first time called a "science."[23] The Roman aristocrat will not dirty his hands in labor, but only close observation and careful study of one's own farm may make agriculture comprehensible. The man who farms his own land is best equipped to learn about agriculture. Columella feared that urban Rome was turning away from its rural roots.

He recounted these rural roots through the legend of Cincinnatus.[24] Cincinnatus was Rome's ablest farmer who possessed the rural values of thrift, hard work, and piety. When crisis arose in the form of an invasion of Rome, the Senate and the people went to Cincinnatus' farm. There they found him plowing his land. Because of his many virtues, the Senate made him dictator, whereupon Cincinnatus commanded the army, defeating the invasion. Putting aside his dictatorial powers, Cincinnatus returned to his farm. From this story, Columella drew the lesson that Rome's greatest military leaders were farmers. Columella believed that Romulus, the founder of Rome, was the village's first farmer. Columella was steeped in the past, when Rome was truly a land of farmers. Nothing was more useful to humans than farming, Columella believed.[25]

Varro believed in the first century BCE that Italy was more intensively and extensively cultivated than any other land.[26] He claimed that agriculture was more prosperous in the northern than in the southern hemisphere. This cannot have been an empirical statement because Varro gave no indication that he had ever observed agriculture in the southern hemisphere. Italian agriculture was more fruitful than Asian farming. Here again, one is not sure of Varro's conception. Given his wealth, it seems possible that he might have observed farming in the Levant, but one must presume that he knew nothing about agriculture in East Asia or probably South Asia for that matter. According to Varro, Italy's warm climate benefited agriculture. Italy surpassed every other land in the production of grapes, spelt, wheat, and olives. As for fruit trees in general, Italy was an "orchard." This boastfulness suggests that agriculture, at least in Italy, was robust in the first century BCE. The contrast with Columella may be significant. A century later, Columella appears to have had none of Varro's enthusiasm. Varro recognized Gaul as an important region for grapes.

Varro warned the farmers to keep goats away from crops, especially grapes and olives, because the animals will eat their fill at his/her expense.[27] Young plants are particularly susceptible to voracious goats. Here one encounters the problem between the agriculturist and the stockman, a conflict that comes to a bad end in Genesis. This conflict may be why the Romans sacrificed goats to Bacchus, the god of wine and all it connoted. Varro observed that Athens forbad the entrance of goats to protect its surrounding olive groves, except the goats that would be sacrificed to the gods once per year. Like Cato, Varro assumed the use of slaves as a matter of course. Varro believed that agriculture was both an art and a science, a judgment that made sense in antiquity. As an art and a science, agriculture teaches one what to plant in each type of soil and how to prepare and cultivate land with the aim of amassing the largest yield. In this sense, then, agriculture was a quantitative and qualitative science. The farmer should have two objectives: profit and pleasure. A farm should thus be a source of sustenance and enjoyment. Attend to profit first in anticipation that pleasure will follow. Be sure to plant fruit trees, among them olives, for both profit and enjoyment. Seek an aesthetic experience because the more attractive a farm is, the higher will be its price upon sale. The application of science to agriculture lessens the risk of farming. In this context, Varro rightly named Theophrastus as the founder of plant sciences. Where the climate is warm, plant oaks upland and almonds and figs in the lowlands. Plant grains on flat terrain and grapevines in the hills. The mountains should contain trees for timber. As a science, agriculture is a progressive activity with yields increasing over time, so that the ancients cannot compare to the moderns of Varro's age. Nut trees should be part of any farm. Vines and olive trees in rows yield better than that in a haphazard planting. The soil determines a farm's fecundity and narrows the choice of what one may plant. A freeholder, a tenant, a sharecropper, and a slave may all labor on a farm.[28] Varro preferred hired labor to slaves, favoring the appointment of an overseer to manage the farm and its labor. The overseer must always be more knowledgeable than the subordinate. He must impart knowledge by both word and example so that the laborer may learn how to accomplish a task.

Good farmers come to their knowledge from experimentation, a point that Columella would likewise emphasize.[29] Here is the core of empirical knowledge of agriculture. The experimentalist will plow at different depths to learn the results of these trials. The farmer should experiment with the number of plowings per year and with other methods of weeding to know what is best.

SOIL DEPLETION AND ITS REMEDIES

The first to investigate the soils and their properties in the West was Theophrastus. The cultivation of the ground increases what plants derive from the soil and how the food tastes.[30] Edible plants serve as food and medicine. Lupines are important in restoring poor soils, although if planted on rich soils, they produce leaves and stems at the expense of pods. Plant legumes very early; otherwise, they will impede summer plowing.[31] Peas are an early legume, though Theophrastus does not mention them in this context. Plant beans before the rain.[32] When it comes, it will nourish bean seedlings. These must have been faba beans. Chickpeas will not germinate if soaked in brine. Theophrastus understood that plants obtain their food from the soil, though he appears to have been unaware of the role of sunlight in plant nourishment.[33] Manure thin soils more and fertile soils less because grains on too fertile a land will produce too much straw and may lodge.[34] Mix light soils with heavy and heavy soils with light, a dictum that portended a law of opposites. These hybrid soils benefit plants more than uniform soils do. Mix depleted soils with clay to restore fertility. Every fourth or fifth year, dig the soil deeply enough to expose the subsoil. Bring subsoil to the surface so that it too has a chance to feed plants. This work must have been arduous. Theophrastus admitted the lack of consensus about whether or how to manure the land. One possibility is to dig it into the ground, possibly with the aid of a plow or perhaps a spade for small plots. Another method is to lay manure atop the soil, letting it decompose. Theophrastus advised against the second method on the grounds that the manure will lose its potency if left to rot. "Pack animals"—perhaps Theophrastus had in mind the donkey—produce the best dung. Next, in value is the manure of ox, sheep, goat, pig, and human. The farmer may manure the land more in wet weather than in dry and less on thin soils.[35] Human dung is too strong for trees and should be reserved for vegetables. Pig manure sweetens pomegranates and almonds. Fruit trees benefit from urine, but Theophrastus does not specify the source. Human dung may be applied to olive trees. As a rule, favor less potent to more potent manure. Just as each plant is suited to a particular soil, so it is suited to a type of manure.

Whatever the crop, Cato advised planting it facing the south to maximize exposure to sunlight.[36] Cato advocated the application of manure on the soil to increase yields. The expectation must have been that farmers as a matter of course kept livestock and so had access to manure. Yet the Roman diet, at least for the masses, contained little meat, suggesting that livestock must have been in shortage. Cato ranked manures, preferring pigeon dung on the garden, grains, olives, and grapes.[37] Goat, sheep, and cattle manure were suitable for fruit trees.

Columella acknowledged that he wrote in a day when soil depletion was claimed to be ubiquitous.[38] The translation speaks of "soil exhaustion," a phrase that reappeared repeatedly in the pages of nineteenth century European and American agricultural articles and books. I prefer the phrase "soil depletion" because it seems more accurate, given that plants absorb elements through their roots. If these elements are not replaced, plants will deplete soils of them. The word "exhaustion" conjures the image of a human or another animal worked to the point of excessive fatigue. Strictly, soil cannot experience fatigue or become tired. Columella did not believe that the problem of soil depletion was insurmountable. If soils were depleted, the fault lay with the carelessness or inadequacy of knowledge about agriculture. Every field of study has experts who train the next generation of practitioners. This is no less true of agriculture than medicine. Although Columella did not use the language, he was attempting to systematize farming, an important step on the road to science. Because agriculture had not been systematized, it was making little progress. The knowledge of agriculture was a type of "wisdom."[39] For the transmission of this wisdom, Columella envisioned the need for schools of agriculture, exactly what the nineteenth century would demand.[40] Columella understood that one could study agriculture on the farm and in the literature.[41] He credited the Greeks, including Greek philosopher Socrates' pupil Xenophon, Aristotle, and Theophrastus. Columella acknowledged Cato as the first Latin stylist to study agriculture and elevated Virgil as the poet of farming. The Roman Senate, according to Columella, so esteemed Carthaginian agriculture that it financed the translation of 28 Punic agricultural treatises into Latin. Yet, observation was the key to understanding agriculture. Long practice fitted men to instruct others in agriculture. It does not suffice merely to read the master farmers, but to gain one's own experience. The master of the farm is akin to the general of the army. In the second volume of his *On Agriculture*, Columella christened agriculture a "science," although he admitted that the field of study had grown so vast that no one person could master every aspect of it.[42]

Columella extolled the value of fallowing land and recommended plowing under leaves and other organic matter to restore soil fertility.[43] Manure was indispensable in this regard. Varro appears to have been among the first to study the effects of various types of manure. In his experience, birds provided the best dung.[44] He judged pigeon dung the finest, although thrushes and blackbirds were also important sources of manure. Human excrement ranked second to that of birds. Third ranked the manure of goats, sheep, and donkeys. Varro thought little of horse manure, although it was suitable for spreading on pasture or grain land. He understood that some land was fertile enough not to need manure. Columella grouped manure into three kinds: bird, human, and cattle, ranking bird dung the best.[45] Within this category, Columella rated dove manure best and the dung of hens second. In contrast, Columella believed the dung of ducks and geese to be inferior. Pigeon dung, human dung, and human urine were all valuable. The fact that Columella resorted to human excrement may lead one to infer that manure from other sources was in shortage. Human urine should be aged 6 months and then applied to fruit trees and grapevines. Human urine was also desirable on olive trees, a point Columella must have added for emphasis because the olive is a fruit tree. After cattle manure, the dung of sheep, goats, and pigs is a poor fertilizer. Where manure is unavailable, plow under leaves and other plant residues. Do not age manure because it loses potency. Spread manure on lands in February. Grain land should be manured again in September in preparation for

October planting. If no bird dung is available, one may resort to goat dung. Frequent applications throughout the year are more efficacious than a single large application. Add more manure to wet soils than to dry. Columella remembered that his uncle, Marcus Columella, did not manure grapevines because the resulting grapes made inferior wine. His uncle judged that human excrement enabled grapevines to produce superior wine. In November, Columella recommended the application of 18 loads of dung per iugerum, a portion going to olive trees.[46] Save pigeon dung for the grapevines, although human urine is also acceptable. Prefer level soil to hilly terrain and loose soil rather than compact.[47] The value of cultivation, and here Columella means the working of the soil, is to pulverize what had been compact soil. Avoid dry, infertile soils. This precept must have been difficult in the Mediterranean Basin where so much land was dry. Consider Syria as an example. Columella warned against dry, infertile soils because they were difficult to work. Even time and effort yielded little. Such land might be too poor even for pasture.

Even where manure was at hand, the ancients understood the value of legumes. Cato accorded special attention to lupine, a legume more important in antiquity than at present.[48] Cato regarded the chief value of lupine in feeding livestock, although humans might also eat it. Lupine had the virtue of tolerating poor sandy soils. Cato urged the reader to plant faba beans, the only beans known to the ancient Mediterranean Basin before the Columbian Exchange, on fertile soil. Strictly, this advice should not have been necessary. The roots of legumes, beans among them, have roots that nitrifying bacteria inhabit. These bacteria convert gaseous nitrogen, oxygen, and hydrogen to create nitrate or ammonium ions, a source of nitrogen that plant roots readily absorb. Because legumes enrich the soil, it should not have been necessary to plant beans on fertile soils. Cato recommended that the farmer plants the legume vetch on newly cleared ground. Vetch would have been suitable as livestock feed or fodder. Lentils should be sown on poor soil, sensible advise for this legume. Cato believed that chickpeas weakened soils, but provided no evidence for this erroneous belief. As a legume, chickpea, as we have seen, would have enriched the soil with nitrogen. Cato faulted barley for depleting soils, but observed that lupine, bean, and vetch enriched the soil.

Varro favored clover and other legumes, except the chickpea, on thin soils.[49] Plow under lupines and beans to enrich the soil. Legumes may enrich the soil to the point that it does not need manure. Plant beans and grains on dry land and asparagus in the shade.

Columella favored a number of legumes, including beans, lentils, peas, chickpeas, lupines, clover, vetch, and "kidney" beans.[50] The last is troubling. The authority K. D. White is no help, although it is clear that the American kidney bean cannot be meant because the Romans had no knowledge of American beans. Columella's peculiar usage must have meant something to him, though today the language is obscure. Columella preferred lupine among the legumes because of its ease of care. He believed lupine improved the soil more than any other legume, terming it "an excellent fertilizer."[51] Lupine may be planted on poor soil in preparation for the establishment of a vineyard or on land to be sown later to grain. Planted in September or October, lupine is a fine food, although livestock are the chief consumers of this legume. If lupine has not germinated by winter, the crop may be lost. Apparently observing a lupine, Columella understood that each seed contained the embryo of a miniature plant. Preferring alkaline soils, lupine does well on poor soils. After lupine, Columella favored the mysterious kidney bean, which he planted on fallowed land or year after year on fertile ground. Peas do well on pulverized soil and should be sown in warm, wet weather. This advice seems odd, given that the pea does well in cool moist weather. Plant the "common bean," apparently the faba bean, on manured land or fallowed land.[52] Plant legumes more deeply than grains by covering them with a layer of soil. Legume roots must penetrate deeply. Columella recommended adding 24 loads of manure per iugerum, plowing the manure under and then planting beans.[53] Beans do not thrive in cold weather. Some farmers plant beans on poor soil to enrich it, though Columella did not. He preferred a grain–fallow system rather than a grain–bean rotation. The worse the land, the more beans one should plant, a maxim that appears to contradict Columella's own practice. Beans do well on heavy soils. Do not plant beans after the winter solstice or in spring, although, curiously, they may be planted in February. Beans store well. Columella recommended the sowing of lentils on poor soil. Prior to sowing, the soil should be repeatedly plowed and dry. One might also plant lentils on fertile ground. Plant lentils at the same time as one manures the land. Lentils may be planted in February. The farmer should sow chickpeas in January or February on fertile, moist ground. In some parts of Italy, chickpea is planted about November 1. The legume is excellent in restoring soil fertility, but Columella admitted to getting only poor yields from the chickpea. He was aware of two varieties of chickpea; the second, *Punicum*, may be sown in March and does well in fertile soils. Plant vetch about September 20 as forage, with a second planting in January or a little later.[54] Plant vetch on fallowed land for best yields with that planting after the sun has evaporated dew from the land. One might interplant forages, clover and vetch, for example. Columella did not think it necessary to hoe lupine.[55] The crop was hardy enough to compete with weeds. In any case, he feared the hoeing lupine might injure its weeds. Columella understood that weeds were deleterious because they competed with crops for water, sunlight, and soil nutrients. Hoe beans thrice during the growing season. Columella observed that beans, peas, and lentils flower about 40 days after germination. Quoting one authority, Columella understood that lupine, bean, vetch, lentil, chickpea, and pea improve soils.[56] So valuable is lupine in this regard that it may take the place of manure. Columella recommended the planting of lupines in vineyards to improve the soil.[57] As a rule, plant lupines early and harvest late.[58]

PLOWING, WEEDING, AND PLANTING

Theophrastus likened germination to giving birth.[59] The soil is pregnant with seeds as a woman is pregnant with child. Planting in spring has the advantage of exposing the seedling to ever lengthening days. Summer brings abundant sunlight. Here Theophrastus appears to acknowledge that sunlight is important in the growth of plants. Plants have an "impulse" to grow. This impulse directs roots to spread underground and shoots to grow above ground. This impulse is really the expression of the genotype. Roots seldom grow below the portion of soil warmed by sunlight, an observation Columella repeated. Because humans tend to plant in spring, they must be receptive to some factor in plants that responds to spring planting. In this sense, the human is the tool of the plant. Where summers are rainy, plant before or after the rains. Columella cites Ethiopia, India, and Egypt as examples, although Egypt is very dry. His breadth of knowledge to include Egypt, India, and Ethiopia is impressive but is not likely empirical. The deeper and stouter the roots, the better the plant will be. The soil must be loose to allow the maximum penetration of roots. As Columella would repeat, Theophrastus advised digging a hole a year before planting, though he cites no empirical evidence for this practice. Theophrastus recommends this procedure because the soil in the hole receives sunlight in summer and experiences the cold of winter. Nature loosens the soil and "warms" it, ideal conditions for germination.[60] Plow in both summer and winter, allowing one to infer that Theophrastus favored multiple plowings.[61] Plowing loosens the soil and eliminates weeds. Soil that receives grain should be barren of weeds. Plow in spring, summer, and before sowing grain in autumn. Where possible, break the soil with mattocks. The people of Thessaly, Theophrastus noted, had a tool with which they dug the soil to a great depth. Those who cultivate the soil well and often find that they save labor by keeping weeds at bay. The opposite is true of slothful farmers. One must be vigilant against weeds because their early germination may ruin the crop.

In the hands of Cato and Columella, the plow was a tool for weeding, so frequent was its use. Indeed, plowing should thus exterminate weeds as to make harrowing unnecessary. This, of course, is a dry farming method used to suppress weeds whose competition for water and nutrients imperiled the crop. Columella recommended plowing in April, July, and September, presumably as a prelude to an autumn planting.[62] The first plowing should be either horizontal or vertical with subsequent plowing at right angles. Manure the land between the first and second plowings at a rate of 18–24 loads of manure per iugerum of land. Plow immediately after manuring the land. Columella understood that manure was a type of "nourishment" for the soil. Columella also appears to have known that the roots emerge from a seed before the shoot.[63] He preferred to plant in dry soil with the seeds awaiting rain. Referring to his own experiences, Columella declared that neither would birds eat nor ants carry away seeds that have been planted in dry soils. Experience dictated that spelt fared better than wheat in dry soils.[64] Columella sowed wheat at 1 2/3 bushels per acre and spelt at twice the density. Admitting that the authorities had no consensus about sowing rates, Columella offered his experience as the guide, noting that climate and soils may require adjustments to these rates. Columella recommended sowing grains in early spring with an additional planting later in the season, unusual advice given Italy's Mediterranean climate. Sow winter wheat and spelt in winter on heavy soils and barley on loose dry soil. Columella recommended the seemingly ubiquitous two-field system with grain one year and fallow the next. Sow more seeds on heavy alkaline solids and more seeds when planted near the shade of trees. Northern latitudes permit spring sowing. Some varieties of wheat, barley, spelt, and beans do well in hot, dry conditions. Columella favored the application of pigeon dung to lands that had not habitually yielded a good harvest. Evincing some knowledge of entomology, Columella understood that some insects, evidently the larvae, chew plant roots. In another case, he saw heredity as a fickle process in which good plants might yield poor offspring and that poor plants will always yield poor progeny. Here, Columella appears not to acknowledge the role of human agency in selecting for better, higher yielding seeds, the basis of plant improvement throughout antiquity. Columella affirmed his line of thought by quoting Virgil. Strictly, this is an appeal to authority rather than to empiricism. Moreover, Columella believed that the third generation of seeds always yielded poorly. Again, it is difficult to find empirical justification for this rule.

Columella appears to have been indifferent toward barley: in one place listing it as only a famine food but in another place praising its utility in both good and bad soils. He recommended planting barley after the second plowing, although whether Columella meant spring or autumn sowing is unclear. Columella observed that barley was less domesticated than wheat because it shattered if harvested late.[65] Land should be fallowed after a barley harvest. One might interplant wheat and barley to create maslin. The two may be sown in January or March. Sow panic grass and millet on thin soils in spring. Hoe repeatedly because these crops compete poorly with weeds. This advice must have been particularly true given a spring sowing. Millet stores longer than other grains and makes an acceptable porridge but not bread.

Columella advised the deep planting of hemp on manured land in March.[66] He noted that commoners often subsisted on turnips. The reality seems to have been, however, that the masses ate chickpeas, although turnips may still have been important. Humans and livestock ate turnips according to Columella. Stockmen in Gaul favored turnips. Columella recommended sowing turnips on pulverized loam, but never on clay. Doing well on moist soils, turnip may be planted after the summer solstice, a period when the Mediterranean Basin would still have been dry. Manure the soil, planting turnips perhaps at the end of August or in early September. Columella appears to have been on the cusp of understanding that organic matter such as leaves rotted in the soil to provide nutrients.[67] Columella expected that farmers would devote a portion of land to pasture in the expectation that they would keep livestock.

THE OLIVE TREE, OTHER FRUIT TREES, AND GRAPEVINE

Theophrastus advised the plant of seeds, and especially trees, when the soil is warm.[68] At such times, the earth desires seeds. A moist, warm, temperate climate favors agriculture. Accordingly, plant trees in spring and autumn, with the planting of trees concentrated in spring. Plant pear and apple trees in spring. The lateness of fall, with its lingering warmth from the Mediterranean summer, makes planting appealing. Dig deep holes for olives and figs because their roots are short.[69] Theophrastus recommends putting straw in the hole before planting to let water collect at the bottom and to cool roots in summer. One wonders whether such practices have an empirical foundation. In this context comes the insight that plants derive food from the sun and water. If one wishes to plant a tree from a cutting, take the cutting from a young tree or one in its prime. Select a cutting from a tree in the same soil in which one wishes to plant the cutting or from a tree on poor soil so that the cutting, now in fertile soil, will benefit from the soil's potency. If the tree faces south, so must the cutting. This advice applies to other orientations. Do not do anything that does not conform to the nature of the tree or its cutting. Select a cutting from a low branch for ease of removal. For grapes and figs, however, take a higher cutting. Plant cuttings deep to encourage root growth. Here again, Theophrastus emphasized that plants feed from the soil. Where the climate is hot, inundate a hole 2 days prior to planting. Apply sand or loose soil to holes that drain poorly. Figs prefer semiaridity. The Laeonian fig, however, tolerates a wetter climate. Theophrastus feared that the wetter the climate, the fuller the figs will be of water. They will not ripen or will rot. Grapes prefer hilly land although some varieties do best on flat terrain. Plant fruit trees at lower levels of a hill. Land that supports pasture or has access to water is suited for grapes. Fertile land is best for olives and figs. Plant fruit trees on loose, moist, light soils. Almonds must have thin soils and plenty of sunshine. Chestnuts prefer shade. Too much wind may stunt trees. Plant pomegranate, myrtle, and bay 9 ft apart. Prune grapevines every year because of rapid growth.[70] Prune fruit trees every second or third year. After pruning grapes and trees, manure the soil. Theophrastus even advised the pruning of roots. Of course, this labor would have been arduous. One wonders whether it had any empirical justification. Theophrastus thought that the pruning of shallow roots would send the remaining roots deeper in the soil. Theophrastus even thought proper pruning of olive and grapevine roots, plants noted for producing comparatively shallow roots. Theophrastus remarked that the larger the roots the better, leading one to wonder why he would ever advise their pruning. Theophrastus noted that trees lost water in hot weather, although he was centuries away from the role of transpiration in the loss of water. Manure grains more than trees because trees need less manure as they produce less fruit in aggregate than grain produces seeds. Too much manure will injure both trees and grains. Apply manure to grapevines no more than every third year. Use potent manure and urine on myrtle and pomegranate trees. As in the case of pomegranate and almond, cultivation may sweeten the fruits.[71] Theophrastus inferred that this sweetening stemmed from changes in the trees' roots. Farmers accomplish this transition by improving the soil, through the application of manure, for example. Irrigation may change flavor as well. In this context, Theophrastus encourages the application of warm water to apple trees. These alterations may cause stone fruits (e.g., peach, apricot, plum, and date) not to produce the pit. Here is the idea, and probably not an empirical one, of seedless fruits before their derivation in modernity. Theophrastus attempted to define sexuality in date palms, but the effort was fruitless because he had no knowledge of the role of flowers in sexual reproduction.

Grapes, figs, and olives should not be grown together. Figs and olives are heavy feeders and may shade grapes. Do not plant almonds with other fruit trees because it is also a heavy feeder, and its roots may entangle the roots of other trees. Apple and pomegranate are good companions because they do not feed voraciously. Some farmers grow grapevines near fig trees for support.[72] Theophrastus noted that red grapes were in cultivation more than white ones, although it is difficult to know whether this practice extended beyond the fourth century BCE Greece. Grapevines need more water than several other crops. Where the climate is wet, plant grapes about October 20. Spade or plow the land deeply in spring so as to incorporate as much air and water into the soil as possible. Theophrastus recommended the application of ashes to fig trees and brine to the date palm, noting that the latter grows on saline soils in Libya and Syria.[73] He was aware that most plants did not tolerate salinity. Here, Theophrastus came close to the insight that saline tolerance must be an adaptation to saline soils. He remarked that the Babylonians added salt to soils on which they grew date palms. From this information, Theophrastus inferred that the roots of date palms must differ from those of many other plants. Date palm roots must have a way of filtering out salt.

The farmer, according to Cato, should make olives and grapes his/her priority, setting the grape above all else.[74] Yet, in his/her list of priorities, the garden comes before the olive trees. Cato understood that the olive (Figure 5.4) is a fruit.[75] Grains rank only sixth among crops. Cato emphasized the importance of keeping an account of grape, olive, and grain yields to determine the productivity of one's land. He regarded grapes and olives as cash crops, and may have had the same view of grains, although wine and olive oil must have yielded the greatest profit. Cato urged the reader to plant olive trees close together (one tree every 25–30 feet), perhaps to maximize the yield of a plot of land being that a densely planted region must yield more than a sparsely planted region.[76] Olives favor heavy "warm" soils, according to Cato. Many varieties must have existed, of which Cato mentioned seven. Some varieties were suited for thin soils. Olives should face west and enjoy full sunlight, wrote Cato. Manure olive trees, presumably at transplantation, allotting one-quarter of one's manure to this job.[77] Cato must have planted at least some olives from seeds, although he recommended the grafting of olives, figs, pears, apples, and grapevines.[78] Land to be transplanted to olive seedlings should be plowed in the days leading up to this event, according to Columella.[79] Once olive seedlings have been transplanted, one may plow so near an established tree to cut its lateral roots in hopes of encouraging the growth of the deeper roots.

Varro advised the apportionment of land for grapevines, gardens, olive trees, pasture, grain, timber, and an orchard. Avoiding the extremes of waterlogged and arid soils, moist ground was best for planting.[80] Clover and cabbage might be grown in a nursery and then transplanted. Plant at lower densities on poor soil and at higher densities on good soil. Be an empiricist in observing what, when, and how your neighbors plant. Apply these lessons on your farm. Varro desired land suitable for fruit trees that will bear two crops per year. The eastern Mediterranean coast near Smyrna, Varro tells his wife, is noted for figs.[81] The Smyrna variety of fig remains a staple in the region. In addition to Smyrna, Consenta is known for apples. Well-cultivated soils produce the best quality of fruits. Varro noted the quality of Spanish wines, although he worried that grapevines may require too much labor for profit, a position that neither Cato nor Columella mentioned. Varro favored heavy warm

FIGURE 5.4   Olive: olive oil was the chief dietary fat in the ancient Mediterranean Basin.

soils for olives, which should face west with full exposure to sunlight.[82] Cypress trees might support olive trees and especially grapevines. Varro believed that grape hybrids yielded early, though how could the Romans have known of hybrids when they did not know that plants were capable of sexual reproduction? The vines exposed to the most sunlight should be harvested first. Grapes from a single variety may be suitable as table and wine grapes, suggesting that distinction need not be drawn. Table grapes may be stored in jars.[83] Store apples in cool, dry environs underlain with straw. Varro writes of the storage of apples in concrete structures, although one doubts that many ordinary farmers had access to concrete. Others stored apples on blankets of wool. Store pomegranates in a layer of sand. Varro recommended boiling of pears, a practice that may seem odd. Why not eat them fresh like apples or table grapes? Others dried pears in the sun, a practice akin to drying figs and dates. Varro noted the practice of buying fresh fruits in Rome, a practice that must have meant that nearby farms specialized in fruits. One may preserve olives in brine, a practice that Cato had mentioned. Age wine one year.

Pick olives where possible rather than shake a tree. Shaking causes olives to hit the ground, where they bruise and yield inferior oil. Harvesters should not wear gloves because these tend to bruise olives. Shaking a tree is inevitable if one is to harvest the highest olives. In this context, it is better to beat braches with a reed rather than a stick because a stick or pole may damage a tree.[84] Varro understood that olives bear heavily in one year and lightly in the next. As with grapes, save a portion of olives for direct consumption with the rest going to oil. Olives must be pressed before fermentation ensues. Fermentation produces rancid oil.

Columella observed that grapes (Figure 5.5) might be grown on thin or heavy soils.[85] Exposure to full sunlight was essential to their success. The best soil should form the nursery, from which grapevines may be transplanted in their third year.[86] Vines find their perfection in grapes.[87] Columella judged the grape among the most important crops in the empire, although they will not tolerate the extreme climate of the northern provinces. In his experience, Columella found that grapevines tolerate a variety of elevations and soils. Grapes came in many varieties, each adapted to a particular soil and climate. His experience appears to have been limited to grapevines planted from seeds. Columella understood that the grape was a type of berry.[88] Warm climates brought the highest yields. Columella observed that grapes did best planted in loose soil, although he did not discount the possibility of planting on compact clays. The soil should be neither too fertile nor too infertile. Dry weather is

FIGURE 5.5 Grapevine: by fermenting grapes, Romans made wine, the chief alcoholic beverage in the ancient Mediterranean Basin.

superior to wet where grapes are concerned.[89] Plant grapes for both table grapes and wine. Farms near a city sell table grapes. Select table grapes for appearance and flavor. Columella knew four varieties of table grapes, two of which stored well in jars. Other farms should concentrate on wine. The vines that yielded the most grapes are most suited for producing wine. One should aim for a grape whose juice is not too fruity. Columella cited the "Aminean varieties" as the best grapes. Columella made clear that Aminean was not a single variety, citing two "sister" varieties among them. Columella preferred small vines for trellising, noting that small vines tended to produce more fruit per unit than large vines. He noted the suitability of Mount Vesuvius near Pompeii for growing grapes, allowing one to date his treatise before the eruption of 79 CE. Columella's emphasis was on wine more than table grapes. Columella followed Cornelius Celsus in placing two categories of wine grapes ahead of table grapes in value. Along with Vesuvius, Pompeii, in general, was a fine region for growing grapes. Where grapevines were unproductive, Columella faulted the farmer for carelessness, greed, or lack of knowledge. Here again, one witnesses Columella's emphasis on practical knowledge, particularly that gained from experience "through the keener eyes of the mind."[90] Columella understood the danger of excessive soil fertility where one wished to grow grapes.[91] Columella called for "experimentation" in viticulture, a principle that would come down to modernity.[92] Agricultural experiment stations promote the value of experimentation to the present. Whereas others claim knowledge from false reasoning, Columella upheld the value of personal experience.[93] He urged the planting of oaks and chestnuts to support grapevines, with planting in November.[94] Columella advised the farmer to plant about 2000 grapevines per iugera.[95] He knew many varieties of grapes in the provinces, although because he had direct knowledge only of Italian grapes, Columella confined himself to what he knew.[27]

Columella judged the olive as "queen of all trees."[96] He recommended the olive because of its easy care. Little effort was necessary to achieve good yields. Here again, Columella confined himself to the 10 varieties of olive trees he had grown. By his account, Posia and Royal yielded the best olives for eating.[97] Orchia and Shuttle were better for food than for oil. Where the climate is warm, plant olives facing the north. Where cool, plant them to the south. A moderate slope is best for the trees. Posia was the most heat-tolerant variety and Sergian the most cold-tolerant. Sandy soils or soils with gravel are fine for olive trees. Plant olives first in the nursery, keeping in mind that they do poorly on alkaline soils. As a rule, land that yields grain is suitable for olives. Columella favored the planting of cuttings taken from the best trees. Here again is an example of human selection, the basis of plant improvement for millennia. Here also, Columella makes no reference to propagation by seeds. Transplant cuttings into the grove in their fifth year in autumn in the anticipation of rain. Where spring is wet, transplant then. Transplant cuttings 60 ft apart where the land has yielded grain in past years. Put manure into holed in which a cutting will be transplanted. Columella recommended an allotment of six pounds of goat dung, though he did not state the number of trees for this amount. Prune olive trees every eighth year to stimulate olive production. Pig or human urine is good for olive trees.

Cato urged his reader to plant apple, quince, pomegranate, pear, and fig trees in alkaline soils. Apples and pears may be planted from seeds.[98] Columella urged the cultivation of "many fruits and vegetables," choosing among the many variet-ies of figs.[99] Fig leaves should feed cattle. Cato suggested that fruit trees, including olives and figs, and elms, pines, and cypress should first be planted in a nursery.[100] Transplantation would occur later, though Cato did not specify a date. He recom-mended against transplanting seedlings when wind and rain threatened. Cato urged the planting of figs, olives, apples, and pear trees and grapevines in spring. In the Mediterranean Basin, spring must have marked the end of winter rains and the onset of dry, hot weather. Columella stated the obvious in warning farmers not to plant figs where the climate is cold.[101] Plant fig trees in a sunny spot. The trees will tolerate thin, rocky soils. Where the climate is cool, plant figs that ripen early to take advan-tage of the short growing season. Several fig varieties are suitable for Italy and North Africa.[102] Plant almond trees about February 1. The soil should be warm and dry. Roots will rot in wet soils. Plant almonds, and perhaps figs as well, from seeds. Plant walnut, pine nut, and chestnut in March. Plant pomegranate no later than April 1, adding human urine to the soil. Plant pear trees in autumn. Columella was familiar with a number of pear and apple varieties. Plant apple and plum trees in winter no later than February 13. Plant mulberry trees from February 13 to June 20. Plant peach trees in autumn. Columella discussed the process of grafting the branches of these trees, but wrote nothing about the choice of rootstock. He implied that grafting was an ancient practice. In late October, transplant olives and other tree seedlings and prune fig trees.[103] Varro favored the grafting of fig branches onto rootstock about June 20.[104] Cherries might be grafted in the middle of winter. Columella implied that fruit trees, among them the olive and date palm and also grapevines, were planted from seeds, although he allows the possibility of vegetative propogation.[105]

GRAINS AND OTHER CROPS

Theophrastus urged the planting of grain on dry land before the rains.[106] The later in the season the sowing, the more seeds should be used. The later the sowing, the less developed the roots will be. Wheat and barley may be planted in rainy, cool cli-mates. Soil that is too poor for barley should be planted to wheat.[107] In this context, Theophrastus puts forward what one might call a theory of climate, asserting that a harsh climate produces "strong" plants, whereas mild climates yield "weak" plants. Some varieties of barley do well on sandy soil. Wheat does better than barley in a wet climate. Wheat outperforms barley on lands devoid of manure. Farmers with a choice should prefer grains to legumes. Rather than rotation, Theophrastus preferred to fal-low soil, noting that wheat yielded better than barley on fallowed land. He suggests that lands with too mild a winter will produce wheat susceptible to rust or lodging.

In this context, Theophrastus was an early writer about plant diseases and pests. Rust was an ancient fungal disease that afflicted the stem and leaves of grains, notably wheat. He observed that a lack of wind caused rust. In truth warm, humid air carries rust fungi, which alight on grain plants. In a more promising vein, Theophrastus asso-ciated rust with the presence of moisture, perhaps dew, on the leaves of wheat and bar-ley plants. Theophrastus observed that Achilles barley was vulnerable to rust, whereas Eteocrithos barley had some level of tolerance. Theophrastus reasoned that the wind helped evaporate the dew, lessening the danger from rust. In the field of entomology,

Theophrastus understood that peas and chickpeas had pests.[108] Grubs assailed wheat, feeding on the roots. Vines and olive trees are subject to insect damage.

Cato recommended the sowing of grains on heavy, fertile soil away from the shade of trees.[109] In this context, Cato distinguished between sowing and planting. One sowed small seeds by broadcasting them over a field. By digging, however, one planted a tree, such as the olive, or a vine such as the grapevine. Under the category of grains, Cato appears to have included millet and panic grass. Varro favored fertile ground for cabbage, winter wheat, spring wheat, and flax.[110] He observed that barley germinated about 7 days after planting.[111] Legumes, millet, and sesame germinated some 4 or 5 days after planting, though beans were slow to germinate. By his observations, Varro understood that roots appear before the shoot. Roots penetrate the ground to the depth that sunlight warms the soil. Varro understood that heliotropes tracked the apparent movement of the sun across the sky. Varro differentiated the grain from the husk, beard, and sheath. Saving the best, largest seeds for next year's planting, Varro demonstrated the potency of human selection. Varro advised against storing grain in a pit, in contrast to the practice of corn growers in pre-Columbian North America. Construct a granary to house grain. Other farmers stored grain in caves, Thrace providing an example. Grain is stored in wells in parts of North Africa and Spain. In North Africa, this practice descends from the Carthaginians of what is today Tunisia. Stored in these ways grain may keep 50 years and millet more than 100 years. These comments appear not to be empirical, but rather the amassed knowledge of two or three generations of humans. Spelt was a winter food. Beans and other legumes store "for a very long time." Varro recommended the storage of legumes in jars.

Columella esteemed the grains as the most important food of humankind.[112] He favored wheat and spelt. He lavished praise on a now-obscure type of wheat. He also favored winter wheat, commenting on its suitability for making bread. Listing four types of spelt, Columella noted that spelt tolerated more rainfall than wheat. He urged the sowing of wheat and spelt in October, though admitting the permissibility of a November sowing with the view that the plants germinate before the onset of winter rains. Columella understood that some other farmers sowed grain only after a period of rainfall. Columella treated millet, barley, and panic grass apart from wheat and spelt. According to Columella, grain land should be plowed to a depth of 2 ft.[113] If grains were no longer important in Italy, to Columella's disappointment, it remained a crop in Egypt and Sudan. Follow turnips with grains to improve yields.[114] Columella believed that turnips should be principally a food for humans, with clover, vetch, barley, and oats reserved for livestock. Barley is an excellent feed for oxen and may be grazed throughout spring. If barley is to feed humans and livestock, harvest a portion in March for human consumption and let livestock graze the rest. Planted in autumn, perhaps October, oats too are suitable for humans and livestock. Livestock should receive an early cutting, whereas humans should eat a fully ripe crop. Hoe wheat, barley, legumes, and spelt when the plants are still immature. Hoe grains before or after the plants flower with the aim of harvesting grain 40 days after flowering. He noted that flax depletes the soil and that sesame tolerates wet soils and so may be planted in October on loam. In this regard, Columella quoted Virgil that oats and flax deplete soils. Columella observed that millet and panic grass likewise deplete soils. Where soil depletion was a problem, he recommended applications of manure and the planting of legumes. This advice would resonate throughout the Middle Ages and early modernity. Columella advised the harrowing of land planted to millets at the end of September.[115] Grain, especially wheat, may be planted in October. One iugerum takes 4–5 modi of two-grained wheat, 9 or 10 modi of emmer wheat, 5 or 6 of barley, 4 or 5 of millets, 8–10 of lupines, 3–4 of peas, 6 of broad beans, 1 of lentil, 8–10 of linseed, 3–4 of vetch, and 4–5 of sesame.

Rape and turnip may be planted near grain fields according to Cato.[116] Cato observed that asparagus did well on wet soils. Cato recommended the manuring of turnips, kohlrabi, and radish.[117] He valued cabbage as food and medicine, believing that the vegetable aided digest and kept one regular.[118] Cato served cabbage with vinegar. Cato planted vegetables and ornamentals for the local market. Varro favored the planting of ornamentals for aesthetic appeal, but he must have been aware that plants with enticing flowers were valuable to urbanites who wanted fresh cut flowers.[119] In this context, Varro noted that violets did well in full sunlight. Columella expected farmers to grow hemp, sesame, and flax.[120] Columella believed that the rosebush was an essential ornamental.[121] Cato and Columella thus witnessed a transition in agriculture from self-sufficiency, which they claimed to prefer, to a market economy. Because Rome was such an important market, the farms near it raised ornamentals rather than grain because the former claimed a higher price. Rome in turn had to find grains elsewhere, Sicily, for example, and later Egypt. Spelt should be sown on alkaline soils. The farmer should plant spring and winter wheat on high ground with full sun exposure.[122] Barley may be cropped year round on newly cleared soil. One expects that manuring must have been necessary to promote continuous cultivation. Varro understood that others favored grain on fertile land. Varro did not limit fertile land to a particular crop, finding rape, turnip, millet, and panic grass acceptable.[123] Beans should be planted at 4 modi per iugerum, wheat at 5, barley at 6, and spelt at 10.[124]

In September or February, Columella advised the planting of cabbage, lettuce, artichoke, coriander, dill, parsnip, and poppy.[125] Where the climate is warm, plant these crops in January. In autumn, plant garlic, onion, and mustard. In February, plant rice, asparagus, leek, and garlic at the end of the month. In March, transplant leeks. In April, plant cucumbers and gourds. Parsley does well in summer, although it must be irrigated. Plant turnips and radishes in August. Plant garlic by the clove and transplant cabbage when it has six leaves. Columella recommended the transplantation of many crops that are not transplanted today. Plant cabbage and lettuce in the garden.[126] At the end of September, plant turnips and vetch on dry soil. Both should feed cattle in the winter.

NOTES

1. Plato, *The Republic*, in *The Great Dialogues of Plato*, trans. W. H. D. Rouse (New York: New American Library, 1956), 312–315.
2. Plato, *Phaedo*, in *The Great Dialogues of Plato*, trans. W. H. D. Rouse (New York: New American Library, 1956), 467–469.
3. Gang Deng, *Development versus Stagnation: Technological Continuity and Agricultural Progress in Pre-Modern China* (Westport, CT: Greenwood Press, 1993), 1–2.

4. Lucius Junius Moderatus Columella, *On Agriculture*, vol. 1, trans. Harrison Boyd Ash (Cambridge, MA: Harvard University Press, 1968), 363.

5. K. D. White, *Roman Farming* (Ithaca, NY: Cornell University Press, 1970),

6. Theophrastus, *De Causis Plantarum*, books III–IX, trans. Benedict Einarson and George K. K. Link (Cambridge, MA: Harvard University Press, 1990), 3.

7. R. W. Sharples, *Theophrastus of Eresus: Sources for His Life, Writings, Thought and Influence*, vol. 5: *Sources on Biology (Human Physiology, Living Creatures, Botany: Texts 328–435* (Leiden: E. J. Brill, 1995), 126–127; "Introduction," in Theophrastus, *Enquiry into Plants*, vol. 1, ed. G. P. Gould (Cambridge: MA: Harvard University Press, 1999), xvi.

8. "Introduction," in Theophrastus, xxii.

9. "Introduction," in Marcus Porcius Cato, *On Agriculture*, trans. William Davis Hooper (Cambridge, MA: Harvard University Press, 1993), ix–xi; K. D. White, *Roman Farming* (Ithaca: Cornell University Press, 1970), 19–20.

10. Cato, *On Agriculture*, 9.

11. "Introduction," in Marcus Terentius Varro, *On Agriculture*, trans. William Davis Hooper (Cambridge, MA: Harvard University Press, 1993), xiv–xvii; White, *Roman Farming*, 22–24.

12. "Introduction," in Lucius Junius Moderatus Columella, *On Agriculture*, vol. 1, trans. Harrison Boyd Ash (Cambridge, MA: Harvard University Press, 1968), ix–xii; White, *Roman Farming*, 26–28.

13. Sian Rees, "Agriculture and Horticulture," in *The Roman World*, vol. 2, ed., John Wacher (London and New York: Routledge, 1987), 481–483.

14. Ibid, 485.

15. Ibid, 498.

16. Theophrastus, *De Causis Plantarum*, vol. 3, 3.

17. Ibid, 5.

18. Ibid, 9–19.

19. Ibid, 141.

20. Cato, *On Agriculture*, 3.

21. Varro, *On Agriculture*, 307–309.

22. Columella, *On Agriculture*, vol. 1, 7.

23. Ibid, 9.

24. Ibid, 13–15.

25. Ibid, 23.

26. Varro, *On Agriculture*, 169–171.

27. Ibid, 179–195.

28. Ibid, 225–227.

29. Columella, *On Agriculture*, vol. 1, 233.

30. Theophrastus, *De Causis Plantarum*, books III–IX, 7–9.

31. Ibid, 153.

32. Ibid, 187.

33. Ibid, 11.

34. Ibid, 147–149.

35. Ibid, 33.

36. Cato, *On Agriculture*, 5, 9.

37. Ibid, 53.

38. Columella, *On Agriculture*, vol. 1, 3, 5.

39. Ibid, 5.

40. Ibid, 7.

41. Ibid, 31, 35–37.

42. Columella, *On Agriculture*, vol. 2, 3.

43. Columella, On Agriculture, vol. 1, 107, 109.
44. Varro, *On Agriculture*, 263, 265.
45. Ibid, 195–205.
46. Columella, *On Agriculture*, vol. 3, 121.
47. Columella, *On Agriculture*, vol. 1, 11–113.
48. Cato, *On Agriculture*, 51–53.
49. Varro, *On Agriculture*, 241–243.
50. Columella, *On Agriculture*, vol. 1, 139.
51. Ibid, 157.
52. Ibid, 159.
53. Ibid, 161–169.
54. Ibid, 175–177.
55. Ibid, 183–185.
56. Ibid, 193.
57. Ibid, 205.
58. Columella, *On Agriculture*, vol. 3, 117.
59. Theophrastus, *De Causis Plantarum*, books III–IX, 23–29.
60. Ibid, 41.
61. Ibid, 153–156.
62. Columella, *On Agriculture*, vol. 1, 131, 135–137.
63. Ibid, 143.
64. Ibid, 145–153.
65. Ibid, 155–157.
66. Ibid, 169–171.
67. Ibid, 207.
68. Theophrastus, *De Causis Plantarum*, books III–IX, 3.
69. Ibid, 31–51.
70. Ibid, 61, 67–77.
71. Ibid, 131–137.
72. Ibid, 87–97.
73. Ibid, 127–131.
74. Cato, *On Agriculture*, 5–7.
75. Ibid, 69.
76. Ibid, 11, 17–19.
77. Ibid, 47.
78. Ibid, 53, 57, 59.
79. Columella, *On Agriculture*, vol. 1, 123.
80. Varro, *On Agriculture*, 275.
81. Ibid, 197–199.
82. Ibid, 245–247.
83. Ibid, 295–299.
84. Ibid, 291.
85. Columella, *On Agriculture*, vol. 1, 19.
86. Ibid, 65.
87. Ibid, 219.
88. Ibid, 231.
89. Ibid, 233–247, 251.
90. Ibid, 273.
91. Ibid, 265.
92. Ibid, 289.
93. Ibid, 363.
94. Ibid, 457.

95. Columella, *On Agriculture*, vol. 2, 25.
96. Ibid, 93.
97. Columella, *On Agriculture*, vol. 1, 73–87.
98. Ibid, 67.
99. Ibid, 21, 47.
100. Cato, *On Agriculture*, 45.
101. Columella, *On Agriculture*, vol. 2, 93.
102. Columella, *On Agriculture*, vol. 1, 95, 101–109.
103. Columella, *On Agriculture*, vol. 3, 125.
104. Varro, *On Agriculture*, 265.
105. Columella, *On Agriculture*, vol. 1, 217.
106. Theophrastus, *De Causis Plantarum*, books III–IX, 149.
107. Ibid, 161–165.
108. Ibid, 169, 173.
109. Cato, *On Agriculture*, 17.
110. Varro, *On Agriculture*, 243.
111. Ibid, 277–295.
112. Columella, *On Agriculture*, vol. 1, 137–143.
113. Ibid, 121–123, 139.
114. Ibid, 165–195.
115. Columella, *On Agriculture*, vol. 3, 121.
116. Cato, *On Agriculture*, 17–19.
117. Ibid, 51.
118. Ibid, 141.
119. Varro, *On Agriculture*, 243.
120. Columella, *On Agriculture*, vol. 1, 139.
121. Ibid, 23.
122. Ibid, 51.
123. Varro, *On Agriculture*, 245.
124. Ibid, 275.
125. Columella, *On Agriculture*, vol. 3, 139–145.
126. Ibid, 21.

6 The Columbian Exchange

BACKGROUND

We remember from Chapter 2 that humans settled the Americas last among the habitable continents. When the climate warmed, the polar ice melted, submerging again the Bering land bridge. The humans in the Americas were isolated for millennia from contact with other people. The initiative now lay with Europe. Europeans had long sought pepper from parts of Asia, especially South and Southeast Asia. Crisis arose in the fifteenth-century CE when Muslim warriors conquered Constantinople (now Istanbul, an important city in Turkey). This conquest allowed the Muslims to close the overland and Mediterranean Sea routes by which Europeans had acquired pepper and other spices. With the Muslims now in command of pepper, they could set the price as high as they wished, squeezing Europe. Portugal on the westernmost edge of Europe felt the price increase most acutely. The Portuguese wondered whether they might be able to circumvent the Muslim monopoly by sailing south. Initial prospects looked grim. Ptolemy, the ancient Greek geographer and astronomer, understood, as all Greek intellectuals did, that the world was a sphere. He correctly positioned Africa between Europe and Asia, though he did not believe that Africa had a southern tip. Rather southern Africa simply circled the globe from north to south in a belt of land. Africa therefore could only be crossed by land.

By the fifteenth century, however, some European intellectuals were willing to challenge Ptolemy, thinking him outdated. Among this new breed of Europeans was Portugal's Prince Henry the Navigator. He wished to test the premise that Africa had no southern tip. The prince sent a series of explorers down the western coast of Africa to determine whether Ptolemy had been right. These voyages posted many obstacles, most of them conceptual. For example, Europeans understood that the farther north one went, the colder was the climate. It seemed reasonable that if one went far enough south, one would encounter boiling seawater and land afire. These Portuguese sailors falsified such beliefs and the expeditions continued. Finally, a ship commanded by Bartholomew Diaz was blown off course. When calm resumed, Diaz realized that he had sailed south and then east. He had rounded the southern tip of Africa, known as the Cape of Good Hope. His men were too frightened to go forward, so Diaz returned to Portugal. Explorer Vasco da Gama completed the voyage, sailing south from Portugal, rounding the southern tip of Africa and sailing northeast to India. Portugal had bypassed the Muslims and at last had their own access to the pepper trade.

Meanwhile, Portugal's success had sharpened its rivalry with Spain. If Portugal had sailed east to find pepper, might it be possible to venture west? On this point, the temptation was to believe Ptolemy, whose map of the world suggested that the distance between Western Europe and East Asia was comparatively short and so feasible to cross by sea. No one, certainly not Ptolemy, had dreamed that two continents stood in Spain's way to East Asia. Italian Spanish explorer Christopher Columbus, a vigorous proponent of a westward crossing, finally secured money from Spain for his voyage. He landed not in India, China, or any other part of Asia, but on an island in the Caribbean Sea. He had come upon the Americas, though he refused to believe that he had never set foot on Asia. The Europeans who followed Columbus came to understand the import of his discovery. The rejoining of the Old and New Worlds by sea—what had begun as a search for pepper—changed the world forever. The Amerindians, never having been exposed to the diseases of the Old World, had no immunities and so died in the millions. With little native labor in the Americas, the European masters of these new lands imported African slaves to toil on the new sugarcane plantations in tropical America. Here, for the first time in history arose the differentiation of people as slave or free based on race, though one should add that some biologists doubt the validity of any notion of race. The racism that grew out of the slave system has complicated the subsequent development of the Americas, particularly the United States, where poor African American communities haunt relations between the whites and blacks. The discovery of the Americas also made possible the transfer of plants and animals from one hemisphere to the other. Pigs, cows, horses, wheat, sugarcane, coffee, and many more came from the Old World to the New, and corn, tomatoes, potatoes, cassava, peanuts, beans, sweet potatoes, and many more came from the New World to the Old. The search for pepper thus shaped the world in ways that are difficult to overstate.

It is impossible to trace the trade of plants between Eastern and Western Hemispheres in detail. I have chosen three plants from each hemisphere: sugarcane, soybeans, and rice from the Old World, and potatoes, sweet potatoes, and corn from the New.

FROM THE OLD WORLD

SUGARCANE

Origin and Diffusion in the Old World

Sugarcane (Figure 6.1) originated in prehistory and is among the oldest cultivated plants. At first, sugarcane was not a crop. Rather, the people of Papua New Guinea used the wild species *Saccharum robustum* to thatch their huts. Sometime before 8000 BCE, *S. robustum* hybridized with the grass *Erianthus arundinaccus* to yield the now familiar Creole cane.[1] In turn, Creole crossed with another wild species, *S. spontaneum*, to yield other sugarcane species including a wild cane variety extant in India.[2]

Around 8000 BCE, sugarcane, probably Creole, spread to Melanesia and Polynesia.[3] Among these islands, sugarcane spread to Fiji, Tonga, Samoa, the Cook Islands, Marquises, and Easter Island. In about 6000 BCE, the people of New Guinea brought Creole to India where it hybridized to yield *S. barberi*, a species adapted to marginal environments. In India, people first learned to crystallize sugar by boiling cane juice. In China, Creole hybridized again, this time yielding *S. sinense*, meaning Chinese cane. Between 4000 and 3000 BCE, sugarcane was cultivated in Indonesia and the Philippines in addition to New Guinea, several Pacific Islands, and India. By 7000

FIGURE 6.1 Sugarcane fueled the plantation economy in tropical America after the Columbian Exchange.

BCE, the people of Southeast Asia, evidently excluding Cambodia, grew sugarcane. By 2000 BCE, sugarcane culture was widespread in China and Cambodia. The latter paid sugarcane as tribute to China. By 100 BCE, sugarcane had reached Africa and perhaps Oman and Arabia. By 600 CE, the people of Iran cultivated sugarcane.[4]

The rise of Islam coincided with the expansion of sugarcane culture. The Koran prohibited the consumption of alcohol, though it permitted the taking of sweet beverages. This pronouncement must have aided sugarcane in its spread. Muslims carried sugarcane from China west to the Atlantic coast. In the seventh century, they spread sugarcane to Syria, Palestine, and Egypt, though one account credits one of Alexander the Great's generals with taking cane from India to Egypt in 325 BCE.[5] By 700 CE, farmers were growing sugarcane in Cyprus, Sicily, and North Africa. In 755, Buddhists took sugarcane from India to Japan. Muslims brought sugarcane to Spain in the eighth century and to Crete and Malta in the ninth century. The crop was tended in lands along the southern shore of the Caspian Sea and in Damascus and Jordan by the tenth century. Around 1100, Crusaders returned to Europe with sugar from the Holy Land, though northern and western Europe were too cold to sustain the plant. Around 1285, Italian adventurer Marco Polo reported the growing of sugarcane in China, India, and east Africa, though he did not observe all these lands firsthand. In the fourteenth century, sugarcane culture shifted from Palestine, Egypt, and Syria to Cyprus, Crete, and the western Mediterranean. By the end of the sixteenth century, sugarcane cultivation in the Mediterranean had faltered as the Little Ice Age made the region inhospitable for the tropical crop. Moreover, cheap sugar from Brazil and the Caribbean undercut the price of Mediterranean sugar.[6]

In 1432, the Portuguese planted sugarcane on Madeira.[7] By 1500, the island, the largest producer in the West, yielded nearly 2000 tons of sugar per year. By 1550, the Portuguese island of Sao Tome off the coast of West Africa had replaced Madeira, yielding more than 2200 tons of sugar per year. In 1788, the British introduced sugarcane to Australia, though it would not be an important crop until the twentieth century. In Australia, planters grew sugarcane in New South Wales and Queensland. Today, sugar is Australia's second most valuable export, wheat being the first.[8]

Sugarcane in the New World

In 1493, Christopher Columbus introduced sugarcane to the Caribbean island of Hispaniola (today Haiti and the Dominican Republic).[9] By 1550, sugar totaled the majority of the island's exports. In 1500, the Portuguese began growing sugarcane in Brazil, with exports to Europe after 1519. Production kept pace with demand for sugar. Despite an increase in production, competition, chiefly from Caribbean plantations, was cutting into Brazil's success. In the seventeenth century, Brazilian sugar lost much of its share of the European market. By 1690, Brazilian sugar accounted for just a few percent of sugar produced worldwide.[10]

Not content to let Spain and Portugal have all the profits from sugarcane, the British, French, Dutch, and Danes carved out sugar estates in the Caribbean islands that Spain could not protect from interlopers. In the seventeenth and eighteenth centuries, the European powers established sugar plantations on Barbados, Jamaica, the Leeward Islands, Martinique, Guadeloupe, Saint Domingue (today Haiti), and the Danish West Indies (today the U.S. Virgin Islands). Landowners planted sugarcane to the exclusion of other crops, creating monoculture on several islands. The most successful planters amassed hundreds of acres. The expense of buying land and slaves and building a mill priced the small farmer out of the enterprise of growing sugarcane. By the eighteenth century, sugar totaled half the value of all French colonial exports and was worth more than all other colonial products in Britain.[11]

Preferring the comforts of home, prosperous landowners returned to the mother country to live as gentry. They entrusted the running of the sugar estates to overseers. Paid little, overseers had few incentives to do a good job. Some cheated landowners. Other allowed the property and mill to dilapidate.

By the eighteenth century, Saint Domingue was the jewel in France's crown. In 1789, on the eve of the French Revolution, the colony produced much of the world's sugar, an amount greater than the yield of the entire British Caribbean. Taking the ideals of the French Revolution to heart, the slaves on Saint Domingue revolted in 1791, destroying the sugar estates.[12] The era of French supremacy in sugar production had passed. Between 1820 and 1824, Haiti averaged only several hundred tons of sugar per year, much less than the production before the French Revolution.[13] Sugarcane retains the stigma of slavery in Haiti. Its people may engage in many occupations, but field hand is not among them.

The failure of sugarcane culture in Haiti did not doom the crop elsewhere. In the nineteenth century, planters in Cuba expanded cane acreage at the expense of coffee, tobacco, and ranchland. Before nationalist Fidel Castro seized power, the United States was the largest importer of Cuban sugar. An avid supporter of the sugar industry, Castro set production quotas though growers never achieved them. In 1980, Cuba produced three-quarters of all Caribbean sugar. Cane occupied millions of acres, two-thirds of the island's arable land. Grenada and Puerto Rico achieved a better balance, growing sugarcane, coffee, and cocoa. Into the twenty-first century, the United States has been the principal purchaser of Puerto Rican sugar.

The perception that the Caribbean has diversified beyond sugarcane is not entirely accurate. Although the islands averaged several million tons of sugar between 1934 and 1938, the amount grew between 1976 and 1980. In the 1980s, Cuba, Barbados, Guadeloupe, and Saint Kitts exported the majority of their sugar, with the United States and Europe being ready buyers. In the twenty-first century, sugarcane remained a leading crop in the Dominican Republic, though exports fell between 1981 and 2003.

In 1750, Jesuit missionaries introduced sugarcane to California and the Louisiana territory.[14] California has never been a large sugar producer. In Louisiana, however, sugar plantations played a leading role in the economy and society. The United States bought the Louisiana territory from France in 1803, opening the plantations to U.S. investment. In Louisiana, planters grew sugarcane along the Mississippi River and in the bayous. The culture of sugarcane along the Mississippi River spurred the building of levees to prevent flooding. The largest planters were among the wealthiest Americans in the antebellum period. In the nineteenth century, sugarcane was Louisiana's principal crop. By 1860, Louisiana produced the vast majority of U.S. sugar.[15] As frost damaged cane, landowners replanted the crop every third year. Since World War II, the trend toward aggregation of land has accelerated. Consequently, the number of plantations has diminished through consolidation. Although Louisiana had tens of thousands of sugar estates in 1957, it counted only several hundred in 1995.

Oxen or horses powered the first mills but in 1822 the first steam mill began to operate in Louisiana. By 1854, more than three-quarters of mills in the state were steam powered. Then a mill cost much to build, a price that barred small farmers from cultivating sugarcane unless they could use a neighbor's mill. In antebellum America, internal markets absorbed all the supply. Attentive to planters' wishes, Congress placed a tariff on sugar to keep cheap Caribbean sugar off the market. Since 1950, the trend toward large plantations has accelerated in the United States. Small and medium farms have declined precipitously.[16]

Hawaii, another region of the United States suitable for sugarcane culture, has the world's largest yield per acre.[17] Yields are so good that landowners may harvest two or three ratoon crops without much loss in productivity. The high wages of sugarcane workers have cut profits in recent decades. Farmers have responded by diversifying agriculture, growing macadamia nuts, ginger, papaya, and cacao in addition to sugarcane. In other cases, landowners, alert to the profits to be made in tourism, have converted sugar estates into resorts.

Since 1920, sugarcane has emerged as an important crop in Florida. Planters grow sugarcane on the rich soils south of Lake Okeechobee, relying on the local African American population and Jamaican immigrants for labor. The embargo on Cuban sugar has benefited Florida growers. Today, the state produces a sizable portion of U.S. sugar and the white crystal is second in value only to citrus in the sunshine state. In Florida, planters fallow a portion of cane land and rotate much of the rest with vegetables or rice. As in Louisiana, large plantations are the rule in Florida with most cane grown on parcels of several hundred acres.[18]

New Opportunities and Problems

In 1768, French sailor Louis Antonin de Bougainville discovered a new variety of sugarcane in Tahiti. This new cultivar, Otecheita, yielded one-third more
sugar than the widespread Creole and soon supplanted it in Java, Burma, Mexico,
the Philippines, Hawaii, Jamaica, the French Caribbean, Puerto Rico, and British
Guiana.[19] Otecheita's tenure was brief. In the nineteenth century, the fungal disease
red rot struck, destroying Otecheita fields in Mauritius, the Caribbean, and South
America.

In the late eighteenth century, farmers planted a new cultivar, Cheribon, in
Louisiana and Georgia. Popular and widely grown, Cheribon remained the standard
cultivar in the United States into the twentieth century. After 1840, Cuban planters
grew Cheribon. In 1796, surgeon James Duncan of the British East India Company
discovered a new variety in Canton, China. Botanist William Roxburgh named it *S.
sinense*. Later, another scientist traced Chinese cane to India, though others insist
that the variety is indigenous to China.[20]

The tendency of farmers to propagate sugarcane vegetatively obscured the obvious fact that it is a sexually reproducing plant. Only in 1888 did amateur scientist John Bovell publicize this finding, leading scientists to breed new varieties. By
hybridizing different species of sugarcane, scientists exploited the phenomenon of
heterosis or hybrid vigor. In Barbados alone, sugarcane hybrids increased sugar production by 76% between 1930 and 1939. In Puerto Rico, hybrids generated more than
$10 million.

Since the nineteenth century, scientists have focused on deriving ethanol from
sugarcane to power automobiles.[21] The plant is a good candidate for this function
because it converts a few percent of sunlight into biomass. Although this may appear
to be a small amount, it is larger than the conversion rate of most other plants. In the
nineteenth century, nascent automakers in the United States used sugarcane ethanol,
and India and Brazil now use it in automobiles. In the United States, ethanol derived
from corn replaced sugarcane ethanol in the twentieth and early twenty-first centuries. In addition to its role in producing ethanol, sugarcane may be used to generate
electricity. Power stations in the tropics burn bagasse, the plant residue that remains
after the extraction of cane juice, to produce electricity. The island of Hawaii gener-
ates a sizable portion of its electricity from burning bagasse. Containing cellulose,
bagasse is also used to make paper and cardboard.

Positive though these aspects are, sugarcane culture depletes soil of nutrients.
Planters in Brazil, reckless in exploiting the soil, moved to new land when they had
exhausted old land by monoculture. Caribbean planters followed this policy until
they ran out of land. The use of manure to fertilize cane lands led planters to keep
livestock. To feed livestock and slaves, planters set aside land for corn, a welcome
practice that diversified agriculture, though soil exhaustion remains troublesome on
many sugarcane estates.

Moreover, the cultivation of sugarcane threatens the environment. In Queensland,
sugarcane growers have denuded hundreds of millions of acres. Without the protective cover of forest, soil erodes, washing into wetlands, rivers, and the ocean. By one
estimate, the growing of sugarcane costs Australia several hundred tons of soil per

hectare per year. In addition to the loss of soil, fertilizer runoff pollutes wetlands, rivers, and the ocean. In the Caribbean, sugarcane growers have removed vast tracts of forest, hastening erosion. In South Florida, sugarcane occupies the majority of land in the Everglades Agricultural Area. This region accounts for more than half of all sugarcane grown in Florida. The runoff of nitrates and phosphates endangers the Everglades, a region of rich biodiversity. Erosion and fertilizer runoff imperil mangrove swamps, sea grass beds, and coral reefs in Florida, the Caribbean, Australia, and Mauritius.

SOYBEAN

Old World Origins

Glycine soja, the ancestor of *G. max*, still grows wild in parts of China, Taiwan, Japan, Korea, and even Russia. *G. soja* is perennial, whereas *G. max* is annual. Compared with grains like barley and wheat in southwestern Asia, the soybean was domesticated late, about 2800 BCE.[22] In 2207 BCE, a Chinese agriculturalist wrote a text on, among other matters, the cultivation of soybeans (Figure 6.2). During the Shang and Chou dynasties (1700–700 BCE), growers spread soybeans throughout China. Like corn, the soybean is adapted to a variety of climates, so it seems plausible that farmers grew soybeans in tropical southern China. Not only did the Chinese eat soybeans, they fed the plants to livestock, and it was the latter practice that would draw the early attention of American stockmen and scientists. Despite their careful treatment of the soybean, the Chinese incorrectly classified the legume a grain. The differences between legumes and grains are so large that the modern reader has difficulty absorbing this error.

In the first-century CE, farmers in Manchuria and Korea began growing soybeans.[23] Thereafter, they spread to Japan, Indonesia, the Philippines, Vietnam, Thailand, Malaysia, Myanmar, Nepal, and India by the fifteenth century. This migration, though not the Columbian Exchange, would be a precursor to it. Throughout Asia, farmers planted soybeans with rice. A nutritious combination, rice supplied carbohydrates and soybeans essential amino and fatty acids. The combination of

FIGURE 6.2 Soybeans have become essential in the corn–soybean–hog triad of the American Midwest.

a grain and a legume was nearly universal. In southwestern Asia, farmers domesticated wheat, barley, peas, lentils, and chickpeas. In the Americas, early farmers domesticated corn and beans. In the seventeenth century, Europe imported soy sauce from Asia, though only in 1712 did German botanist Englebert Kaempfer introduce the soybean to Europe.[24] Again this was not the Columbian Exchange but marked an important step in that direction. As in Asia, the primary purpose of the soybean in early modern Europe was to feed livestock, especially chickens.

The Columbian Exchange and the Americas

When the Americas finally imported the soybean, interest in it grew rapidly among America's scientific elite. In 1765, British East India Company sailor Samuel Bowen gave soybeans to surveyor Henry Yonge, who planted them as a curiosity on his Georgia farm.[25] The next year, Bowen planted soybeans on his farm near Savannah, Georgia, making sauce and noodles. In 1769, the American Philosophical Society, an early promoter of agriculture and science, took interest in the soybean. The next year, one of its members, Benjamin Franklin, then in Britain, sent soybeans to John Bertram, perhaps America's most accomplished botanist at that time. Bertram planted them in a garden near Philadelphia. In 1804, physician James Mease reported the cultivation of soybeans throughout Pennsylvania. The botanical garden in Cambridge, Massachusetts, raised soybean in 1829, though farmers were reluctant to grow them, thinking of soybeans as a specialty crop with a limited market.[26]

Less well documented, the soybean also reached the Americas by crossing the Pacific Ocean during the eighteenth century.[27] The source of these soybeans was likely Japan, Korea, and China. The idea that the United States could obtain soybeans directly from Asia interested the U.S. Department of Agriculture (USDA) almost from its origin in 1862. In 1898 came a decisive moment in the Columbian Exchange. The USDA dispatched scientists William Morse and Charles Piper to Japan and China to collect soybean varieties. This action marked the federal government's commitment to advancing the cause of the soybean. Morse and Piper returned with more than 400 varieties. With small pods and seeds, these soybeans were suitable for feeding livestock. The USDA had committed itself to the view that the soybean was a forage crop, not foreseeing the explosion in demand after 1940 for soybean oil and meal.[28]

Meanwhile, soybeans established its presence in the Midwest. In 1851, farmers began cultivating soybeans in Illinois, from where they spread to Iowa and Ohio the next year, and thereafter to the rest of the nascent Corn Belt.[29] Northwestern Ohio would emerge as an important region. The heavy, wet clay soils in northwestern Ohio necessitated a late spring planting of corn, causing a diminution in yields. Farmers found that they could plant soybeans late with no reduction in yields. Consequently, northwestern Ohio has emerged as a region of soybean monoculture.

At the same time, Morse and Piper's varieties found favor in the South. The variety Mammoth Yellow, which appears to have predated the USDA's efforts, was the most important early variety in the South. In 1915, the boll weevil, having spread from Mexico to the United States, destroyed much of the cotton crop, causing a shortage of cottonseed oil. Southern farmers turned to soybeans as a source of oil. In the 1920s, the tractor spread throughout the United States, replacing the horse as a source of traction. Without horses, farmers no longer needed to cultivate oats as feed. In many cases, they transitioned from oats to soybeans.

The Great Depression, harmful in so many respects, benefited soybeans.[30] The drought of 1934 hurt corn growers, who fastened ever tighter on soybeans, which were more drought-tolerant than corn. Moreover, the New Deal, hoping to increase farm income by raising commodity prices, limited acreage of several crops, including corn. Because soybeans were not among these crops, farmers in what was becoming the Corn and Soybean Belt of the Midwest grew soybeans on land that might otherwise have gone to corn. Between 1934 and 1939, soybean acreage increased to 40% in the United States, with the Midwest and South reaping the greatest benefits from soybeans.[31]

By the end of the Great Depression, the tide was turning away from the use of soybeans as forage. New chemical techniques eased the extraction of meal and oil from the bean. The soybean now took off a crop that farmers raised for beans rather than for foliage. This meant that farmers wanted soybeans that yielded large pods and beans. The USDA collection, which had by then grown to more than 2500 varieties, posed a problem because most varieties had small pods and beans. In an effort to breed new, large-seeded varieties, scientists discarded all, but about 140 varieties from this collection. These 140 varieties became the backbone of the soybean-breeding program in the United States. Mukden from Manchuria would become particularly important to soybean breeders because of several desirable traits.

Between 1940 and 1944, U.S. soybean production increased from 78 to 192 million bushels.[32] By 1944, Americans consumed 72% of the soybean crop, most of it as oil. Illinois, Indiana, Ohio, Iowa, and Missouri grew bumper crops. In 1956, the United States emerged as the world's largest soybean producer and exporter, a position that it has not relinquished. In 1973, soybeans replaced corn and wheat as the leading source of farm income in the United States. By 1979, the United States harvested more than 1 billion bushels of soybeans. In 1990, Illinois was the leading producer of soybeans in the United States, with Iowa, Minnesota, Indiana, Ohio, and Missouri not far behind.[33] Today, the United States harvests more soybeans than all Asian countries combined. In this context, soybeans have emerged as the miracle crop. Its oil goes to margarine, cooking oil, shortening, salad dressing, paint, and plastic. Soy protein is an ingredient in infant formulas, soymilk, and vegetable burgers. Soybeans are found in hospital meals, vegetarian foods, sports drinks, pizza toppings, seafood, diet aids, frozen foods, canned goods of all kinds, dietary supplements, cookies, pet food, peanut butter, chicken nuggets, and ravioli. Billions of people worldwide have consumed soybean products without recognizing their presence as an ingredient.

The New World of Genetic Engineering

Genetic engineering is likely to shape the soybean's traits for decades to come. Chemical company DuPont has bioengineered a variety, with oil that is healthier than oil from traditional varieties. Agrochemical giant Monsanto aims to engineer nematode-resistant soybeans. Because nematodes are the worst pest of soybeans, this breakthrough would mark an important triumph. In the twenty-first century, farmers worldwide, especially in the United States, have transitioned rapidly to genetically engineered (GE) soybeans. Already Monsanto and DuPont own the rights to more than 1000 GE cultivars. At the moment, Monsanto has positioned itself to become the leading seller of the herbicide Roundup® and Roundup-resistant (known as Roundup Ready or RR) soybeans.[34] Roundup, a broad-spectrum herbicide, kills whatever plant it contacts. This poses problems for farmers, who must be careful not to get Roundup on a crop. Monsanto may have solved this problem in the short term by bioengineering soybeans, as well as other crops, resistant to Roundup, freeing growers to use Roundup without reservations. RR soybeans appear to be only a stopgap measure. Already weeds have evolved resistance to Roundup so that it is much less effective against them. Even so, RR soybeans have remade agriculture throughout the Americas. Argentina plants RR soybeans on virtually all soybean acreage.[35] By 2006, Brazil planted RR soybeans on 40% of acreage.

Because Monsanto owns these soybeans, its lawyers vigorously protect the company's intellectual property rights, suing several farmers for allegedly keeping soybeans to plant next year. This aggressiveness has tarnished Monsanto's image. It appears to be a corporate bully harassing the farmers it claims to help.

RICE

Origins

Rice (Figure 6.3) has sunk roots deep in time. The ancestor of rice likely inhabited the supercontinent Gondwana 130 million years ago.[36] When the continents separated, the proto rice evolved separately in what became Africa and Asia, though scientists do not know which continent first possessed fully evolved rice. The decisive moment, whether in Africa or Asia, came when humans, likely women, saved seeds from year to year, planting them in swamps near their homes. The presence of inundation from an early date must have lessened if not eliminated the burden of weeding. The first farmers also kept livestock from eating the rice. Over millennia, these farmers selected rice that yielded abundant, large seeds. These plants were shorter than wild varieties, demonstrating the ability of rice to put its biomass into seeds rather than stems and foliage. The species native to Asia is *Oryza sativa.*

FIGURE 6.3   Rice was a plantation staple in antebellum South Carolina and Georgia.

O. glaberrina is the African indigene. Farmers worldwide prefer Asian rice because of its higher yield and resistance to shattering during milling.

Asia

Scientists disagree about where and when in Asia humans domesticated rice. One thesis holds that humans domesticated rice in the Yellow River valley about 8000 BCE, an impressive date.[37] Another thesis points to northeastern India or perhaps the border between India and China about 7000 BCE.[38] A third account identifies China's Yangtze River valley as the cradle of rice culture about 7000 BCE. Another hypothesis puts the first cultivation of rice in India about 6500 BCE. Yet another posits a later date: 4000 BCE. By 5000 BCE, it is possible that rice was grown in China, Southeast Asia, and India.[39] It is tempting to think that both dryland and paddy rice were grown then. Dryland rice, which subsisted on rainwater, was popular in Thailand, Laos, Vietnam, and southwestern China. The introduction of paddy rice may have been a later development. In Asia, rice is cultivated from India and Sri Lanka in the west to Southeast Asia and the islands of South Asia in the southeast to China and Japan in the north. Into the twenty-first century, dryland rice prevails in about half of India and parts of Laos and Borneo. The rest of Asia relies on paddy rice. India rather than China has the largest area planted to rice.[40] The reliance on a single crop, as the Irish Potato Famine proved, can cause catastrophe. Famine has plagued India and China, killing millions and destabilizing government. Today, Thailand is the world's largest rice exporter.[41]

Africa

Legend holds that the rain god gave Africa rice in the mythic past. One should note the differences between African and Asian rice. African rice comes in greater diversity than Asian rice, opening the possibility that rice has had a longer tenure in Africa than Asia. Asian rice is fairly uniform. The brown husk, though nutritionally valuable, may be removed to reveal a long white grain. African too has a white grain, but it is short and a subspecies of *O. glaberrina*. In contrast to Asia, Africa also has a long red grain, which too is a separate subspecies of *O. glaberrina*.

Scholars have proposed the Niger River valley as the center of rice culture in West Africa, where farmers likely domesticated rice between 4500 and 3000 BCE.[42] Another thesis puts the origin of rice cultivation in West Africa about 1500 BCE.[43] From the Niger River valley, rice spread to Senegal between 1700 and 800 BCE and to East Africa in the ninth century CE. An independent center of domestication may have been the Guinea coast between 1000 and 1200 CE.

From an early date, the history of rice in Africa is intertwined with the history of rice in Asia and with the trans-Indian Ocean trade that linked Africa and Asia. As early as 1000 BCE, sailors may have brought Asian varieties to Madagascar, from where they spread to East Africa. If this is true, from an early date, African farmers grew *O. sativa* and *O. glaberrina*. Alternately, East Africa may have imported rice from South Asia, surely India. It is also possible that Arabs, among the great traders of the Middle Ages, brought Asian rice to Africa between the sixth and eleventh centuries. It is also possible that Africa did not grow Asian varieties until the Portuguese planted them in West Africa in the fifteenth century, on the cusp of the slave trade.

Because Asian rice does not compete well against African weeds, one must question how widely Africa planted Asian varieties. Yet where these varieties were grown in swamps, Asian rice yielded well. By all accounts, Asian varieties spread slowly. Only in the nineteenth century did Asian rice spread from Mozambique to the Congo.

Rice and the Columbian Exchange

At a quick glance, one might think it easy to determine when and from where the Americans acquired rice. If the grain was red or short and white, then it came from Africa. If it was long and white, however, the matter becomes contentious because this rice, having originated in Asia, may have reached the Americas from either Asia or Africa. One fact that no one disputes is that the Americas acquired rice from across the Atlantic rather than Pacific Oceans. This fact should favor an African introduction.

Yet controversy remains. About 1570, a Portuguese writer observed the culture of rice in Brazil, though he claimed not to know how Brazil acquired rice.[44] One tradition holds that a slave woman smuggled rice into Brazil in her hair. Given the importance of women in rice culture and the enormous role that slaves played in teaching whites how to grow rice, this story is attractive even if fanciful. In 1579, a Spanish writer noted that rice culture extended to the Gulf coast of Mexico. According to one account, slaves in South Carolina grew rice as early as 1680.[45] Another version holds that South Carolina did not plant rice until 1685 and that this rice came from Madagascar. If this account is true, and if Asian rice was grown on the island, then this rice may have been African or Asian. If, on the other hand, rice came to South Carolina aboard a slave ship, perhaps as leftover provision, then it cannot have originated in Madagascar. If the source of rice was West Africa, as this thesis holds, and if one recalls the importance of the African species to West Africa, then Africa should have been the conduit through which the Americas received rice. Yet, there is always a possibility that if Portugal owned the slave ship and given the country's familiarity with Asian rice, then Asia may have been the ultimate source of American rice. The matter awaits additional research. Although one commonly thinks of the Columbian Exchange as a European–American affair, in the case of rice, Africa or perhaps Asia, and not Europe, was the link to the Americas.

Like sugarcane in tropical America, rice in South Carolina and later Georgia cemented the oppressive relationship between whites and Africans. The near extirpation of the Amerindians and the reluctance of whites to work someone else's land conspired to prompt Europeans to compel Africans to work the rice estates.[46] By 1698, slaves imported from the Caribbean grew rice in South Carolina, though the colony turned to Africa to meet its needs in the eighteenth century. A majority of slaves imported to South Carolina came from rice-growing regions of Africa. Slavers understood the value of these slaves, announcing in newspapers the intention to sell slaves skilled in growing rice.[47] As rice production soared in South Carolina and spread to Georgia, so did the number of slaves. As in Africa and Asia, women did much of the work in the Americas. Slaves imported not only the technologies of cultivation, but also the culture of rice production. For example, their field songs originated in Africa.

By 1690, rice had become a cash crop. In the eighteenth century, rice made South Carolina the wealthiest plantation economy in North America. Planters in South

Carolina and Georgia sent their rice to Britain, which re-exported it to the continent and India. Demand was strong in Europe, where it was used to make beer and paper. Catholics ate rice with fish on meatless Fridays and during Lent. By 1763, rice was British North America's chief export to Britain.

Until 1720, South Carolinians depended on rainfall to sustain their crop and so must have grown dryland varieties. Around 1720, they began using water from ponds and streams to flood their fields.[48] About 1750, planter McKewn Johnstone, probably using the knowledge of his slaves, introduced the tidal flow system to South Carolina.[49] He used a gate to capture water from a river. The gate opened to receive water that the tide pushed upriver and closed when the tide receded to prevent water from flowing back downriver. In the nineteenth and twentieth centuries, however, South Carolina's decline was swift. By 1900, South Carolina and Georgia produced little rice and none by 1929.[50] Competition from Cuba, South America, and the American Southwest undermined the two.

By the mid-nineteenth century, lands along the Mississippi River grew rice.[51] Production moved into western Louisiana and Texas. By 1903, these states produced nearly the entire rice crop in the United States. During the twentieth century, rice migrated west to California and north to Arkansas and Missouri. Today, Louisiana, Mississippi, Missouri, Texas, California, Florida, and Arkansas are the major producers. In Louisiana, farmers rotate rice with soybeans in the same way that the Midwest rotates corn and soybeans. In the upper Mississippi River valley, 2 years of soybeans follow rice. California rotates rice with sunflowers, beans, sugar beets, vegetables, or tomatoes. Florida rotates rice with sugarcane or vegetables.[52]

FROM THE NEW WORLD

POTATO

A New World Indigene

The ancestor of the potato was probably native to the American Southwest, Mexico, and Guatemala. The potato (Figure 6.4) arose in tropical South America but is not a tropical crop. Grown in the cool highlands of the Andes Mountains, the potato is a crop of temperate regions. The prehistoric peoples of the Andes may have domesticated the potato about 6000 BCE.[53] During the next 2000 years, the people on the border between modern Peru and Bolivia began selecting potatoes for size and flavor. At the same time, they domesticated other tubers, though only the potato is a world staple. From these early people, the Inca learned to grow potatoes. In the Americas, the potato has migrated to the temperate zone of North America. From Washington State to Maine, Americans grow potatoes, though Idaho received acclaim for the celebrated Idaho potato. Most farmers grow variants of American botanist Luther Burbank's Russet Burbank potato.

The First Columbian Exchange

Spanish conquistador Francisco Pizarro, invading South America in 1532, initiated Europe's relationship with the potato.[54] In 1537, the Spanish encountered Amerindians growing potatoes. Curious, they sampled the food, finding it palatable.

© Smithsonian Institution

FIGURE 6.4   The potato remains a staple of the temperate zones.

Yet, their interest in finding gold and silver led them to undervalue the potato. Even when the Spanish turned to farming, they focused on introducing Old World crops to the Americas rather than developing the potential of indigenous crops.

How and when the potato reached Europe remains a matter of debate. According to one thesis, English adventurer Francis Drake found the potato on an island near Chile in 1578, but he may not have returned to England with it.[55] Botanist Carolus Clusius visited Drake, who never mentioned the potato, making him an unlikely conduit between the Americas and Europe. Another thesis holds that Drake gave the potato to a friend in 1580. Supposing it to be akin to grains, he ate the seeds, which were unpalatable. So disgusted was the man, that he ordered his gardener to burn the plants. By chance, a potato emerged unscathed and baked. Finding the tuber agreeable, the gardener rescued the rest from destruction. Alternatively, the Spanish may have used potatoes as provision for their ships. Leftover potatoes might have been planted in sixteenth-century Spain.[56] Once established, probably in northern Spain, the potato may have spread thereafter throughout northern Europe. Yet, another tradition holds that the Spanish brought the potato to the Canary Islands in 1567, only later transplanting it in Spain. Still another thesis credits Sir Walter Raleigh with introducing the potato to Europe. Tradition holds that he served a plate of boiled potatoes to Queen Elizabeth I, but historians hold that Raleigh never visited regions of the Americas where potatoes grew. How then could he have brought the tuber to Europe? Economist Earl Hamilton discovered from a search of archival records that a hospital in Seville bought potatoes for its patients in 1573, establishing this date as the latest when the potato could have been cultivated in Spain.[57] Geneticist Radcliffe Salaman believes that Spain grew the potato as early as 1570.[58] This date might have marked the year of transfer from New to Old World.

European botanists were interested in the potato from an early date. In 1596, botanist Gaspard Baubin named the potato *Solanum tuberonum*, a designation that

Swedish naturalist Carl Linnaeus retained in the eighteenth century.[59] In 1597, English herbalist John Gerard published a text with a description and illustration of a potato.[60] Gerard may have grown potatoes in his garden. If so, he must have had firsthand knowledge about them. Yet, his knowledge was far from complete. He claimed to have gotten his potatoes from Virginia, but the colony grew none and was climatically unsuited to the potato. Only in 1601 did Carolus Clusius pinpoint the potato's origin in the Andes Mountains.[61] As early as 1588, he had acquired potatoes from Belgium, though it is not clear as to how they came to be in Belgium. Clusius remarked that Germany and Italy grew potatoes, though Mediterranean Italy is unsuited to the potato. Better must have been the highlands of the Apennine Mountains.

In the seventeenth century, many Europeans believed the potato to be an aphrodisiac. They appreciated that wherever the potato grew, population increased. The potato must have made people eager for sex. Eighteenth-century herbalist William Salman speculated that the potato increased the number of sperm in men. Treading a more prudent path, physician Tobias Venner in 1622 advocated the potato's nutritional value.

Not everyone admired the potato. In the early eighteenth century, clergymen warned against it, claiming that it could not be wholesome if the bible did not mention it. Of course, the biblical writers, ignorant about the existence of the Americans, could not have known of the potato. Into the nineteenth century, some Europeans were slow to acknowledge the value of the potato. Because the potato took strange shapes, some people feared that it might cause leprosy, among the most feared diseases of that time.[62] As early as the seventeenth century, Burgundy, a region in France, outlawed the potato because of its putative connection to leprosy.

Yet, the potato transcended superstition. As early as 1640, it was a staple in Belgium. Because the potato yielded so much food per plant, the poor who had subsisted on bread found that they could more easily afford potatoes. Scientists and economists were beginning to understand the danger of this phenomenon. If population increased because of the potato, the destruction of the tuber would imperil the masses.

Europe's royalty were slow to grasp this reality. In 1744, King Frederick the Great distributed seed potatoes to Prussian farmers.[63] (Prussia is today part of Germany.) Austria, Russia, and France, witnessing Prussia's success, encouraged their farmers to adopt the potato. In 1765, Russian tsarina Catherine the Great followed Frederick's example.[64] By the 1770s, the potato was a staple in what would become Germany, Central Europe, and Russia. French soldier Antoine-Augustin Parmantier, captured by the Prussians, spent 3 years in jail, eating little more than potatoes.[65] Upon regaining his freedom, Parmantier credited the potato with his survival. Returning to France, he crusaded on behalf of the tuber, even winning an audience with French nobles. He presented King Louis XVI and Queen Marie Antoinette with a bouquet of potato flowers, which quickly became fashionable. Marie wore them in her hair and the king pinned them on his lapel.

Europe's Last Subsistence Crisis and the Second Columbian Exchange

No one needed to convince the Irish to grow potatoes. The climate and soil favored the crop. Equally important was the social and economic system. The English thought the Irish an inferior type of human and exploited them as they did Africans in the Americas. Like the sugarcane plantations in the Caribbean, English landlords lived in style in England, taking no interest in the welfare of the Irish. Charged high rent, the Irish tenant could hope to pay only by planting most of the land to grain or converting it into pasture. The tenant could afford to set aside only two or three acres for his family. On this land, he planted potatoes. The potato was therefore the only food that stood between the masses and starvation.

Flawed as this system was, it worked initially. Two or three acres were enough to feed even a large family, with potatoes left over to feed a pig. By one account, the potato enabled the Irish tenant to eat better than his counterpart on the continent, who subsisted on bread. One historian estimated that the average adult laborer ate 14 pounds of potatoes per day.[66] Supplementing this fare with cabbage, turnips, and milk when possible, he consumed some 5000 calories per day, a high figure by current standards.

Nourished on potatoes, Ireland's population rose from 1.5 million in 1600 to 2 million in 1700, to 5 million in 1800, and to 8.5 million in 1845.[67] The first inkling that this system might collapse came in 1740 and 1741, when a frigid winter killed the potato crop.[68] More than 100,000 Irish starved and they might have fared worse but for the willingness of English landlords to let the Irish keep some of their grain. Between 1800 and 1845, 114 commissions reported that Ireland's masses were on the verge of starvation. Yet, Britain refused to reform the system. In the early nineteenth century came the final dagger that would burst the population balloon. The introduction of the Lumper variety yielded well in Ireland but was in all other matters an inferior potato with poor taste and fewer nutrients than other varieties. By planting the Lumper everywhere in Ireland, the Irish unwittingly adopted a genetically uniform crop. What killed one plant had the potential to kill all. From the United States came a pathogen in 1845.[69] Here was a second manifestation of the Columbian Exchange. Scientists long classified the pathogen as a fungus, though recent consensus has formed around the idea that the pathogen is really a water mold. The mold afflicted all Europe. Ireland, battered by heavy spring rains, was particularly vulnerable, and between 1845 and 1849, the mold eliminated the potato crop, causing 1 million Irish to starve.[70] Another 1.5 million escaped death by fleeing Ireland. Most came to the United States and Canada. The potato famine thus caused a portion of the human migration that was part of the Columbian Exchange.

A Third Columbian Exchange?

The rise of plant biotechnology during the last quarter of the twentieth century affected the potato. Monsanto and other multinational companies took the lead in this development. In the 1990s, Monsanto bioengineered a type of potato resistant to several insects. Inexplicably the potato sold poorly in the United States and Europe, leading Monsanto to withdraw it. Monsanto's latest potential success has been to bioengineer a type of potato that will yield starch suitable for industrial extraction. The company has approached the European Union for permission to sell the potato. This cultivar would appear to be well suited to Europe, where some 75% of all potatoes go to industrial processes. Yet, Europe has never been entirely comfortable with biotechnology and has yet to grant Monsanto permission to sell the potato. If Europe relents, the moment will mark the third transfer of the potato or a potato pathogen from the Americas to Europe.

SWEET POTATO

New World Origins

The genus *Ipomoea* has 400 species, most of them diploids.[71] That is, each plant has two pairs of chromosomes in every cell. Many other organisms, including humans, are also diploids. The sweet potato (Figure 6.5), however, is a hexaploid, containing six pairs of chromosomes in each cell. Scientists continue to wonder how a diploid ancestor gained so many more chromosomes over the millennia, though the sweet potato is not alone in exhibiting polyploidy.

An American indigene, the sweet potato may have originated between the Yucatan Peninsula and the Orinoco River in Venezuela and Colombia.[72] In addition, Panama, northern South America, and the Caribbean have been proposed as regions of origin. Central America has the greatest diversity of sweet potato varieties, and according to Russian agronomist Nikolai Vavilov's logic, should be the site of origin.[73] Yet, wild relatives in the Caribbean have focused scientists' attention there. The remains of the sweet potato date to 600 BCE in a Peruvian cave, but scientists do not know whether the people of the Andes Mountains domesticated the plant by then. Given the plant's suitability for warm climates, the Andes Mountains may not be a promising place to seek the origins of the sweet potato. The Amerindians of the tropics probably domesticated the sweet potato at least 5000 years ago. In pre-Columbian America, the Maya grew the sweet potato in Central America and the Inca cultivated the root in Peru, though the focus must have been the lowlands.

Prelude to the Columbian Exchange

The sweet potato appears to have migrated from the Americas even before the Columbian Exchange.[74] One account holds that farmers grew the sweet potato in New Zealand as early as 1000 CE, though how it crossed the Pacific Ocean is unclear. Perhaps Polynesians acted as intermediaries. Another thesis holds that New Zealand acquired the sweet potato directly from Peru, though this line of thinking appears to raise more questions than it answers. From Central America, another line of thought holds, Oceania received the sweet potato. This thesis insists that humans were not

FIGURE 6.5   An important root crop, the sweet potato is a worldwide staple throughout the tropics.

involved in this transfer, making it unclear, again, how the sweet potato crossed the Pacific. The sweet potato of Papua New Guinea differs from that of Peru, disqualifying Peru as the source of transmission.

The Columbian Exchange

Christopher Columbus quickly appreciated the value of the sweet potato, returning from his first voyage with it.[75] About 1600, Portugal, a latecomer to the Columbian Exchange in this instance, planted the sweet potato in West Africa, Southeast Asia, and the East Indies. Later, the plant migrated to India, China, and Japan, though not everyone supports a late date for China.

One story holds that Chinese businessman Zhenlong Chen brought the sweet potato from Luzon in the Philippines to Fujian, China, in the sixteenth century.[76] This story leaves unclear how and when the Philippines received the sweet potato if Portugal did not plant it until 1600. In fact, it is not clear that Portugal visited the Philippines then. The idea that ocean currents carried it across the Pacific may be implausible. According to this account, the date of transmission is unknown because the merchants of Zhengzhon, the port of Fujian, kept this information secret. Chen's son presented the sweet potato to the governor of Fujian, and either father or son introduced sweet potato to Zhejiang, Shandong, and Henan provinces. The sweet potato must have been in cultivation before 1594, because Fujian's governor, familiar with the root, ordered peasants to plant it to avert famine that year. One account holds that Burma (now Myanmar) introduced the sweet potato to Tali, China, in 1563, though again we have the problem of determining from where and when Myanmar acquired the sweet potato.[77] Again the date is too early to invoke the Portuguese. Another account notes that the sweet potato spread from Vietnam to Dongyuan, China, in 1582. In this case, it may be possible to invoke the Portuguese as the transmitters of the sweet potato to Vietnam. Another possibility is that India spread the sweet potato to China, but from where and when did India acquire the plant? From the south, the sweet potato spread to Quanzhou, Putian, and Changle counties.

In the 1950s and 1960s, the Chinese increased sweet potato production to feed a burgeoning population. In 1961, acreage peaked. Thereafter, the Green Revolution doubtless hampered the sweet potato. The Green Revolution focused, perhaps myopically, on grains at the expense of roots, tubers, and legumes. By 1985, many Chinese farmers had switched to high-yielding varieties (HYVs) of rice, wheat, and corn. Doubtless, the hybrids from the United States were at the core of this transition, which marks a Columbian Exchange that undermined part of the original Columbian Exchange. In China, sweet potato ranks fifth behind rice, wheat, corn, and soybeans. Being a Chinese indigene, one is not surprised to find soybeans on this list. Still one would be wrong to forecast the demise of the sweet potato in China. It is still from Hainan in the south to Inner Mongolia in the north and from Zhejiang in the east to Tibet in the west. The lands along the Yellow and Yangtze Rivers are the center of sweet potato culture. The leading producers are Sichuan, Hanan, Chengzing, Anhui, Guangdong, and Shindong provinces. Stockmen feed sweet potato plants to pigs, cows, and goats.

In India, sweet potato ranks third among tubers and roots, trailing potato and cassava, all being imports of the Columbian Exchange.[78] Curiously, the transmission of sweet potato from the Americas to India appears to be unknown. The sweet potato is cultivated everywhere in the subcontinent except Jammu, Kashmir, Himachal, Pradesh, and Sikkim. Compared with other nations, India ranks fifth in yield per acre, eighth in aggregate production, and twelfth in total hectarage to sweet potatoes. Not surprisingly, the sweet potato is rainfed in Orissa, West Bengal, Utter Pradesh, Bihar, and Jharkhand. These states produce the majority of India's sweet potatoes and hold the majority of acreage to the crop. India does not engage in sweet potato monoculture. Most farmers cultivate only a few acres to sweet potato and the rest to other crops, particularly since the Green Revolution, though in Bihar farmers cultivate sweet potatoes year round. Elsewhere sweet potatoes are grown in the rainy season between September and January and in summer. The yield in Bihar is modest, and the poor soils of Orissa and Jharkhand do not fare better.

As incomes have increased in India, as in the United States, people have eaten fewer sweet potatoes. Accordingly, many farmers have switched from sweet potatoes to grain, a trend, again, that the Green Revolution has reinforced. In India, the majority of sweet potatoes are eaten fresh. Indians roast, bake, and boil the root. Many Indians believe that sweet potatoes and milk are an adequate diet but that sweet potatoes alone are unwholesome. In Bihar, farmers rotate sweet potatoes with corn, wheat, and onions. In Orissa, sweet potatoes follow corn or rice. In West Bengal, farmers rotate moong, taro, and sweet potatoes. In Andhra Pradesh, sweet potatoes follow corn and precede vegetables. In Chhattisgarh, Uttar Pradesh, and Maharashtra farmers grow sweet potatoes and cowpeas.

Portugal brought sweet potato to Africa during the slave trade. Farmers first grew sweet potato in West Africa about 1520.[79] By the end of the seventeenth century, West Africa was an important region for the cultivation of the root. These farmers grew sweet potatoes with cassava and yams, making the region rich in root crops. The sweet potato is not, however, the most important crop in West Africa, as it must compete against yams, cassava, rice, cowpea, cocoyam, and acha. Nonetheless, West and East Africa account for the majority of Africa's sweet potatoes. In Ghana, the root ranks third in acreage, trailing only cassava and yams. The people of Ghana boil, fry, or roast sweet potatoes.

The people of sub-Saharan Africa adopted sweet potato from an early date, often planting it over yams. So long has the sweet potato been in cultivation that some Africans wrongly assume it to be a native crop. In 2007, sub-Saharan Africa yielded nearly 13 million metric tons of sweet potatoes.[80] The poor grow sweet potato, a subsistence crop, on small plots. In this context, the African reliance on sweet potato as a food of the masses resembles Ireland's dependence on the potato in the nineteenth century. Even without fertilizers, small farmers have managed to increase yields. Africa grows sweet potato from sea level to 7500 feet. Sweet potato is important in Africa because it demands less labor and fertilizers than rice or other grains. This is particularly true because Green Revolution rice demands heavy inputs of fertilizers. The sweet potato has gained adherents where diseases of cassava and banana obviated cultivation of these crops. The farmland near Lake Victoria has arisen as an important region of sweet potato culture. In this region, farmers plant sweet potato and cassava. Where cassava fails, farmers replant sweet potatoes. Where farm size has decreased, farmers have increased the proportion of land to sweet potato to derive maximum calories and nutrients from the soil.

Although they have taken time, government policies favor the sweet potato in Africa. At the end of colonialism, governments subsidized corn growers as an odd way to encourage subsistence farming. Corn is a high-input crop that is not ideally suited for subsistence agriculture. These subsidies were too expensive and in the early twenty-first century, governments, especially in southern Africa, ended them, leading corn growers to transition to sweet potato. The AIDS epidemic, depopulating the countryside, has prompted single-parent households seeking to minimize labor and costs to plant sweet potato. Many Africans regard sweet potato as a famine food. The Romans used the phrase "famine food" as a disparagement, though Africans do not seem to hold this view.

In sub-Saharan Africa, 23 countries produce almost all the continent's sweet potatoes.[81] Uganda and Nigeria total one-third of all sub-Saharan Africa's harvest. The most densely populated regions of these nations produce the most sweet potatoes. Uganda, Rwanda, Burundi, and Malawi harvest hundreds of pounds of sweet potato per person. The lowest producers are Congo, Ethiopia, South Africa, Cote d'Ivoire, Niger, and Burkina Faso. Not as productive as the United States and China, sub-Saharan Africa yields 1.8 tons of sweet potatoes per acre. South Africa, using irrigation and fertilizers, harvest a higher yield. In eastern and southern Africa, sweet potato is often a secondary crop to corn, banana, and plantain. In parts of Africa, sweet potato follows a cash crop like peanuts. Ugandan farmers dry the harvest, so it may last 5 months. Ugandans grind sweet potato to make porridge known as *atapa*. Throughout Africa, women tend sweet potatoes, giving rise to the association among the root, women, and poverty.

Corn

American Origins

The debate over the origins of corn raged for centuries. In the 1970s, American botanist Paul C. Mangelsdorf identified southern Mexico as the cradle of corn culture, a judgment that has withstood scrutiny.[82] Pre-Columbian peoples from Canada to the southern tip of South America grew corn (Figure 6.6). It was the staple of millions of Amerindians. Perhaps, recognizing that corn was not sufficiently nutritious to be consumed alone, the Native Americas grew corn with beans and squash, the "three sisters." Corn provided starch and some vitamins and minerals, beans

FIGURE 6.6 An American grass, corn has sustained humans and livestock for centuries.

supplied protein, and squash was the source of additional vitamins. Together, these three were a nourishing combination. As elsewhere, the Amerindians combined a grain and a legume by eating corn and beans. The Europeans who settled the Americas established a corn-livestock complex, and corn became the staple of the U.S. livestock industry, a position it retains today. Corn's intensive cultivation in the Midwest gave the region a new name, the Corn Belt. Many farmers in this region grow nothing but corn and soybeans, an Asian indigene, in rotation. Soybeans are now the legume of choice with corn, whereas the Amerindians had once favored beans. As corn's importance grew, so did scientific interest in the plant. In the early twentieth century, American agronomists, borrowing ideas from British naturalist Charles Darwin and German monk and founder of genetics Gregor Mendel, developed varieties of hybrid corn that yielded far more corn than did traditional varieties. Indeed corn was the first plant from which scientists tapped the potential of heterosis or hybrid vigor.[83] In some cases, these new hybrids were resistant to insects and diseases. The hybrids of U.S. Department of Agriculture (USDA) breeder Glenn H. Stringfield were among the first resistant to the once menacing European Corn Borer (see Chapter 8). Biotechnology companies, including the American agrochemical giant Monsanto, have bioengineered corn with several valuable traits, including again resistance to the European Corn Borer. This type of biotech corn, known as Bt corn, was widely popular in the 1990s and was among the first successes of biotechnology (see Chapter 10).

Corn, Asia, and the Columbian Exchange

In October 1492, Italian Spanish explorer Christopher Columbus encountered corn in the Caribbean island of Hispaniola (now Haiti and the Dominican Republic). A trip to Cuba that November established corn's presence there as well. Columbus likely brought corn back to Spain in 1493.[84] Before 1700, farmers in Europe, Asia, Africa, and several Pacific islands grew corn. China cultivated corn by about 1550, though the date may have been as late as 1578.[85] The Chinese were well aware of the New World origins of the plant, betraying nothing of the confusion that would surround corn in Europe. In 1585, Spanish visitors to China confirmed the presence of corn. Corn quickly became a staple. It may have reached China's east coast via the Pacific Ocean. There may also have been an overland introduction from India, a circumstance that appears to confirm an earlier date of cultivation in India. Burma (today Myanmar) may have been an intermediary between India and China, confirming again an early date of cultivation in Burma. The Portuguese may have introduced corn and peanuts, both American crops, to China. The Chinese called corn "wheat of the western barbarians," a recognition that their western neighbors grew corn and that the Chinese had little respect for their neighbors.[86] The decision to refer to corn as a variant of wheat is interesting and may signal China's understanding that the two were related. Both are grasses, as we noted earlier. Bangladesh and Tibet were also early adopters, growing corn during the summer monsoon. Indeed, corn tolerates as many as 400 inches of rain per year. Today's hybrids can do well with as few as 10 inches or rain per year. Corn may not have been the first choice. The people of Asia appear to have preferred rice, but where it could not be grown, corn was the substitute. India's northern region was quick to adopt corn. In China, the poor and ethnic groups that the Chinese did not accept as full-fledged compatriots were the first to adopt corn. The Chinese ate young corn fresh and used flour to make a kind of cornbread and porridge. They also distilled corn into whiskey, an American practice transplanted in China. As a rule, the Chinese tended to plant corn on land that was not suited to rice or even peanuts. Corn had an advantage over rice in being less labor-intensive. Corn also impressed the Chinese with its geographic range. The rise of China's population after 1700 may have stemmed from the cultivation of corn and other American crops, including peanuts, as we have seen, sweet potatoes, and cassava.

Corn, Africa, and the Columbian Exchange

Corn was an important provision aboard slave ships.[87] Africans began to grow corn about 1530.[88] Portugal had probably introduced corn to Africa, though the Portuguese word for corn was generic for other grains, making it difficult to know whether the Portuguese meant corn or another grain. The first unambiguous reference to corn in Africa dates back to 1550.[89] West Africa was the original center of African corn culture. Africans paired corn with millet, an old staple. The African reference to corn as "Portuguese grain," appears to confirm Portugal as the source of corn's introduction to Africa. Some Africans did not like corn, finding it suitable only for pigs, an attitude that sunk roots with the stockmen who populated the Americas. Portugal probably acted to spur corn culture as a cheap food for slaves. On Portuguese slave ships, slaves ate two meals per day, one of corn and the other of beans, almost certainly those of American origin. Not an especially nutritious food, corn may have contributed to the poor health of overworked and undernourished slaves. By the seventeenth century, corn had spread throughout Africa's western coast. The average slave consumed about two pounds of corn per day aboard ship. Africans had no illusions about the voyage west, referring to slave ships as "tombs." All told, corn was the indispensable food of the slave trade. Africans turned to corn because it matured early and yielded more abundantly than millet or sorghum. Corn also out-yielded the African subspecies of rice.

The momentum toward independence allowed Africa to chart its own, if perilous, course. Rapid population growth since about 1850 must have depended in part on corn and cassava.[90] Between 1934 and 1938, Africa actually exported corn. South Africa was then the continent's leading exporter. Africa must then have been self-sufficient. Part of these exports may be explained by the fact that Africans were not avid corn consumers. They appear to have preferred cassava and sweet potatoes, another American crop. These surpluses vanished after World War II, as Africa became a net importer. By 1960, Africa needed 2.4 million tons of grain imports to feed itself.[91] By then, Africans preferred to import wheat. Corn was losing its status. Even in the 1960s, however, Africans grew corn for export. Had Africans fed this corn to livestock, they might have increased the amount of meat or pork in the diet, though they could not of course raise livestock in areas of sleeping sickness and the tsetse fly. Africa also lagged well behind the United States in planting varieties of high-yielding hybrid corn. The United States emerged as the world's granary exporting corn and other grains to Africa, perhaps contributing to the continent's reliance of food imports.

Corn, Europe, and the Columbian Exchange

Italy may have grown corn as early as 1494.[92] European botanists grew corn in greenhouses as early as 1539.[93] Not all scientists were smitten. German naturalist Jerome Bock thought corn a "strange grain." Bock believed, however, that corn might have medicinal value. He apparently did not conceive of its use as food and feed. Some Europeans, ignorant of corn's origins, thought that it had arrived from Asia, a mistake that Italian naturalist Petrus Matthiolus corrected in 1565. Indeed, Spain had earlier amassed literature that corn was an American indigene. In northern Europe, the error of origins persisted. Indeed, for many years, the people of northern Europe had no direct knowledge of corn because they did not grow it. In 1597, English herbalist John Gerard, who may have planted corn in his garden, pointed to corn's nutritional inadequacy as a bar to the plant's widespread acceptance. Corn made an inferior bread that was difficult to digest according to Gerard. He noted that only inferior people like the Amerindians actually ate corn. In 1640, English naturalist John Parkinson opposed Gerard's ideas. Corn was an acceptable food in the Americas and Europe, noted Parkinson. The fact that Christians ate corn proved it a gift from God. He noted that people and livestock ate corn in the seventeenth-century Europe. In the eighteenth century, Linnaeus took an interest in corn, though French naturalist George-Louis Buffon derided any plant or person from the Americas as inferior.[94] Antoine-Augustin Parmentier, who had done so much to promote the potato, did the same for corn in the eighteenth-century France. Parmentier and other agronomists were not content to describe corn. Others had already done that. They wanted to determine the best methods for cultivating corn, an approach to science that resonated with the American breeders who would develop hybrid corn in the twentieth century.

In Europe, corn ranked third in tonnage in the twentieth century, trailing only wheat and barley.[95] Corn represented about 20% of Europe's grain harvest. By the 1980s, Europe produced one-eighth of the world's corn crop. Like the United States, Europe grows corn for livestock. All nations in Europe, except those in Scandinavia, grow corn. The former Yugoslavia, Romania, and France rank in the top 10 among corn producers. Hungary, Portugal, Romania, and the former Yugoslavia grew more corn than wheat.

NOTES

1. Sanjida O'Connell, *Sugar: The Grass that Changed the World* (London: Virgin Books, 2004), 6.
2. J. E. Bowen and D. L. Anderson, "Sugar-Cane Cropping Systems," in *Field Crop Ecosystems*, ed. C. J. Pearson (Amsterdam: Elsevier, 1992), 150; Peter Macinnis, *Bittersweet: The Story of Sugar* (Crows Nest, Australia: Allen & Unwin, 2002), 143.
3. O'Connell, *Sugar*, 8.
4. Bowen and Anderson, "Sugar-Cane Cropping Systems," 150; Macinnis, *Bittersweet*, 4, 9–10.
5. O'Connell, *Sugar*, 12–13.
6. Macinnis, *Bittersweet*, 14–23.
7. Sidney W. Mintz, *Sweetness and Power: The Place of Sugar in Modern History* (New York: Penguin Books, 1985), 31.

 8. O'Connell, *Sugar*, 26.
 9. Mintz, *Sweetness and Power*, 32–33.
10. Macinnis, *Bittersweet*, 30–38.
11. O'Connell, *Sugar*, 34–37.
12. Elizabeth Abbott, *Sugar: A Bittersweet History* (London and New York: Duckworth Overlook, 2009), 212.
13. Mintz, *Sweetness and Power*, 69.
14. O'Connell, *Sugar*, 74.
15. Ibid, 75.
16. Bowen and Anderson, "Sugar-Cane Cropping Systems," 148.
17. Ibid, 151.
18. Ibid, 148.
19. O'Connell, *Sugar*, 73–74.
20. Macinnis, *Bittersweet*, 144.
21. Bowen and Anderson, "Sugar-Cane Cropping Systems," 149.
22. C. Wayne Smith, *Crop Production: Evolution, History, and Technology* (New York: Wiley, 1995), 350.
23. J. G. Vaughan and C. Geissler, *The New Oxford Book of Food Plants* (New York: Oxford University Press, 1997), 28.
24. Smith, *Crop Production*, 351.
25. Ibid, 352.
26. Ibid, 352.
27. Ibid, 353.
28. Vaughan and Geissler, *New Oxford Book*, xvi.
29. A. H. Probst and Robert W. Judd, "Origin, United States History and Development, and World Distribution," in *Soybeans: Improvement, Production, and Uses*, ed. Billy E. Caldwell, Robert W. Howell, Robert W. Judd, and Herbert W. Johnson (Madison: American Society of Agronomy, 1973), 11.
30. Smith, *Crop Production*, 356–357.
31. Ibid, 356.
32. Ibid, 356.
33. John H. Martin, Richard P. Waldren and David L. Stamp, *Principles of Field Crop Production*, 4th ed (Upper Saddle River, NJ: Pearson/Prentice Hall, 2006), 613.
34. Wayne A. Parrott and Thomas E. Clemente, "Transgenic Soybean," in *Soybeans: Improvement, Production, and Uses*, 3d ed. eds. H. Roger Boerma and James E. Specht (Madison: WI: American Society of Agronomy, 2004), 265–268.
35. Christine M. Du Bois and Ivan Sergio Freire De Sousa, "Genetically Engineered Soy," in *The World of Soy*, eds. Christine M. Du Bois, Chee-Beng Tan and Sidney W. Mintz (Urbana and Chicago: University of Illinois Press, 2008), 74–75.
36. Te-Tzu Chang, "Origin, Domestication, and Diversification," in *Rice: Origin, History, Technology, and Production*, eds. C. Wayne Smith and Robert H. Dilday (Hoboken, NJ: Wiley, 2003), 4.
37. Smith, *Crop Production*, 221.
38. Martin, Waldren and Stamp, *Principles of Field Crop Production*, 471.
39. Dong Hanfei, "Upland Rice Systems," *Field Crop Ecosystems*, ed. C. J. Pearson (Amsterdam: Elsevier, 1992), 192.
40. Ibid, 183.
41. Henry C. Dethloff, "American Rice Industry: Historical Overview of Production and Marketing," in *Rice: Origin, History, Technology, and Production*, eds. C. Wayne Smith and Robert H. Dilday (Hoboken, NJ: Wiley, 2003), 69.
42. Te-Tzu Chang, "Origin, Domestication, and Diversification," 12.
43. Dong Hanfei, "Upland Rice Systems," in *Field Crop Ecosystems*, 142–143.

44. Judith Carney, "Out of Africa: Colonial Rice History in the Black Atlantic," in *Colonial Botany: Science, Commerce, and Politics in the Early Modern World*, eds. Londa Schiebinger and Claudia Swan (Philadelphia: University of Pennsylvania Press, 2005), 218–219.
45. Dethloff, "American Rice Industry," 69.
46. Carney, "Out of Africa," 204–206.
47. Carney, Ibid, 217.
48. Henry C. Dethloff, *A History of the American Rice Industry, 1685–1985* (College Station: Texas A & M University Press, 1988), 19.
49. Smith, *Crop Production*, 232.
50. Dethloff, *History of the American Rice Industry*, 46.
51. Martin, Waldren and Stamp, *Principles of Field Crop Production*, 471.
52. Joe E. Street and Patrick K. Bollich, "Rice Production," in *Rice: Origin, History, Technology, and Production*, eds. C. Wayne Smith and Robert H. Dilday (Hoboken, NJ: Wiley, 2003), 289–290.
53. Vincent E. Rubatzky and Mas Yamaguchi, *World Vegetables: Principles, Production, and Nutritive Values*, 2d ed (New York: Chapman and Hall, 1997), 105.
54. John Reader, *Potato: A History of the Propitious Esculent* (New Haven: Yale University Press, 2009), 10.
55. Reader, *Potato*, 86–87; Milton Meltzer, *The Amazing Potato: A Story in which the Incas, Conquistadors, Marie Antoinette, Thomas Jefferson, Wars, Famines, Immigrants, and French Fries All Play a Part* (New York: HarperCollinsPublishers, 1992), 24.
56. Sylvia A. Johnson, *Tomatoes, Potatoes, Corn, and Beans: How the Foods of the Americas Changed Eating around the World* (New York: Atheneum Books, 1997), 68.
57. Redcliffe N. Salaman, *The History and Social Influence of the Potato* (Cambridge, UK: Cambridge University Press, 1949, reprinted 1970), 143.
58. Ibid, 68–69.
59. Ibid, 96.
60. Meltzer, *The Amazing Potato*, 23–24.
61. Reader, *Potato*, 84–85.
62. Johnson, *Tomatoes, Potatoes, Corn, and Beans*, 72.
63. Meltzer, *The Amazing Potato*, 31.
64. Johnson, *Tomatoes, Potatoes, Corn, and Beans*, 77.
65. Meltzer, *The Amazing Potato*, 29.
66. "Glossary," in *The Prendergast Letters: Correspondence from Famine-Era Ireland, 1840–1850*, ed. Shelley Barber (Amherst and Boston: University of Massachusetts Press, 2006), 193.
67. James S. Donnelly, Jr. *The Great Irish Potato Famine* (Gloucestershire: Sutton Publishing, 2001), 4.
68. Reader, *Potato*, 156–157.
69. Reader, *Potato*, 194.
70. Donnelly, *The Great Irish Potato Famine*, 178; John Percival, *The Great Famine: Ireland's Potato Famine, 1845–1851* (New York: Viewer Books, 1995), 171; Gail L. Schumann and Cleora J. D'Arcy, *Hungry Planet: Stories of Plant Diseases* (St. Paul, MN: American Phytopathological Society, 2012), 12.
71. Rubatzky and Yamaguchi, *World Vegetables*, 130.
72. Gad Lobenstein, "Origin, Distribution and Economic Importance," in *The Sweetpotato* (Dordrect, Netherlands: Springer, 2009), 9; J. G. Vaughan and C. Geissler, *The New Oxford Book of Food Plants* (New York: Oxford University Press, 1997), 192.
73. I. C. Onwueme, *The Tropical Tuber Crops: Yams, Cassava, Sweet Potatoes, and Cocoyams* (New York: Wiley, 1978), 167.

74. Lobenstein, "Origins, Distribution and Economic Importance," 9; Vaughan and Geissler, *New Oxford Handbook*, 192; Rubatzky and Yamaguchi, *World Vegetables*, 130.
75. Vaughan and Geissler, *New Oxford Handbook*, 192.
76. L. Zhang, Q. Wang, Q. Liu, and Q. Wang, "Sweetpotato in China," in *The Sweetpotato* (Dordrect, Netherlands: Springer, 2009), 325.
77. Ibid, 326.
78. S. Edison, et al, "Sweetpotato in the Indian Sub-Continent," in *The Sweetpotato* (Dordrect, Netherlands: Springer, 2009), 391.
79. M. Akoroda, "Sweetpotato in West Africa," in *The Sweetpotato* (Dordrect, Netherlands: Springer, 2009), 442.
80. J. Low, et al, "Sweetpotato in Sub-Saharan Africa," in *The Sweetpotato* (Dordrect, Netherlands: Springer, 2009), 359.
81. Ibid, 363.
82. Paul C. Mangelsdorf, *Corn: Its Origins, Evolution, and Improvement* (Cambridge, MA: The Belknap Press of Harvard University Press, 1974), 14; Betty Fussell, *The Story of Corn* (New York: Knopf, 1992), 16; Johnson, *Tomatoes, Potatoes, Corn, and Beans*, 10.
83. Martin, Waldren and Stamp, *Principles of Field Crop Production*, 318–321.
84. Arturo Warman, *Corn and Capitalism: How a Botanical Bastard Grew to Global Dominance*, trans. Nancy L. Westrate (Chapel Hill and London: University of North Carolina Press, 2003), 37.
85. Ibid, 39.
86. Ibid, 40.
87. Johnson, *Tomatoes, Potatoes, Corn, and Beans*, 22.
88. Warman, *Corn and Capitalism*, 60.
89. Ibid, 61.
90. Ibid, 82–83.
91. Ibid, 84.
92. Ibid, 98.
93. Ibid, 100.
94. Ibid, 102.
95. Ibid, 97.

7 The Rise of the Land Grant Complex

OVERVIEW

Late-eighteenth-century Americans were heirs to the Enlightenment, which stressed that humans could improve society by acquiring and applying knowledge of nature. Here pioneers faced the daunting task of building a nation out of deciduous and coniferous forests and the prairie. This task required the practical talents of the civil engineer, the surveyor, the navigator, the mason, and the farmer. Nowhere was the zeal for practical knowledge more evident than in the northeast, where a society of small farmers, craftsmen, and artisans concentrated, in good Puritan fashion, on earning a living.

In Philadelphia, Benjamin Franklin (Figure 7.1) summarized the practical outlook of Americans by asserting that all knowledge should "tend to increase the power of man over matter, and multiply the conveniences or pleasures of life."[1] Franklin applied this vision to agriculture in 1769 by establishing the Committee on Husbandry and American Improvement within the American Philosophical Society. A more important manifestation of this concern for agricultural knowledge emerged from the Philadelphia Society for Promoting Agriculture, which John Bordley, a former member of the Governor's Council in Maryland, founded in 1785.[2] That Bordley received enthusiastic endorsement from Philadelphia's leading gentry confirmed their belief that science could improve agriculture. This was of course not a new idea. We have traced it to the Greeks, Romans, and Chinese. As the first agricultural society in the United States, the Philadelphia Society for Promoting Agriculture set the tone for subsequent societies by defining itself as an institution to help farmers increase production. To this end, it advocated crop rotations that incorporated legumes, manuring, deep plowing, draining damp soils, and selective breeding of plants and livestock. Its practical character appealed to farmers and by 1858 nearly 1000 agricultural societies existed throughout the United States.[3]

Beyond their function as forums for the exchange of practical information, these societies began agricultural experimentation in the United States. Initially, local farmers devised experiments on topics of immediate interest. A corn grower might compare the yields of several varieties or various planting depths of a single variety. The rigor and thoroughness of such tests naturally varied among farmers who, after all, were not scientists. To farmers, scientists were people who confined themselves to laboratory investigations of academic questions that provided nothing useful for agriculture. The esoteric pursuit of science, therefore, violated the practical ethos of agricultural institutions. Equally important, agricultural science did not exist as an organized body of knowledge at the beginning of the nineteenth century. Although physics had by then provided powerful mathematical descriptions

FIGURE 7.1 A founding father, Benjamin Franklin understood the importance of linking education with a thorough grounding in the sciences.

of physical phenomena, it excluded the living world from its compass. Natural history was then a taxonomic system rather than a science that could describe plant and animal growth in physical terms. Even had the members of agricultural institutions welcomed academic scientists into their midst, the latter possessed no special knowledge of agriculture. In large measure, agriculture in the West still depended on the treatises of Greece and Rome.

A nontheoretical approach typified not only agricultural institutions, but also the agricultural periodicals, the first of which a Baltimore journalist founded in 1819.[4] By 1830, roughly 30 farm periodicals existed with a total circulation of more than 100,000. By the Civil War, more than 400 journals existed, most of which were edited by journalists who augmented the pen with the plow and who considered themselves "practical farmers."[5] As practical people, they espoused "scientific farming," which entailed the standard practices of crop rotation, manuring, draining wet soils, deep plowing, and plant breeding. They reported experiments that illustrated the value of one or more of these practices.

As important as these movements were, they involved private rather than public initiatives and were only local in scope. In 1819, the New York state legislature took the first tentative step toward creating a public institution for agricultural science by founding the New York State Board of Agriculture.[6] It apparently stimulated little interest among farmers and when its charter expired in 1825 the legislature did not renew it. The New Hampshire legislature followed New York's example, establishing an agricultural board in 1820, but it held few meetings, inspired no more interest than the effort in New York had, and became defunct. The Ohio legislature founded the first permanent state board of agriculture in 1846.

Larger in scope than state efforts were ideas that Congress should sponsor agricultural education and create a national bureau of agriculture. Although the first U.S. president George Washington had advocated these ideas in 1797, Congress maintained a strict interpretation of the Constitution, which allowed it to promote science

Justus Liebig.

FIGURE 7.2   From the work of Justus von Liebig arose the agricultural experiment stations, first in Britain and the German states and then in the United States. (From Shutterstock.)

only by awarding patents. In 1839, Henry L. Ellsworth, the commissioner of patents, persuaded Congress to grant the Patent Office $1000 to distribute seeds to farmers and to collect agricultural statistics, but this was still far from Washington's vision.[7]

The great stimulus for federal funding of agricultural science and education came not from within the United States, but from Europe. In 1840, Justus von Liebig (Figure 7.2), the celebrated German chemist, announced that plants absorbed nitrogen through their roots in the form of ammonia and that farmers could increase crop yields by adding ammoniacal fertilizers to their soils.[8] Liebig had made an important advance, but he was wrong to suggest the direct uptake of ammonia from the soil. Roots absorb nitrogen in the form of nitrate or ammonium ions. Benjamin Silliman Jr., a Yale University chemist and editor of the *American Journal of Science* immediately praised Liebig, and in 1850 John P. Norton, a former pupil of Liebig, urged the state legislatures to create a system of agricultural colleges complete with experimental farms.[9] In the next few years, nothing happened. If the states would not act, perhaps Congress would fill the void, but the problem, then as now, was obstructionism. The Democrats were then the party of conservatism, the South, slavery, and states rights. It was the party of "no." As long as they retained their place in both houses of Congress, nothing would happen. Almost overnight, the Civil War changed these calculations. The succession of much of the South and with it the exodus of Democrats from the House and Senate left the new Republican Party free to act. The Republican Party was then the party of economic growth, of individual initiative, and upward mobility. Because agriculture was such an important part of the economy the Republicans, with President Abraham Lincoln's support, acted quickly, in 1862 creating the U.S. Department of Agriculture and passing the Morrill Act, which allotted public land to the states. The states sold this land to create an agricultural and mechanical college in each state, an action that appears to have gone beyond what even George Washington had hoped to achieve. Significantly, these achievements balanced agricultural education and experimentation. The Morrill Act required the land grant colleges to maintain a farm on which to conduct experiments, and Congress empowered the USDA to oversee "practical and scientific experiments."[10] The American West was not entirely happy with the Morrill Act because it gave federal land in the West to populous eastern states. Because the West had fewer people, it received less land. To be precise, the formula for the Morrill Act gave 30,000 acres of land per congressperson and senator in each state, so that New York, for example, received the most land. In 1890, Congress passed a second Morrill Act under the same terms but aimed this land at the southern states to encourage the establishment of land grant colleges for African Americans.

MOVEMENT TOWARD FEDERAL FUNDING OF THE AGRICULTURAL SCIENCES

The first European colonist in the Americas labored to change the agricultural landscape, a development treated in Chapter 6. This work had by the seventeenth century a scientific character in North America, with white colonists selecting the best plants, whose seeds would be sown next year. They experimented with different crops on different soils. The challenge was immense as the colonists aimed to plant Old World cultigens in the Americas and to adopt important American staples like corn, potatoes, sweet potatoes, cassava, beans, peanuts, and cacao. The tropical nature of some of these plants precluded their planting in what would become the United States, though citrus fruits would prove their value in South Florida, Arizona, Texas, and southern California and sugarcane in Louisiana. Massachusetts, Virginia, and other colonies planted new crops on a vast scale in a trial-and-error attempt to determine what would grow where. Colonial assemblies employed plant collectors to traverse Central and South America and the Caribbean for new crops or new varieties of an established crop. South Carolina and Georgia learned the hard way that citrus trees could not withstand a hard frost.

In the eighteenth century, eminent American botanist John Bartram established a research garden near Philadelphia.[11] American polymath and friend Benjamin Franklin sent Bartram seeds from France, and George Washington visited him to learn of his latest experiments. Bartram could boast an acre that yielded 36 bushels of wheat. Under his son William, the garden grew as it became more commercial. William raised a variety of ornamentals and shrubs for sale rather than for an overt scientific purpose. At the same time, William imported novelties from Europe, Asia, and Africa to attract the discerning eye. At its apex, the garden held more than 4000 species of plants. By 1825, William could count 116 varieties of apple trees, 108 of pears, 54 of cherries, 50 of plums, 16 of apricots, 74 of peaches, and 255 of the ornamental geranium. Sorghum was among the novelties that debuted in the Bartram garden.[12]

Agricultural practices and experimentation were still close to Roman traditions, as Americans experimented with different manures. Others tried legumes and plowed their land at different depths. Much of this work may have been ad hoc even if motivated by the Roman texts and under the direction of a single gentleman farmer. There appears to have been little effort to coordinate experiments. No institution then devoted its sole attention to the agricultural sciences. Some gentlemen farmers published the results of their experiments, though it is difficult to know how widely they were read.

About 1760, George Washington (Figure 7.3) turned increasing attention to agricultural experimentation.[13] Determined to diversify, he planted several varieties of wheat so that he would not rely solely on tobacco for cash. Washington soaked wheat seeds in brine to discover whether the process might deter pathogens, though he had little success protecting wheat from rust or the Hessian fly. Washington planted alfalfa to restore soil fertility and to feed livestock. Washington read widely about agriculture and, if he knew Latin, one would not be surprised to learn that he read the Roman texts. This may not have been true given that he referenced a large number of English titles. Washington experimented with clover, rye, spelt, and several vegetables not widely grown in Virginia. He mixed compost with horse, sheep, and cow dung, and worked the mixture into the soil to determine which was best. Washington mixed compost into clay soils, possibly to make them easier to work. Like Varro and Columella, Washington was interested in increasing soil fertility, a resolve that would become a guiding principle of the early agricultural experiment stations. Washington plowed deeply enough to bring subsoil to the surface and plowed under green manure. Rotating crops, he planted several species of fruit trees and ornamental shrubs to test their suitability to Virginia's soils and climate.

Thomas Jefferson shared this vision. More than Washington, he promoted all branches of science, particularly those with practical value like the agricultural sciences. Jefferson once remarked that one could do no greater good for his or her country than to introduce a new and useful plant, a vision that would guide the USDA.[14]

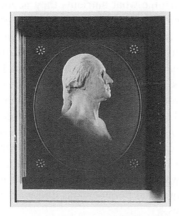

FIGURE 7.3   The first U.S. president, George Washington favored the creation of a national university to train students in the principles of agriculture.

Jefferson grew innumerable varieties of old and new world crops in his estate. He understood early that the tomato was not toxic, as folklore held. He imported several fruit trees from Europe and Asia. His vision for agriculture was deeply democratic, at least for whites. He opposed the idea that people should amass vast tracts of land. Rather the United States should remain, what Jefferson envisioned the Roman Republic to have been, a nation of small farmers. Nothing promoted democracy more than the free ownership of land.

In some ways sympathetic to the views of Jefferson, seventh U.S. president Andrew Jackson lauded farmers as the "chosen people of God." He must have been more sympathetic to the agricultural sciences than his biographers appear to have appreciated, because in 1835 he appointed Connecticut advocate of the sciences, Henry L. Ellsworth, to a post in the U.S. State Department. With grand visions, Ellsworth convinced Congress to appoint him the next year as the first Commissioner of Patents. He came to these duties as the author of articles on agriculture and the agricultural sciences and so was eager to use the Patent Office to promote the agricultural sciences. He had been for several years the author of an advice column, in a Connecticut newspaper, aimed at practical farmers. A lawyer by profession, Ellsworth, appears to have been an agricultural advocate by avocation. He was among a handful of agriculturists with the foresight to envision the Midwest as a corn empire. Under his direction, the Patent Office distributed free seeds throughout the United States. He believed that the federal government should distribute free agricultural literature to anyone who wanted the latest science. Through Ellsworth's efforts, Congress granted the Patent Office one clerk and two assistants with an annual budget of $1000 to send seeds to all who requested them. An additional $1000 followed to enable the Patent Office to collect and publish agricultural statistics throughout the United States.

Never satisfied with his duties, Ellsworth sought still greater advancement, asking Congress to establish in the Patent Office, a national laboratory for the study of agricultural chemistry, a branch of science that the early agricultural experiment stations would pursue. The prime purpose of this laboratory, Ellsworth hoped, would be to analyze soils to determine what nutrients they needed. This request came on the heels of Liebig's well-publicized work on agricultural chemistry. Given the popularity and importance of such research, it seems surprising that Congress was slow to act, though Ellsworth lived long enough to see the fulfillment of his objectives. In 1842 came a new request. Ellsworth wanted federal land set aside for experiments with crops, manures, and the new fertilizers on the market. He hoped that the Smithsonian Institution would inaugurate a series of lectures to communicate the findings of the agricultural sciences to farmers. He believed that every farmer should grasp the essentials of chemistry and so conduct his or her own experiments. Perhaps peculiarly, Ellsworth advocated the extension of sugar from cornstalks at a moment when America was awash in sugar from Louisiana, the Caribbean, and other parts of tropical America. He wanted the federal government to fund attempts to breed new varieties of wheat, corn, and potatoes and find ways to combat plant diseases, especially those of potatoes.

By 1847, two years after Ellsworth's retirement, the Patent Office distributed more than 60,000 seed packets.[15] The French minister of agriculture and commerce augmented this total by sending the Patent Office additional seeds for distribution. In 1848, Congress authorized the Patent Office to perform chemical analyses of foods, not quite what Ellsworth had wanted, but it was a start. The office hired people to analyze the contents of various types of wheat flour and of sugar from various sugarcane varieties in Louisiana. The results cannot have been surprising, for sugarcane, whatever the variety, yields the sugar sucrose. The same is true of sugar beets.

This work did not occur in a vacuum. As early as 1817, both the House and Senate considered a bill to create an agricultural department. Despite its failure, the idea remained in circulation. Similar bills followed in 1824 and 1825, and presidents John Quincy Adams, a gifted scientist, Andrew Jackson, and James K. Polk favored the idea of a department of agriculture. Perhaps because of Liebig's renown, the 1840s witnessed a series of debates about the desirability of such a department. As elsewhere, southern Democrats were intransigent. Only with the formation of the Republican Party in the 1850s, and its rise to prominence in the election of 1860 would the U.S. Department of Agriculture become reality.

THE LAND GRANT COLLEGES

The Morrill Act owed much to Vermont congressman Justin Smith Morrill (Figure 7.4), the architect of these laws. He put forth his proposals in 1857, drawing on a long tradition of government support for education. What was unique about Morrill's proposals was the insistence that education should produce utilitarian results for the masses of practical farmers and craftsmen. Morrill did not envision the new colleges as bastions of the liberal arts. He noted that land grants had supported the creation of Princeton University, Columbia University, Brown University,

FIGURE 7.4 Justin Morrill spearheaded the drive to create a national network of land grant colleges, at which training in agricultural and mechanical subjects should be the priority.

and Dartmouth College. In this context, it appears to be curious that the Constitution neither permits nor prohibits Congress from funding education, yet Congress did act. In 1787, the Northwest Ordinance authorized the nascent states in the Northwest Territory to fund education through the sale of land grants. In 1802, the Academy at West Point was an early beneficiary of Congressional funding for education. In the 1840s, on the heels of Liebig's celebrated publication, Jonathan B. Turner, professor at Illinois College, emerged as an advocate for agricultural education. Higher education must keep abreast of the times. A focus solely on the liberal arts did not truly advance economic opportunity. Colleges, extant and new, should concentrate on practical subjects like agriculture. Turner impressed several congressmen, but no one acted until Justin Morrill put forth the 1857 bill. Journalist Horace Greeley, the National Agricultural Society president Marshall O. Wilder, and the editors of farm periodicals supported Morrill. The argument was that agriculture was so important to the economy that Congress and the states could not fail to support it. Despite the opposition of Democrats, Morrill managed to shepherd his bill through the House with Ohio Senator Benjamin Wade guiding it through the Senate. As a senator 30 years earlier, James Buchanan supported agricultural education but then as president he vetoed the bill. The United States Agricultural Society denied the president honorary membership but the damage had been done. We have seen that everything changed with succession.

In 1862, just three U.S. public colleges taught agriculture: Michigan State University, Pennsylvania State University, and the University of Maryland.[16] In September 1862, Iowa became the first state to accept a land grant, creating Iowa Agricultural College (now Iowa State University).[17] Vermont and Connecticut acted later that year. In 1863, 14 more states joined the roster. By the end of the Civil War, 21 states had established agricultural and mechanical colleges. As was true of Pennsylvania, Michigan, and Maryland, Minnesota, Wisconsin, and Missouri allotted the proceeds from the sale of their land script to existing colleges. In some states, private colleges and universities accepted the land grant: MIT, Brown University in Rhode Island, and Cornell University in New York. Delaware, Illinois, Kentucky, and California created new colleges and universities. By 1870, all 37 states had accepted the land grant, though not all had created land grant colleges.[18] For instance, the Ohio legislature created The Ohio Agricultural and Mechanical College in 1873 though it did not matriculate the first student until 1878. By then, the college had mushroomed into The Ohio State University, a change that angered farmers. Elsewhere states struggled. In 1876, Delaware College had seven professors and 40 students. Illinois Agricultural College had just four faculty and 41 students. Louisiana State University had just three students, yet employed five professors. In 1877, not a single student enrolled in New Hampshire. Between 1865 and 1873, the University of Maryland graduated just eight students as it wore out five presidents. By these standards, Cornell and the University of Wisconsin were light years ahead of the pack. At the University of California at Berkeley, as at Ohio State, critics charged that the university neglected agriculture in favor of a classical education. The two were elitist rather than democratic.

Today, the land grant universities enroll 3 million students and graduate 500,000 men and women each year.[19] The land grant universities were among the earliest to enroll women and African Americans. Yet, scientist Neil E. Harl fears that these universities have scaled back their presence in the classroom and extension to immerse themselves in research. This is a bit ironic given that the early critics of Ohio State faulted it for doing too little agricultural research. As the number of farmers has declined, the land grant colleges have had to adjust. Colleges of agriculture enroll less than 10% of students at the land grant universities. These universities today emphasize engineering, technological fields, pharmacy, nursing and other health sciences, management, and the liberal arts. Land grant universities cannot remain viable if they concentrate only on agriculture. Instead, the universities must find new ways to partner with the EPA, the National Institutes of Health, and the Department of Commerce.

As important as these acts were as a declaration of the federal government's responsibility to ensure agriculture's vitality, their creation may have been premature given the rudimentary character of the agricultural sciences. No one yet knew, we have seen, that plant roots absorb nutrients as ions. The control of insects depended largely on preventing their reaching plants, as was the idea behind coating a shallow trench around cornfields with tar to trap chinch bugs. The control of diseases, as in Roman times, depended on isolating and often eliminating diseased plants. The lack of an extensive body of scientific knowledge about agriculture led land grant colleges to emphasize liberal arts rather than agricultural subjects. In effect, they emulated their more prestigious kin in the east in an effort to become more like Harvard and less like a cow college. By the 1870s, this emphasis angered farmers, who had initially hoped that these colleges would mold students into capable farmers. Instead, they felt that land grant colleges had betrayed them by abandoning agricultural subjects for Latin and Greek. Consequently, farmers scoffed at the claims of administrators and faculty that land grant colleges were the proper institutions to conduct agricultural experiments. They pointed to the low enrollments in agricultural classes and to the low esteem in which faculty and students held agricultural courses. Worse, few graduates of land grant colleges returned to farming, leading farmers to regard these colleges as antifarm.

Inequality is a serious problem. The historically black colleges and universities that have funded agricultural research have lagged behind the mostly white institutions that trace their lineage to the first Morrill Act. Not until 1966 did black land grant universities receive direct federal patronage, whereas the mostly white institutions traced their federal funding to the Hatch Act of 1887.[20] In 2006 and 2007, 18 black land grant universities received more than $72 million in federal funds. Much research naturally focuses on the needs of minority farmers. By then, the federal government had instituted a matching program where it funding agricultural research with $1 for each $1 of state funding. State funding, with this incentive, had begun to rise as early as 2000.

The emphasis is on co-operative research among Tuskegee University, Alabama A&M University, Alcorn State University, Florida A&M University, Fort Valley State University, North Carolina A&T University, Southern University and A&M College, and South Carolina State University, all historically black universities. These institutions created the Southern Food System Educational Consortium to focus on soil improvement, crop yields, and increasing incomes for poor farmers.

This alliance trains teachers to specialize in the sciences and mathematics in public schools. These universities train farmers to adopt organic farming practices in order to produce high-value vegetables and specialty crops. This consortium also emphasizes the importance of nutrition and health.[21]

A second partnership unites the USDA, Alabama A&M University, Alcorn State University, Florida A&M University, Fort Valley State University, Langston University, North Carolina A&T University, Prairie View A&M University, South Carolina State University, Tennessee State University, Tuskegee University, and the University of Arkansas at Pine Bluff to create the Southern Agricultural Biotechnology Consortium for Underserved Communities. The consortium funds plant and animal biotechnology at universities that might not otherwise have sufficient laboratory facilities for this research. The consortium also seeks to gage public sentiments regarding biotechnology, noting that many southerners, and Americans, in general, believe that Congress should mandate the labeling of biotech foods.[22]

The University of Georgia, North Carolina State University, Tuskegee University, Alabama A&M University, Alcorn State University, and the University of Arkansas at Pine Bluff created the Black Belt Regional Commission to combat poverty. The commission has concentrated on long-term rural poverty, regions where poverty has not improved over the last 20 years. The commission has partnered with health-care providers, employment agencies, public schools, and public transportation to improve opportunities for the poor.[23]

Research is one facet of agricultural research. The initial impetus was on applied research, which had been the aim of the Hatch Act. In 1906, however, Congress passed the Adams Act to award additional federal funds for basic science. The distinction between the two is fairly easy to draw. Applied science aims to create practical technologies. The USDA's mass production of penicillin in the 1940s for widespread use was applied science. Basic science does not seek an immediate utilitarian end. The measurement of the speed of light is basic science because it has no immediate application. Other areas are not so easy to pigeonhole. Is the collection of exotic varieties of rice in Africa basic or applied science? If the intent is to give plant breeders access to novel genes to incorporate into elite rice cultivars, the science appears to me to be applied, even though the application may not be immediate.

For research, particularly applied, to be relevant to farmers, some agency must communicate this science to them. For this purpose, the USDA hired Iowa Agricultural College professor Seaman A. Knapp to travel the American South to speak before groups of farmers.[24] In this respect, Knapp may have been the first extension agent, though others hold that renowned agricultural chemist George Washington Carver deserves this distinction. At any rate, Knapp was the USDA's first Special Agent for the Promotion of Agriculture in the South. Attuned to the visual impact of good farming practices, Knapp established demonstration farmers throughout the South so that farmers could evaluate new crop varieties, methods of water conservation, and other scientific matters. To this end, Knapp in 1903 rented 70 acres in Terrell, Texas, to create a sizable demonstration farm. Here he showed farmers ways to combat plant pests. As word of his success spread, Knapp found himself a correspondent with farmers nationwide eager to learn of other scientific advances. Knapp's guiding principle was to learn firsthand what farmers needed and then deliver the science to meet these needs.

Following Knapp's example in Terrell, Smith County, Texas, thanks to a USDA grant, hired its first extension agent in 1906. As extension spread to other areas of the United States, Congress in 1914 passed the Smith Lever Act, which empowered each land grant college to establish a Cooperative Extension Service as the voice of applied research at the colleges and experiment stations. The sole function of extension was to bring the benefits of science directly to farmers. It is useful to remember that the experiment stations had in an informal way pursued extension, though some scientists resisted time away from the laboratory and experiment plot and so welcomed the Smith Lever Act as assurance that they could concentrate on research. Congress, the states, and counties all help fund extension.

Today, 51 land grant universities, one in each state and Puerto Rico, trace their lineage to the Morrill Act of 1862.[25] An additional 16 states created land grant universities with the proceeds from land sales granted by the Morrill Act of 1890. These are the historically black land grant universities. A seventeenth university, Tuskegee University, also accepted a land grant in 1890. Some community colleges, many attended by Native Americans, receive federal funding for the teaching of agricultural subjects. Guam and the U.S. Virgin Islands have their own land grant universities. The land grant universities, as a generality, are an outgrowth of the empirical sciences of Aristotle, Theophrastus, and the Roman writers. More immediately, the land grant universities took as their model the German research universities that had such prestige in the nineteenth century. These universities taught a number of agricultural scientists including several American agricultural chemists. This was the moment when practical knowledge trumped the classics. Applied science was the route to success. Faculty at the land grant universities were to teach, conduct research leading to publication, and serve the community. Today research dominates with publication the measure of success. Faculty aim to spend more time in the laboratory and less in the classroom or extension. The researcher has more status than the extension agent. Extension agents resent this classification and often view their work as fulfilling a core mission at their university. As a rule, basic science carries more value than applied research.

Since the 1960s, researchers have come increasingly to rely on outside grants. This circumstance may give scientists greater freedom in the choice of research than the founders of the land grant complex had intended. Since the 1990s, the focus has turned to cellular and molecular biology in an attempt to improve agriculture, horticulture, and forestry. There remains an emphasis on improving yields to keep pace with the global population. But this may not be the dominant emphasis. Competing for funding is research to preserve and improve wetlands, forests, and rangeland. Genetics and biotechnology are demanding more funding, as is the project to map the genomes of numerous crops. Money invested in biotechnology expects in return plants with more nutrients, resistance to pests, drought, and other environmental stresses. Research seeks to make agriculture sustainable and decrease the use of pesticides and herbicides. Environmental quality, including safe water, depends on such strides. An important aim of both public and private research is to derive plants resistant or at least tolerant of insects, nematodes, and pathogens and better able to compete against weeds. In a sense, scientists are still fighting the problems that had plagued Cato the Elder 2300 years ago. Scientists aim to rejuvenate plant agriculture so that it remains competitive worldwide. Americans and others around the world are concerned not just about abundance and low prices, but also about how plant production affects the environment.

Since about 1970, the rise of integrated pest management has provided farmers a tool to reduce pesticide use. New efforts seek to use the methods of biological control that American entomologist Charles Valentine Riley pioneered in the nineteenth century. In the realm of biological control, researchers seek the latest generation of predators, parasites, competitors, pathogens, attractants, repellants, insect sterility, and pest-resistant plants to combat pests. In addition, research is ever more precise about quantifying the best cultural practices to limit the use of agrochemicals. New crop rotations, multicropping, tillage, or minimum or no tillage must all be part of the farmer's arsenal. Minimum and no till, of course, may increase pest populations.

Texas A&M University has recommended that a plant sciences department spend about 25% of its budget on biological control, 20% on tillage methods including minimum and no till, 17% on the development of new pesticides, fungicides, herbicides, and other agrochemicals, and 38% on developing an "integrated system" of plant protection.[26] In a different context, one might divide research into quadrants, with one-fourth going to plant population biology and ecology, another quarter to plant functional mechanisms, the third quarter to rhizosphere biology and ecology, and the last quarter to the relationship among plants and soil microbes. One may need as well a way of funding research on minor crops, a recommendation that may lead one to infer that the United States spends too much money on corn, soybeans, cotton, and other major crops. Basic research needs additional funds because it is a key to improving plant productivity. Research should include the study of population biology and genetics, soil microbiology, plant physiology, the interface between roots and soil, water conservation, macro- and micronutrients, and soil microbes.

TOWARD THE STATE AGRICULTURAL EXPERIMENT STATIONS

Farmers created a pragmatic and informal alternative to the land grant colleges. In 1863, the Massachusetts Board of Agriculture sponsored a series of practical lectures on farming.[27] By 1876, eight other states had followed Massachusetts by initiating lectures under the name of Farmers' Institutes. These lectures—delivered at locations throughout a state by farmers, urbanites who dabbled in farming, and ultimately agricultural scientists—were the nineteenth century's most successful answer to the problem of teaching farmers the lessons of the agricultural sciences. While farmers were establishing institutes, USDA bureaucrat Oliver H. Kelley, founded the Grange in 1867.[28] Initially, a social organization peripheral to the agricultural science movement, the Grange entered the mainstream of this movement during the next 20 years, helping to fashion the Hatch Act of 1887, which established a national network of agricultural experiment stations.

These stations, however, originated in Europe rather than the United States. In England, where agricultural improvement had long been an avocation of the gentry, John B. Lawes, an eccentric and wealthy developer of patent medicines, established a small greenhouse at Rothamstead in 1837.[29] During the 1840s, he expanded it into an agricultural experiment station complete with a barn for livestock feeding trials, experimental plots in which to compare crop varieties, crop rotations and fertilizers, and laboratories for fertilizer analyses. Meanwhile, Liebig's work had precipitated the rapid growth of a fertilizer industry throughout the German states. Germany was not then unified. The need to ensure the quality of fertilizers prompted Saxony, in 1851, to create a station, which was originally little more than a laboratory for analyzing fertilizers.[30] The German states had established 11 stations by 1857, 30 by 1867, and 70 by 1875. By 1874, the stations were also conducting feeding and breeding trials with cattle, field experiments with fertilizers, experiments on tobacco, grapes, wine making, silk production, and tests of seed germination rates. By then, it no longer made sense to distinguish between the Rothamstead example of an experiment station as primarily a farm for crop and livestock experiments and the German ideal of a station as a chemical laboratory. Instead, by the mid-1870s, these extremes had merged into a comprehensive model that combined experiments in the field and barn with chemical analyses.

As in the German states, the growth of a fertilizer industry in the United States exposed farmers to unscrupulous dealers. Because farmers generally had at best only a poor knowledge of chemistry and no way of analyzing fertilizers, they did not realize the extent of fraud until the 1850s, when Samuel W. Johnson, professor of chemistry at the Yale Analytical Laboratory, began to publicize fertilizer analyses.[31] Like several of his peers, Johnson had studied under Liebig and had come to appreciate the value of the German stations as watchdogs of the fertilizer industry. With support from Wilbur O. Atwater, the chemistry professor at Wesleyan University, and local farm organizations, Johnson lobbied the Connecticut legislature to create America's first agricultural experiment station, in 1875.[32] Although this was an important moment, the Connecticut station did not typify many of its successors because it initially spent too much time analyzing fertilizers and not enough applying science to help farmers increase productivity. The Connecticut station therefore reflected the German ideal of the early 1850s rather than the comprehensive model of the 1870s. By 1880, North Carolina, California, and New Jersey had established a station. In that year, however, none of these stations had the resources to conduct a broad range of experiments, and consequently they too reflected the outmoded German model.

As early as 1882, Seaman Knapp argued that Congress should finance the creation of a network of experiment stations in the United States. By early 1886, those who kept abreast of the issue recognized that Congress would eventually enact Knapp's proposal in the form of the Hatch bill, which would provide an annual appropriation of $15,000 to each state and territory to create or maintain an agricultural experiment station. Since the Morrill Act of 1862 had designated agricultural experimentation as a function of the land grant colleges, the Hatch bill envisioned an experiment station as part of each land grant college. The trustees of each college, then, would administer Hatch funds.

But matters did not go smoothly everywhere. In Ohio, farmers and farm leaders believed that the Ohio State University had squandered its opportunities to help farmers. So far as they knew the university performed no agricultural experiments of any kind. Hatch funds should not go to such a place but rather to the Ohio Agricultural Experiment Station, which had been independent from Ohio State University since the former's origin in 1882. With deep and distinguished roots in Ohio, the secretary of the National Grange, Joseph H. Brigham departed for Washington, DC, with the aim of convincing Congress to award Hatch funds to the agricultural experiment stations themselves wherever they were independent from a state's land grant college. In this case, Ohio and Massachusetts would be the beneficiaries of such language. With Ohio and Massachusetts united, Congress amended the Hatch bill to stipulate that wherever an experiment station was independent of its land grant college, it and not the college would receive Hatch funds. Congress passed and President Grover Cleveland signed the Hatch Act on March 2, 1887. Yet, Congress had been short-sighted. The act had not authorized the treasury to disperse these monies in 1887, with the result that the Ohio Agricultural Experiment Station, bereft of state and federal support, was alive in name only during 1887 and would not receive funding until 1888.[33]

A VARIETY OF PLANT SCIENCES AT PUBLIC AND PRIVATE INSTITUTIONS

The remaining chapters focus wholly or partly on the central achievements of the plant sciences. Here it is enough to overview the great diversity of achievements in the plant sciences, with particular reference to the land grant complex as well as some private initiatives. Focusing on the land grant universities and agricultural experiment stations, journalist Ron Smith judges the notable achievement of the agricultural sciences to be the production of more food on fewer hectares and with less labor.[34] Much of these gains have been possible through mechanization, satellite technology, and greater pest and disease control, though there are different ways of achieving the last two goals. Focusing on the achievements of the land grant universities and agricultural experiment stations in the American South, Smith credits Texas A&M University and its experiment station with producing better crop varieties not just for the South but also for the Midwest. The Texas land grant complex has bred new varieties of cotton, sorghum, grapefruit, and onion. Research has made more efficient use of irrigation to support these and other crops. Research also focuses on reducing the runoff of fertilizers. The Texas complex has bred half the wheat varieties grown in Texas. Current projects include research to improve biofuels and to breed better varieties of ornamentals, turfgrass, harvesting technology, and the technologies of food storage and transportation.

In the 1940s, the Mississippi State University Agricultural Experiment Station began research on improving soil fertility that remains vital worldwide. This research pioneered the use of anhydrous ammonium, a nitrogenous fertilizer first available to U.S. farmers in 1947. Melissa Mixon, associate vice president of the Mississippi Agricultural Experiment Station, believes the development of anhydrous ammonia to be among the top 10 achievements in the history of the agricultural sciences. The Mississippi Station also pioneered research in controlling the boll weevil throughout the South and the genetic engineering of cotton resistant to the insect. The station helped farmers convert to minimum tillage. Currently, the station works to develop biofuels from woody plants. According to Mixon, the agricultural sciences have improved "the quality of life of all Mississippians."[35]

California almond growers rely on the agricultural sciences for new and higher-yielding varieties (HYVs). The sciences have taught these growers to sharpen the use and timing of fertilizers. At the same time, the nature of the agricultural sciences appears to be in flux, with increasing engagement with environmental and social issues and a broad commitment to all aspects of the economy. Research focuses on nutrition, food safety, pest control, floriculture, and forestry. Despite this trend, the University of California at Davis, part of California's land grant system, is proud to have contributed to the research necessary to feed the world, a tall order with populations at or above 7 billion people worldwide. According to Neal Van Allen, dean of the college of agriculture at UC Davis, "One of the great success stories of the twentieth century was the ability of agriculture to keep ahead of the food needs of a very rapidly growing population on earth. Much of the credit for this success must be given to the land grant system of higher education, research and extension that was established in the United States."[36] The engines of research, the agricultural experiment stations, made breakthroughs in plant breeding, soil management, pest control, and mechanization. To remain effective, the land grant complex must have a greater commitment of funding and other resources. UC Davis has pioneered new varieties of tomatoes and harvesting methods. The plant sciences have made possible the growth of a $1 billion tomato industry in California per year. UC Davis has bred half the strawberry varieties grown worldwide as well as new grape varieties for the wine industry. California's food industry is competitive worldwide because it has benefited from the agricultural sciences.

The University of Georgia and its agricultural experiment station emphasizes increasing crop yields to keep food abundant. Research focuses both on improving soil fertility and on conserving soils for future generations. The University of Georgia land grant complex focuses on both basic and applied research, a tradition that dates to the Adams Act of 1906. The University of Georgia and its experiment station aim to breed better varieties of cotton, peanuts, forage grasses, grains, soybeans, turfgrass, tobacco, and vegetables. This system emphasizes interdisciplinary research among agronomists, entomologists, agricultural engineers, agricultural economists, food scientists, and plant pathologists. Many other agricultural experiment stations work in interdisciplinary teams as well. Among other activities, these teams work to find the precise amount and kind of nutrients for each crop in a particular location. Based on this work, the University of Georgia land grant system encourages farmers to add only that much fertilizer and water to the soil, so they need not worry about the cost and hazard of excess. This work appears to have some basis in Liebig's Law of the Minimum. The University of Georgia and its experiment station use aerial surveillance to identify broods of pests and to estimate yields. This system is investigating the potentials of sorghum, napier grass, and Bermuda grass as biofuels. The University of Georgia and its experiment station have the nation's largest staff of plant breeders, many of whom focus their research on cotton, peanuts, soybeans, wheat, turfgrass, ornamentals, blueberries, pecans, and plants suitable for biofuels. Thanks to the land grant complex in Georgia and elsewhere, the agricultural sciences have made farming more efficient.

According to scientists K. E. Woeste, S. B. Blanche, K. A. Moldenhauer, and C. D. Nelson, plant breeders have made "extraordinary contributions to the economic expansion of the United States."[37] Plant breeding has contributed to an increase in the population and to economic growth in general. The Amerindians had selected plants but were unfamiliar with the science of plant breeding (Chapter 4). Early European inhabitants of North America likewise selected plants but lacked knowledge of plant breeding. As Chapter 6 made clear, people throughout the world, including those in North America, imported crops from other parts of the world, a process that immigration hastened. Again Americans selected the most promising plants from a new population. Cross-pollination of varieties of corn, probably without human aid, created the dent corns that would be so important to the rise of the Midwest as the Corn Belt. As the next chapter makes clear, plant breeders developed new varieties resistant to pests or pathogens. These innovations reduced the risk of farming. Plants that yielded more biomass more reliably enriched American farmers and so contributed to rural development, including an increase in land values.

The New Deal of the 1930s invested in plant breeding as one strategy of many to improve rural lives that had been beset by hardship since the end of World War I. Plant breeding has stabilized agriculture, improved nutrition, and lowered food prices. Plant breeding made possible urbanization and industrialization. Plant breeding also enlarged the meaning of intellectual property rights and contributed to the rise of agribusiness. One can debate the merits of a system in which large farms have come to dominate American agriculture and to increase income inequality. The very success of plant breeding has led to obesity and diabetes. Will plant breeders help solve these problems? Plant breeding has also been at the forefront of developing plants for conversion into biofuels. One wonders as well whether plant breeding can help alleviate the problems of global warming and climate change. Perhaps plant breeders might derive perennials that capture carbon dioxide year round, as possible only for the subtropics and tropics. Perhaps one might breed trees and grasses that absorb extra carbon dioxide. Plant breeding may be an impetus behind the reforestation of denuded land. Plant breeding might tap into the organic movement by deriving new varieties of crops suited to organic agriculture. The organic model is tied to sustainability.[38]

Since 1958, the Agricultural Research Service (ARS) has conducted the research of the USDA. It too has pursued a diversity of sciences. Even before 1958, the USDA, as the next chapter makes clear, conducted research in breeding pest- and pathogen-resistant corn hybrids. This effort spilled over into the ARS. The agency works to keep new pests and plant pathogens out of the United States. The service is working to determine the effect of increasing concentrations of carbon dioxide on the growth of weeds, crops, tree crops, and other plants. The ARS is attempting to answer two basic questions. First, how will climate change affect agriculture in the United States and worldwide?[39] Second, how will climate change affect the population of plant pathogens and pests? The ARS has amassed among the world's most extensive seed banks to provide genetic resources to combat a pest or pathogen. Chapter 9 makes clear the value of these resources. To this end, the ARS maintains the National Plant Germplasm System, which attracts plant scientists worldwide. These resources aided the Green Revolution by helping scientists increase yields, nutrition, and tolerance of stresses, including insects and diseases. Other efforts include the U.S. National

Seed Storage Laboratory in Fort Collins, Colorado, which holds more than 1 million seeds, each with a unique genotype.[40] Another seed bank, the ARS Photosynthesis Research Laboratory in Urbana, Illinois, aims to help scientists draw on exotic germplasm to engineer plants more efficient at photosynthesis. Some of this research focuses on the enzyme rubisco, which directs the capture of carbon dioxide through stoma. Yet, rubisco, like many enzymes, is so ancient that it evolved when earth harbored little oxygen in the atmosphere. Consequently, it is not foolproof in differentiating carbon dioxide from oxygen. Consider the soybean as an example of this phenomenon. Four out of five gaseous molecules that enter the stoma of soybean plants will be oxygen rather than carbon dioxide. If soybean could be engineered to be more efficient, five out of five gaseous molecules will be carbon dioxide. The soybean, more effective at absorbing carbon dioxide, might in some small way mitigate the human production of carbon dioxide. Scientists wonder whether they might be able to replace rubisco with an enzyme from green algae to achieve such a soybean plant.[41] Like other members of the land grant complex, the ARS works to derive better biofuels.

With worldwide population forecast to be between 8 and 10 billion people by 2050, the ARS, in deference to its history, continues to apply science to increase crop yields.[42] U.S. crop yields increase by about 1% per year. Because humans rely on plants to sustain populations, the plant sciences must remain the focus of the agricultural sciences. The challenge of the plant sciences may be to increase crop yields in an environment of higher temperatures, greater fluctuations in rainfall, and cropping on marginal lands. As we have seen, the more efficient a crop is at photosynthesis, the better it should meet these conditions. The emphasis must fall upon the derivation of crop varieties that better tolerate environmental stresses. The ARS National Center for Agricultural Utilization Research in Peoria, Illinois, works to expand the markets for proteins, fatty acids, and other components of corn, soybeans, and other oilseed crops. This center was the first worldwide to mass produce penicillin about 1943. It was also the first to convert soybean oil into ink. Presently, the center aims to make plant-based cosmetics, lubricants, pharmaceuticals, baked goods, ice cream, and hamburger.[43]

The ARS Southern Regional Research Center in New Orleans, Louisiana, focuses on corn, cotton, rice, peanuts, and other regional crops. It is attempting to eliminate *Aspergillus* fungi carcinogens from cottonseeds, peanuts, corn kernels, and the seeds of other grains. Other species of *Aspergillus* act as pesticides. Scientists aim to encourage their growth and spread. The U.S. Environmental Protection Agency has authorized farmers to use this natural pesticide. Growers in Arizona have used the fungi to protect cotton. The ARS Beltsville Agricultural Research Center Vegetable Laboratory in Beltsville, Maryland, focuses on the old menace *Phytophthora infestans*, the water mold that caused the Irish Potato Famine in the 1840s. The water mold has mutated, giving rise to new and dangerous species. Accordingly, Beltsville scientists aim to breed potatoes resistant to these new threats. Peru and the United States have suffered from these new sports. The germplasm bank in Peru may contain sources of resistance, drawing U.S. scientists to Peru to study the collection in hopes of finding a resistant line to backcross into elite potato cultivars. For this reason, Peru has attracted scientists from New York, Washington, Oregon, Michigan, and other states where the potato is an important crop. Beltsville scientists are working to breed tomatoes, potatoes, peppers, eggplants, and several vegetables for enhanced nutrition. A new line of tomatoes is rich in beta-carotene, the precursor of vitamin A. These tomatoes have a long shelf life and more lycopene, an antioxidant.[44]

The ARS Western Regional Research Center's Processing Chemistry and Engineering Research Unit in Albany, California, has perfected a technique for growing cherry tomato plants through tissue culture in what is little more than a Petrie dish. At the moment, it is unclear what the practical implications might be, but these tomatoes have 10 times more lycopene than current elite varieties, furthering the aim of breeding high lycopene tomatoes.[45]

Even before the creation of the ARS, the USDA was aware of the value of genetic diversity among crop plants. For this reason, it made two bold moves in 1898. First, it dispatched two scientists, Charles Piper and William Morse, to China and Japan to collect new soybean varieties (see Chapter 6). This was a prescient move at a time when the soybean was yet to reach its status as a major crop in the Americas. Because soybeans were then used as forage, the size of the pods and seeds did not matter. For this reason, Piper and Morse returned with small podded soybeans. By the 1940s, the transition of the soybean plant from fodder to seed led scientists to revisit Piper and Morse's collection, among which were several large seeded varieties. These formed the foundation of all modern soybean cultivars in the United States. One of these varieties, Mukden, was important as a source resistant to *Phytophthora* root rot, a disease that might have threatened the soybean industry throughout the Midwest had not scientists quickly used Mukden to breed resistant varieties. Second, the USDA sent scientists to Russia to collect new varieties of wheat. The collection included durum and hard red winter wheats. The durum wheats remain important to farmers in Minnesota and North Dakota. The wheat collection is the foundation of a seed bank with some 500,000 unique genotypes. This collection is part of the U.S. National Plant Germplasm System, which has about 25 gene banks nationwide. The ARS National Seed Storage Laboratory alone has more than 327,000 seeds, each with a unique genotype.[46] This laboratory collects seeds worldwide, which are stored at −160°C and may remain viable for centuries. The transcription of a seed's DNA will allow scientists to evaluate it without the need for planting it. These collections are indispensable because the environment is not constant. New pests and pathogens are always evolving. Genetic novelty is essential to combat these hazards. An example of this success occurred in the 1990s. In 1986, the Russian wheat aphid entered Texas. By 1892, it had infected half the West's irrigated winter wheat, 14% of spring wheat, and one-third of barley. The immediate response was to screen the germplasm at the ARS laboratories. Accordingly, Colorado State University released the first aphid-resistant wheat, Halt. Rice germplasm centers in the United States and China fuel research in the breeding and genetics of the grass.[47]

An interesting use of the plant sciences has been to improve conditions for crops in the semiarid West. This research allows comparison with the dry farming empiricism of the Romans (Chapter 5). In both the West and Rome, dryland farming was and is the focus. In the West, this area stretches from the Canadian border to southern Texas. One encounters a range of soils: silty loam, silty clay, and sandy loam. Rain falls in summer, whereas winters are dry and cold. This model does not perfectly fit

Roman circumstance with a hot dry summer and a cool rainy winter. Drought is a danger, as the Dust Bowl made clear. Recurring droughts plagued the early 1950s, the 1960s, the 1970s, and the 2000s.[48] Precipitation is erratic, doubling one year only to halve the next. In the West, as in Rome, it was and is important to identify methods to minimize crop failure. The answer in Rome was to plow the land many times to eliminate weeds, the primary competitors for water and nutrients. The land grant complex has taken a different approach, as early as the 1980s advocating a reduction or abandonment of plowing. The first case is minimum tillage and the second no tillage. By not turning over the soil with the plow, no tillage prohibits water from evaporating from this soil and so is a water conservation technique. Minimum and no till are important on the Great Plains, whose chief crop is wheat, an opportunistic crop that absorbs water from newly planted fallowed land in winter and spring. It is interesting to note that the West has retained the practice of fallowing that is evident in the Roman texts. When planted early in spring, wheat matures before the onset of the hot weather and aridity of late summer.

In wheat's early history in North America, the colonists in the east had planted it in monoculture. The movement west after the American Revolution found more fertile lands in the Midwest and ultimately the Great Plains, but here monoculture failed, leading to the practice of cropping and then fallowing land. The fallowed land stored moisture. In the southern Great Plains, farmers planted wheat in October with the harvest in July. A fallow of 3 months ensued, followed by another crop of wheat. This system failed, leading to a fallow of 15 months, producing one crop every 2 years. The southern Great Plains favored winter wheat and the northern Great Plains spring wheat. In the northern plains, farmers planted spring wheat in April with a harvest in July. A 21-month fallow ensued. The use of fallow had drawbacks including soil erosion and a decrease in organic matter in the soil. At best, fallowed land captures 40% of rainwater and retention may be as low as 15%. The comparatively high percentage corresponds to minimum or no tillage. Since the late twentieth century, the movement has been toward no tillage, a technique that encourages the early germination of weeds and so increases the use of herbicides. Because the land is not plowed, weeds are not plowed under. Scientists recommend crop rotation along with no tillage. In 2002, in the Great Plains, about one-sixth of the land is not tilled and one-third is minimum tilled.[49] No till faces the barrier that yields are not as high as on plowed land because of competition from weeds. Absent tillage weedy grasses are difficult to control. In the southern Great Plains, two crops are grown every 3 years, with corn or sorghum being the summer crop. To its benefit, no tillage makes possible the use of land year round. In Montana and North Dakota, for example, the practice of fallowing land decreased from 1.2 million hectares in 1990 to 400,000 hectares in 2010.[50] No till works because the less one disturbs the land, the less evaporation one causes. Now farmers have replaced wheat and fallow with wheat and legume or wheat and oilseed crops. In 2010, legumes and oilseed crops occupied 14% of farmland in Montana and North Dakota. Since 2008, farmers have been more apt to plant legumes than oilseed crops. This transition has decreased erosion and increased soil fertility and organic matter in it. In no till spring, wheat, legumes, and oilseed crops all efficiently use water. In no till spring, wheat monoculture yields 25% more wheat than wheat and fallow in Montana.

In parts of Canada and Montana, pea or lentil in rotation with wheat in no tillage yields more wheat than monoculture or wheat and fallow. In Montana, the removal of pea or lentil from rotation reduces wheat yields 25%. One USDA study reported a 5% diminution in yields on no tilled land in Colorado and Nebraska in 1990 but a 20% increase on no tilled land in these states in 2010. No till appears to permit more intensive farming with grain and legume rotations.[51]

In the central Great Plains, farmers plant winter wheat in autumn, corn after the wheat harvest, and fallow the land after the corn harvest. Another rotation may be winter wheat, corn, proso millet, and fallow. No till supports corn, sorghum, winter wheat, foxtail millet, and sunflower. Corn had benefited from no till, as Colorado went from less than 10,000 ha of corn in 1986 to 96,000 ha in 2010. Between these years, the total area in Colorado planted to crops increased by 208,000 ha. No till has made it possible for farmers in Colorado and Nebraska to increase incomes by as much as 40%. In no till, corn and sorghum may be grown in rotation 2 years with the third to fallow. The use of more crops adds more crop residues and thus organic matter to the soil. Yet, these residues may harbor and increase pest populations. Wheat stem sawflies have become a burden to farmers on the northern plains. By 2007, the sawfly had spread south into the central plains.[52]

The southern Great Plains have been tardy in adopting minimum or no till. The USDA estimates that farmers no till fewer than 5% of cropland in northwestern Texas and southwestern Kansas. The lag may be due to the investment in cattle rather than crops and because forage crops grow slowly in no tilled land. Cattle also compact the soil leading stockmen to plow the land to lighten it. The southern Great Plains demonstrates the rule that the drier the climate, the greater the temptation to fallow land because the amount of water in the soil is the limiting factor in plant agriculture. The amount of water in the soil at planting best predicts yields. Rain or its absence in May and June are also predictors. The best correlation of soil water with yield holds for proso millet and *Phaseolus* beans. Long season crops, sunflower, corn, and sorghum, need moisture throughout the growing season.

In northeastern Montana, scientists are experimenting with using lentils and peas as green manure. Legume rotations, however, are less successful in the southern Great Plains. No till land may be compact and more vulnerable to runoff from a sudden storm. No till has made strides in the northern plains where more than one-quarter of farmland is not tilled. In the central plains, about 20% of farmland is not tilled.[53]

Public and private research has made U.S. agriculture the world's envy. American agriculture is productive and competitive in a global economy. As production has increased, however, the farm population has dwindled so that less than 2% of Americans farm. Left to its own devices, the free market seldom devotes sufficient money to the agricultural sciences and to science, in general. For this reason, the land grant complex has been and is necessary. Government has gone beyond the traditional funding of the sciences to protecting the intellectual property rights of what plant breeders and biotechnologists produce. To this, public sector has arisen a growing private initiative. Private firms, such as Monsanto, spend about $6 billion on agricultural research per year. The public investment in agricultural research totals about $4.3 billion per year and another $1.8 billion in extension.[54]

Since 1958, the ARS has been the research arm of the USDA. Funding for research and extension grew most rapidly during the 1960s and 1970s. Thereafter, public financing has grown at a slow pace, even halting during the 1980s. The 1990s have not seen a generous rebound, with public funding rising only 0.12% per year. In the twenty-first century, public funding has increased for research but not for extension. Neither have private funds for the agricultural sciences been abundant. Overall funding for the agricultural experiment stations is increasing faster than for the USDA. Less money goes to boost crop yields and more to health and nutrition, environmentalism and sustainability, natural resources, and biofuels. One might argue that the United States is underinvesting in farm productivity. This may be a misguided trend because every $1 invested in productivity yields more than $1 in increased crop yields. The U.S. retrenchment comes at a time when China, India, and Brazil are all increasing public funding of the agricultural sciences.[55]

These trends have brought the land grant complex far from its origins. Between 1862 and 1914, an increase in domestic and foreign demand for food spurred federal and state funding of research aimed at raising crop yields. The land grant complex believed that an increase in yields per acre was the best way to boost farm income. This mindset continued into the twentieth century. Between 1914 and 1950, research and extension grew rapidly, as the trend in investment makes clear. In 1910 the states and Congress spent $9 million on agricultural research. In 1930, the figure stood at $46 million. At the end of World War II, funding had surpassed $70 million per year. The USDA was then the prime mover, conducting 60% of total U.S. agricultural research. The emphasis was on developing HYVs of crops, mechanizing agriculture at a rapid pace and advancing the chemistry of pest and weed control.[56]

In hindsight, some scientists perceive the period between 1950 and 1990 as the Golden Era of public agricultural research.[57] This assessment may go too far, for we have seen the diminution of funding in the 1980s. This does not negate the fact that public funding rose between 1950 and 1980. In 1950, the Congress, the states, and localities spent $136 million on agricultural research, in 1960 $310 million, and in 1980 $1.7 billion. By 1980, however, the USDA was in danger of becoming superfluous. With a budget of $500 million, the USDA conducted just 30% of public agricultural research. Robust state funding had allowed the land grant universities and agricultural experiment stations to shoulder a larger share of research. Nonetheless, the USDA remained important in funding extension.

Other changes were afoot. As early as 1965, the USDA began to award grants to universities that had not been part of the land grant charter.[58] Since the 1990s, agricultural research, particularly plant biotechnology, has become more costly so that federal and state dollars do not go as far as one might hope. Meanwhile, the federal deficit and the growth in payments for social security, medicare, and medicaid have squeezed money that might have otherwise funded science at all levels. Hatch Act funds actually decreased in the 1990s and extension took an even harder hit. The land grant universities retooled, hiring experts in cellular and molecular biology and seeking grants from public and private sources outside the land grant system. In this era of austerity, the USDA refused to fund research that duplicated what other institutions had already done. A wealth gap widened, with elite land grant universities

172   Plants and People

like the University of Wisconsin and the University of California system growing in
wealth and prestige and the small and regional institutions languishing. Across the
board, extension has been in retreat.

Will old interests and initiative remain viable? Plant breeders, plant pathologists,
entomologists, and others defined success as increasing the yields of corn, soybeans,
cotton, or other important crops. Research attempted to derive drought-tolerant crops,
allowing farmers to skimp on irrigation. Pathogen- and pest-resistant crops served
a similar purpose, reducing the need for pesticides. The expansion of consumerism
in the United States, not a new phenomenon, led scientists to focus not merely on
crops, but also on turfgrass and ornamentals. At a time when Americans expect more
from science, many people are more than a generation removed from farming and
no longer understand why investment in the agricultural sciences remains important.
Land grant universities once looked to the children of farmers to fill enrollment
but now must turn elsewhere for students, many of whom do not have an interest in
agriculture. Where the taxpayer had once funded virtually all agricultural research,
the large farmers pay for this research by buying Monsanto's Roundup Ready corn,
soybeans, sugar beets, or some other crops.

NOTES

1. Benjamin Franklin quoted in Alexandra Oleson, "Introduction: To Build a New
   Intellectual Order," in *The Pursuit of Knowledge in the Early American Republic:
   American Scientific and Learned Societies from Colonial Times to the Civil War*, eds.
   Alexandra Oleson and Sanborn C. Brown (Baltimore: Johns Hopkins University Press,
   1976), xvi.
2. Simon Baatz, *Venerate the Plough: A History of the Philadelphia Society for Promoting
   Agriculture, 1785–1985* (Philadelphia: Philadelphia Society for Promoting Agriculture,
   1985), 3–4.
3. Margaret W. Rossiter, "The Organization of Agricultural Improvement in the
   United States, 1785–1865," in *The Pursuit of Knowledge in the Early American
   Republic: American Scientific Societies from Colonial Times to the Civil War*, eds.
   Alexandra Oleson and Sanborn C. Brown (Baltimore: Johns Hopkins University
   Press, 1976), 291.
4. Donald B. Marti, *To Improve the Soil and the Mind: Agricultural Societies, Journals,
   and Schools in the Northeastern States, 1791–1865* (Ann Arbor, MI: Published for the
   Agricultural History Society and the Dept. of Communications Arts, New York State
   College of Agriculture and Life Sciences, Cornell University by University Microfilms
   International, 1979), 126.
5. Albert L. Demaree, "The Farm Journals, Their Editors, and Their Public," *Agricultural
   History* 15 (July 1941): 183.
6. Alfred C. True, *History of Agricultural Education in the United States, 1785–1923*
   (Washington, DC: U.S. Department of Agriculture Miscellaneous Publications 36,
   1926), 24–28.
7. Alfred C. True, *History of Agricultural Experimentation and Research, 1607–1925,
   Including a History of the United States Department of Agriculture* (New York:
   Johnson Reprint, 1970), 25.
8. Charles A. Browne, "Liebig and the Law of the Minimum," in *Liebig and after Liebig*,
   ed. Forest Ray Moulton (Washington, DC: American Association for the Advancement
   of Science, 1942), 71–82.

9. H. C. Knoblauch, et al., *State Agricultural Experiment Stations: A History of Research Policy and Procedures* (Washington, DC: U.S. Department of Agriculture Miscellaneous Publication 904, 1962), 12.
10. True, *History of Agricultural Experimentation and Research*, 34–40; Margaret W. Rossiter, "The Organization of the Agricultural Sciences," in *The Organization of Knowledge in Modern America, 1860–1920*, ed. Alexandra Oleson and John Voss (Baltimore: Johns Hopkins University Press, 1979), 213.
11. True, *History of Agricultural Experimentation and Research*, 3.
12. Ibid, 4.
13. Ibid, 5.
14. Ibid, 6.
15. Ibid, 27.
16. Dennis W. Johnson, *The Laws that Shaped America: Fifteen Acts of Congress and Their Lasting Impact* (New York and London: Routledge, 2009), 98.
17. Ibid, 99.
18. Ibid, 100–101.
19. Ibid, 104.
20. Walter A. Hill, "Enhancing Small and Minority Farm Profitability and Rural Community Viability through New Partnerships," in *Perspectives on 21st Century Agriculture: A Tribute to Walter J. Armbruster*, eds. Ronald D. Knutson, Sharron D. Knutson, and David P. Ernstes (Oak Brook, IL: Farm Foundation, 2007), 47–48.
21. Ibid, 49.
22. Ibid, 49–50.
23. Ibid.
24. Edward G. Smith and Roland D. Smith, "Will Extension Be Relevant in the 21st Century," in *Perspectives on 21st Century Agriculture: A Tribute to Walter J. Armbruster*, eds. Ronald D. Knutson, Sharron D. Knutson, and David P. Ernstes (Oak Brook, IL: Farm Foundation, 2007), 57.
25. George R. McDowell, *Land-Grant Universities and Extension into the 21st Century: Renegotiating or Abandoning a Social Contract* (Ames: Iowa State University Press, 2001), 5.
26. *Opportunities to Meet Changing Needs: Research on Food, Agriculture, and Natural Resources* (College Station: Texas A & M University Press, 1994), 77.
27. True, *History of Agricultural Extension*, 9–10.
28. D. Sven Nordin, *Rich Harvest: A History of the Grange, 1867–1900* (Jackson: Mississippi State University Press, 1974), 11.
29. E. John Russell, "Rothamstead and Its Experiment Station," *Agricultural History* 16 (1942): 161–183.
30. Margaret W. Rossiter, *The Emergence of Agricultural Science: Justus Liebig and the Americans, 1840–1880* (New Haven: Yale University Press, 1975), 130; Samuel W. Johnson, "The Agricultural Experiment Stations of Europe," *Annual Report of the Sheffield Scientific School of Yale College* 10 (1874–1875): 21.
31. Rossiter, *Emergence of Agricultural Science*, 151–156; Charles A. Browne, "Agricultural Chemistry," *Journal of the American Chemical Society* 48 (September 1926): 179.
32. Rossiter, *Emergence of Agricultural Science*, 157–170; Alan I. Marcus, "From State Chemistry to State Science: The Transformation of the Idea of the Agricultural Experiment Station, 1875–1887," in *The Agricultural Scientific Enterprise: A System in Transition*, eds. Lawrence Busch and William B. Lacy (Boulder, CO: Westview Press, 1986), 3–5.
33. Congress, Senate, 49th Congress, 2d sess., *Congressional Record* (17 January 1887), 726; *Sixth Annual Report of the Ohio Agricultural Experiment Station for 1887* (Columbus, OH: Myers Brothers, 1888), 7.

34. Ron Smith, "Agricultural Experiment Stations Provide Catalyst for 150 Years of Innovations in Agriculture," *Southwest Farm Press* 36 (2009), 8.
35. Ibid, 9.
36. Ibid.
37. K. E. Woeste, S. B. Blanche, K. A. Moldenhauer, and C. D. Nelson, "Plant Breeding and Rural Development in the United States," *Crop Science* 50 (2010): 1625.
38. Ibid, 1628.
39. Jan Suszkiw, "Crop Plants: At the Root of Civilization," *Agricultural Research* (December 1999): 2.
40. Ibid, 10.
41. Ibid, 10–11.
42. Ibid, 10.
43. Ibid, 11.
44. Ibid, 11–12.
45. Ibid, 12.
46. Ibid, 14.
47. Ibid, 14; Hong Ma, Kang Chong, and Xing-Wang Deng, "Rice Research: Past, Present, and Future," Journal of Integrative *Plant Biology* 49 (2007): 729–730.
48. Neil Hanson, Brett Allen, R. Louis Baumhardt, and Drew Lyon, "Research Achievements and Adoption of No-Till, Dryland Cropping in the Semi Arid U.S. Great Plains," *Field Crops Research* 132 (2012): 196.
49. Ibid, 197.
50. Ibid, 198.
51. Ibid, 199.
52. Ibid.
53. Ibid, 202.
54. Julian M. Alston and Philip G. Pardey, "U.S. Agricultural Research and Technology Policy for the 21st Century," in *Perspectives on 21st Century Agriculture: A Tribute to Walter J. Armbruster*, eds. Ronald D. Knutson, Sharron D. Knutson, and David P. Ernstes (Oak Brook, IL: Farm Foundation, 2007), 67.
55. Kenneth R. Farrell, John E. Lee, Duane C. Acker, and Ronald D. Knutson, "Agricultural Research and Extension Policy in Retrospect: Implications for the Future," in *Perspectives on 21st Century Agriculture: A Tribute to Walter J. Armbruster*, eds. Ronald D. Knutson, Sharron D. Knutson, and David P. Ernstes (Oak Brook, IL: Farm Foundation, 2007), 74–76.
56. Ibid, 72–73.
57. Ibid, 73.
58. Ibid.

8 The Hybrid Corn Revolution

OVERVIEW

The development of hybrid corn was arguably the most important achievement of applied science in the twentieth century. The powerful, new technique of crossing inbred lines of corn to induce heterosis not only enabled breeders to derive high-yielding corn, but also to derive insect- and disease-resistant corn. The case of the European Corn Borer is illustrative. An unpleasant product of the Columbian Exchange, the borer entered the United States in 1917, migrating west. By 1921, it had inhabited western Ohio, then the gateway to the Corn Belt. Alarmed, the U.S. Department of Agriculture assigned corn breeder Glen Herbert Stringfield to Toledo, Ohio, to derive inbred lines and hybrids resistant to the insect in hope of stopping it in its tracks. Stringfield immediately sought an alliance with the Ohio Agricultural Experiment Station in Wooster. From the outset, then, the battle against the borer was a cooperative venture. The success of hybrid corn and its resistance to the European Corn Borer was thus largely an achievement of the land grant complex.[1]

In 1933, farmers in the Corn Belt planted 1% of acreage to hybrids. In 1943, they planted 90% of acreage to hybrids. In 1933, the corn yield in the United States averaged 22.6 bushels per acre. In 1965, it more than doubled to 55 bushels per acre. Equally remarkable, in 1936, Ohio corn growers planted hybrids on less than 2% of acreage. In 1948, they planted hybrids on 98% of acreage. In 1936, Ohio's corn yield averaged 33 bushels per acre, while in 1965 it nearly doubled to 74 bushels per acre.[2]

IMPROVEMENT OF CORN THROUGH BREEDING

After 1921, the European Corn Borer spread westward through northwestern Ohio, southeastern Michigan, Ontario, Canada, and beyond to all the states bordering the Great Lakes.

The improvement of soil fertility, the nineteenth century's answer to what plagued farmers, was but one means of raising crop yields. A parallel development was the improvement of crops through breeding techniques, and corn provided the exemplar. The science of corn breeding owes its origins to Charles Darwin, more famous as the originator of the theory of evolution by natural selection. Darwin was no practical breeder, but he was interested in the effects of crossing different varieties of corn. In 1871, he observed that intervarietal crosses of corn produced more vigorous offspring than intravarietal crosses, crosses between plants of the same variety, which in turn were more vigorous than inbreds.[3]

At Michigan Agricultural College (now Michigan State University), William J. Beal, who had studied under the renowned botanist Asa Gray at Harvard, followed Darwin's work with interest. Beal had likely known of Darwin's research because Darwin and Gray had been confidants. Unlike Darwin, Beal was intensely practical, and in 1877 he began to cross corn varieties solely in hopes of increasing their yields. He achieved small increases, but his results were not impressive enough to galvanize the attention of agronomists at other agricultural colleges and experiment stations.

The temptation among corn breeders to trace a linear descent from Beal to modern corn breeders is therefore an oversimplification. In reality, at the turn of the century, corn breeders followed the example of Illinois agronomist Cyril George Hopkins rather than of Beal. Between 1896 and 1899, Hopkins discovered a method to modify the percentage of oil and protein in corn. He planted seeds from one ear of corn per row, with each row comprising one of four characteristics: high oil, low oil, high protein, or low protein. By detasseling alternate rows and allowing open pollination, he was able to trace the female pedigree of each plant and to determine the female progenitors' contribution to oil and protein content in corn. He thereby aligned corn-breeding techniques more closely with livestock breeding practices in which breeders commonly evaluated cattle and hogs by the female pedigree. By using recurrent selection among the female rows, Hopkins announced in 1899 that he had increased oil content from 4.7% to 6.1% and protein from 10.9% to 12.3%. He also suggested the possibility of using this method, the ear-to-row technique, to increase corn yields.[4]

This suggestion impressed agronomists throughout the Corn Belt, who quickly adopted the technique. Their initial enthusiasm in 1899 faded during the 1910s because the ear-to-row method, when repeated over numerous generations, reduced rather than increased yields as the offspring were increasingly crosses of genetically similar individuals. The effect was akin to inbreeding, which reduced yields, as corn breeders since Darwin had recognized. During the 1910s, breeders largely abandoned Hopkins' method. By then, breeders were beginning to appreciate that yield is not a trait that depends on some fraction of genes that one could concentrate, as is the case with oil and protein content, but is rather a measure of vigor that depends on the entire genotype, specifically its degree of diversity and the complementary coupling of alleles. The genotype is the sum of all genes in an organism. Genes come in pairs. Each member of a pair is an allele.[5]

As the language of this explanation suggests, the impetus for modern corn breeding came not from Hopkins, but rather from the rediscovery of Austrian monk Gregor Mendel's laws in 1900. Briefly Mendel (Figure 8.1) had determined in the 1860s from experiments with peas that physical traits are inherited in pairs of discrete units, or, we have seen, alleles. The male and female parent each contributes half of each pair of alleles. If, in a heterozygous pair of alleles, one allele determines the expression of a trait, then that allele is dominant. A heterozygous coupling is one in which two different genes, or alleles, occupy that pair of alleles on a chromosome. For example, the allele for brown is dominant over that for blue in human eye color. The allele for the hidden trait in the heterozygous pair, blue in this case, is recessive. The recessive trait will only be expressed when it is in a homozygous coupling. A homozygous coupling is one in which the same gene or allele occupies both loci in the pair.[6]

FIGURE 8.1 Obscure during his life, Gregor Mendel's experiments with pea plants created the science of genetics.

Although the transmission of traits and their expression, particularly in humans, is more complex than Mendel had recognized, his understanding of heredity enabled scientists to revolutionize corn breeding. Geneticist George Harrison Shull at the Carnegie Institute was the first to recognize the potential of Mendelian principles to increase corn yields. Around 1907, Shull had begun to inbreed corn in an attempt to apply the theory of pure lines of Danish botanist Wilhelm Johannsen. After failing to derive exceptionally large and small beans by crossing beans of a normal self-pollinating variety, Johannsen concluded that a self-pollinating variety must be a pure type of nearly identical genotypes, which necessarily prohibited individuals from deviating markedly from the mean. The same is true of peas, which also self-pollinate, Mendel had known.

Unlike beans and peas, corn has widely separated anthers and stigmas, which make it a cross-pollinating plant. In any cornfield, the silk of each ear will likely receive pollen from many different corn plants, making each ear a complex of pure types. Shull realized by 1908 that in inbreeding corn he had isolated the pure types from a variety of corn. Consider as an example a variety of corn with the heterozygous alleles B and b. Half the progeny in a selfed line will retain the heterozygous pairing of Bb, and the other half will be homozygous in equal proportions of BB and bb. Remember, a heterozygous coupling involves two different genes or alleles, whereas a homozygous coupling is a doubling of the same gene or allele. After 5–7 generations of inbreeding, the descendants of a variety will be segregated into one of several pure types with all individuals in a pure type clustering near a single genotype.[7]

By 1908, Shull (Figure 8.2) realized that the isolation of pure lines would make possible the wide crosses necessary to achieve heterosis or hybrid vigor, a phenomenon that humans had long exploited by crossing the horse and donkey to obtain the mule, but which no one had systematically applied to increase crop yields. On one level, then, heterosis is the result of overdominance that is of marked heterozygosity. The overdominance theory states that the pair Bb is superior to either BB or bb. By separating corn varieties into pure types, breeders could imagine each pure line as a variety of peas. By crossing pure lines of corn, as Mendel had peas, breeders could induce heterosis.

Although the cross of two inbred lines of corn, the single cross, induced heterosis in offspring, it was not then commercially feasible because the inbred lines did not produce large enough ears from which to obtain a large seed stock. At the Connecticut Agricultural Experiment Station, geneticist Edward M. East assigned Donald F. Jones, then a graduate student, the task of developing a practical technique for obtaining large stocks of hybrid seed. In 1917, Jones solved the problem of limited seed stocks with the double cross (Figure 8.3), that is, a cross of two single cross lines. By providing the breeder with abundant seeds, the double cross made hybrid corn possible.

Beyond this practical work, Jones explained heterosis by postulating that intervarietal crosses, that is, the crossbreeding of two plants that are members of different varieties of a species, that produced high-yielding offspring must combine the best dominant traits of both parents so that advantageous alleles partially or completely masked less advantageous alleles. Moreover, advantageous alleles must be numerous enough that several of corn's 10 pairs of chromosomes contain multiple favorable

FIGURE 8.2   George Harrison Shull described but did not make the first hybrid crosses of corn.

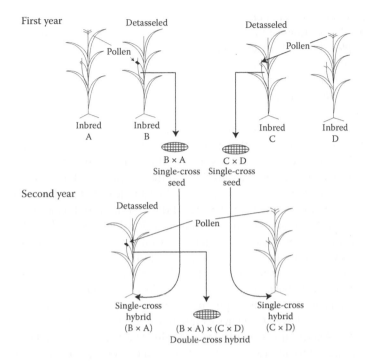

FIGURE 8.3 Donald F. Jones devised the first practical method for hybridizing corn on a commercial scale.

genes, making the fortuitous arrangement of genes that produced a high-yielding hybrid virtually impossible to duplicate by allowing hybrids to cross-pollinate on their own. Farmers could not therefore save hybrid seed for the result would be a diminution rather than an increase in yield.[8]

RISE OF THE EUROPEAN CORN BORER

The potential for increasing corn yields through the double cross coincided with the potential for enormous losses with the arrival of the European Corn Borer in Massachusetts in 1917. The adult female is a brown moth that deposits clusters of 15–20 eggs on the underside of corn leaves, usually in two broods, one in late May or early June and the second in mid-August. Larvae hatch in roughly 7 days and burrow into corn plants, where they mature through five instars or stages. The borer winters as a full-grown larvae, then pupates in early May. Although it feeds primarily on corn, the borer also devours some 200 species of plants, including oats.[9]

From Massachusetts, the borer spread to New York and Pennsylvania in 1919, Ontario, Canada, in 1920, Ohio in 1921, and Indiana in 1926. Thereafter, it spread more slowly, reaching Illinois only in 1939. Still, by 1952, it had reached 37 of the 48 contiguous states.[10]

Larvae of the first brood tunnel into leaf whorls, where they feed on emerging leaves and tassels, reducing yield by 3% or 4% per plant for each borer that reaches maturity. Even two borers per plant can reduce yields enough to diminish the farmer's income. Infestations of more than 20 borers per plant are severe and have caused nearly complete losses. Larvae of the second brood eat pollen accumulations on leaf axils, sheaths, and collars, all different parts of a corn plant, and tunnel into stalks. Because they infest corn after pollination and therefore after yield is largely set, they reduce yield only when infestation is severe enough to break stalks, thereby impairing mechanical harvesting.[11]

By 1922, the European Corn Borer (Figure 8.4) had become a serious pest, damaging between 5% and 20% of the corn grown in New England. In areas of the greatest infestation, corn growers lost nearly their entire crop. In 1924, growers in Kent and Essex counties, Ontario, lost 25% of their corn to the borer. By 1926, damage was severe enough to prompt one entomologist to fear that the borer would soon eliminate corn culture in the United States if the pest increased unabated.[12]

The serious threat that it posed to corn made critical the need to limit borer populations. Insecticides were then ineffective against the borer, leaving state legislatures and Congress with little alternative but to compel farmers to eliminate corn residues to prevent borers from wintering in them. In 1925, Ohio enacted the first law to require farmers to plow under, burn, or shred all cornstalks by May 1 of the following year. In 1926, the Canadian Parliament required all corn growers to eliminate residues after harvest. Congress followed Canada's example in February 1927 by granting the U.S. Department of Agriculture $10 million to coordinate a campaign with the experiment stations and state agricultural departments to reduce borer populations. In order to share in the federal largess, states needed to legislate that farmers eliminate corn stalks sometime before a new generation of borers appeared in May. In Republican fashion, this left the onus of legislating compliance to the states. In March 1927, New York, Pennsylvania, Indiana, and Michigan joined Ohio by passing similar laws.[13]

This federal and state effort failed to limit borer populations. In 1928, a USDA entomologist announced that the borer had more than 200 host plants, making the elimination of corn residues inconsequential because borers would overwinter elsewhere. Legislators, seemingly oblivious to this news, continued to hound farmers to destroy corn residues.[14]

Far more successful than the cleanup campaigns were the programs to breed borer-resistant hybrids. Resistance is the ability of a host to deter or defeat infestation

FIGURE 8.4 The European Corn Borer was once arguably the most dangerous pest of corn.

from an insect or pathogen. The highest level of resistance requires that the host possess antibiosis to the insect or pathogen. In 1924, the USDA established the first research center to screen inbreds for borer resistance and then combine resistant inbreds into double cross hybrids at Toledo, Ohio. By decade's end, it added a center at Lafayette, Indiana, and later one at Ankeny, Iowa.[15]

The earliest evidence for insect resistance in a crop came in 1792 when a New York farmer reported that Underhill, a variety of wheat, was resistant to the Hessian fly. The genetic basis of resistance was only discovered in 1905, when British plant breeder Rowland H. Biffen crossed American club wheat, a variety resistant to the fungus *Puccinia glumarum*, with Michigan bronze, a susceptible variety, finding all offspring susceptible to the fungus. By interbreeding the offspring, however, he obtained three-fourths of the second generation susceptible to the fungus and one-fourth resistant, a 3:1 ratio that corresponded to Mendelian segregation for a single recessive gene. Thereafter, plant breeders rapidly broadened the concept of resistance as a hereditary trait to include insect resistance.[16]

Resistance to the European Corn Borer depended on the derivation of resistant inbreds. In a double cross hybrid, at least three of the inbreds must be resistant for the hybrid to possess high resistance. Inbreds facilitated the search for resistance because, as homozygous lines, they revealed traits hidden in heterozygous lines. Consider again the hypothetical case of alleles B and b, in which B is dominant and b recessive. The recessive b will only express itself in the homozygous coupling bb. Inbreeding thereby allowed breeders more readily to identify advantageous traits such as a strong stalk, drought tolerance, or resistance to a disease or insect.[17]

Yet, the search for resistant inbreds was complicated by the fact that the borer hatched two broods per summer, which attacked different plant structures at different developmental stages. Resistance to both broods involved different complexes of genes. During the 1960s, breeders found that at least six genes confer a high level of resistance to the first brood, and at least seven genes confer resistance to the second brood. Unfortunately, the polygenetic nature of resistance increased the difficulty of combining all of the genes into a double cross hybrid. Not until 1979, for example, did breeders at the Iowa Agricultural Experiment Station derive inbreds with high resistance to both broods. Inbreds highly resistant to either brood have often lodged, that is, fallen down, more frequently and yielded less-than-elite inbreds. The polygenetic nature of borer resistance has made difficult the task of backcrossing resistance from a resistant inbred to a susceptible but otherwise elite inbred.[18]

To complicate the task of deriving borer-resistant lines, breeders initially could rarely acquire large enough borer populations with which to test lines for resistance. Instead, beginning in 1933, entomologists and breeders collected borer-infested corn stalks in the fall to use as a source of borers in spring. Not until 1965 did Wilbur Guthrie, an entomologist at the Corn Research Laboratory in Ankeny, derive a wheat germ diet suitable for raising borers in the laboratory. This diet finally enabled entomologists to raise large populations of both broods.[19]

The difficulty of breeding borer-resistant hybrids has led breeders to settle for tolerant hybrids.[20] Tolerance is the ability of a crop to withstand moderate infestations of an insect or pathogen with only slight yield reduction. Double cross hybrids that have a pedigree with two borer-resistant inbreds usually possess moderate tolerance.[21] Tolerance has been a byproduct of yield trials, which all experiment stations in the Corn Belt have conducted since the 1930s.[22] In these trials, breeders have compared the yields of hybrids in the field, where hybrids have faced normal environmental stresses including the borer. Over several years, the hybrids that consistently yielded well did so partly because they were tolerant to the borer. By the 1960s, most hybrids, whether double or single cross, possessed tolerance along with high yield and a strong stalk.[23]

THE EUROPEAN CORN BORER SPREADS TO THE GATEWAY TO THE CORN BELT

As one of the three research centers for the national program to breed borer-resistant hybrids, the OAES exemplified the close cooperation between the USDA and experiment stations. The choice of Ohio was logical because southern and western Ohio were fertile regions for corn culture in the Corn Belt and because Ohio was the first state in the Corn Belt to encounter the borer. The borer was the first significant example of the parasitic insects and diseases that plagued Ohio's agriculture during this century and that compelled the OAES to emphasize insect and disease control to a significant degree. The borer exacerbated the spread of diseases in Ohio because the holes it left provided access to fungi, especially stalk rot. Frederick D. Richey, a USDA corn breeder once characterized Ohio, with its profusion of insects and diseases, as "the armpit of the Corn Belt."[24]

In Ohio, the borer first appeared in the counties bordering Lake Erie in August 1921. That year, infestations were low, reaching only 2% or 3% of corn plants in Ashtabula county, and inflicted no appreciable damage to the corn crop. Although the legislature authorized the OAES and the Ohio Department of Agriculture to monitor the borer, it allocated no funds for this purpose until 1923. Extension agents encouraged farmers to limit borer populations by cutting cornstalks close to the ground at harvest and burning or shredding them to prevent borers from wintering in them. Reports of severe damage in Canada in 1924 and infestations as high as 40% in Ohio prompted the legislature in 1925 to mandate that farmers cut corn within 2 in. of the ground and plow under, burn, or shred all corn stocks by May 1 of the following year.[25]

This law had little immediate effect on borer populations. By 1927, the Ohio Department of Agriculture estimated that the borer had reduced corn yields 20% throughout Ohio, and that year the legislature appropriated $200,000 to the department to enforce the law of 1925.[26]

Just as the national cleanup campaign had failed to limit borer populations, it likewise failed in Ohio. Enforcement of the law was lax. As late as March 1928, the department charged four men, only two of whom were full time, to enforce compliance in all of Ohio's 88 counties. Farmers in eastern Ohio, where they usually grew more wheat than corn, generally ignored the law. Because few borers infested eastern Ohio, farmers believed they had no reason to obey the law. In contrast to eastern Ohio, the sheer scale of corn culture in western and southern Ohio made compliance burdensome, leading some farmers to ignore the law. Finally, as we have seen, the law could have little effect given that the borer had more than 200 host plants.[27]

The failure of the cleanup campaign made essential the breeding of resistant hybrids. In 1924, the USDA assigned Glen Stringfield (Figure 8.5) to the research center in Toledo to initiate a program to breed borer-resistant hybrids at the OAES and Toledo research center. A native Nebraskan, he had earned a B.S. in agronomy from the University of Nebraska in 1924. In both aptitude and training, Stringfield was a corn breeder, and that year he joined the USDA while working toward an M.S. in agronomy. One entomologist recalled Stringfield as the most adept corn breeder in the Corn Belt. He had an uncanny ability to recognize the slightest manifestations of insect and disease damage in corn. This acuity served him well in Ohio, where he seldom had enough money to hire a technician to assist him. Consequently, he maintained a small breeding plot of 3 or 4 acres, whereas the national average was 6 or 7 acres. Although such a small plot might have hampered others, Stringfield used it to his advantage by concentrating his work on borer resistance and discarding plants that had even the slightest damage.[28] In pursuing his work, Stringfield approached the agricultural sciences more narrowly than had the previous generation of scientists who had maintained close ties to farmers through lectures at Farmers' Institutes and copious correspondence with them. Instead, he concentrated on research and left to extension agents the transmission of practical knowledge to farmers. Unlike Charles Embre Thorne, the first full time director of the OAES, Stringfield sometimes pursued knowledge even when it had no immediate practical application.[29]

For example, in a series of experiments published in 1950, Stringfield discovered cases in which second-generation progeny of two inbred lines obtained by backcrossing, a technique in which a breeder crosses offspring with one of their parents, the first generation to either parent yielded more than second-generation offspring obtained by selfing the first generation. According to the overdominance theory, the yields of both have been equal because they contained the same degree of heterozygosity. Because overdominance could not adequately explain the difference

FIGURE 8.5 USDA corn breeder Glen Herbert Stringfield developed the first inbreds and hybrids resistant to the European Corn Borer.

in yields, Stringfield posited either that the second generation obtained by backcross had a more even distribution of favorable dominant genes than the selfed lines or that it had a more complementary arrangement of favorable genes than the selfed lines. The second explanation corroborated Donald Jones' explanation of heterosis as a function of the degree to which favorable genes are arrayed.[30]

As this example suggests, Stringfield brought to the OAES an expertise in corn breeding that it had previously lacked. In 1905, OAES agronomist Carlos Grant Williams had begun to tinker with Cyril Hopkins' ear-to-row method, though like other agronomists he had not succeeded in raising yields. The absence of publications on the method after 1917 suggests that he may have abandoned it around this time. Williams' interest in corn breeding should not disguise the fact that he was a generalist, not a professional corn breeder. In fact, Williams was better known as a wheat rather than corn breeder.[31]

Immediately upon his arrival at the OAES, Stringfield began the laborious process of screening inbreds for resistance to the borer's first brood. He concentrated only on first brood resistance because the second brood did not appear in Ohio until the late 1930s. Thereafter, it never reached damaging frequencies because Ohio's growing seasons were too short to sustain appreciable second brood populations. A contrast to Ohio is Iowa, where high populations of both broods have compelled corn breeders to seek both first and second brood resistance, an achievement that they finally realized in 1979. Still, Stringfield faced an arduous task because he needed to identify inbreds that combined high levels of first brood resistance with high yield and strong stalks.[32]

His first success was Oh51A, which he derived in 1932 and released to seed companies in 1940. The inbred, which was resistant to four of the borer's five instars, was the first highly resistant inbred in the Corn Belt. Stringfield had derived Oh51A by inbreeding progeny from a backcrossing program between a variety from Carlos Williams' experiments and an old, open-pollinated variety, Oh51, as the recurrent parent, that is, the parent to which offspring were backcrossed. In addition to broad resistance to the first brood, Oh51A possessed excellent root structure and a strong stalk. Unfortunately, its husks were tight and short, which exposed kernels to predatory birds.[33]

Stringfield used Oh51A to breed OhioK24, an early maturing hybrid developing at roughly 95 days between planting and mid-silk, which he released in 1942. He had derived OhioK24 in 1934 by crossing the offspring from the cross between Oh51A and a moderately borer-resistant line with the offspring from a cross between two early maturing inbreds. With two borer-resistant inbreds in its pedigree, OhioK24 possessed intermediate resistance to the first brood. In trials at the Toledo center in 1941, OhioK24 yielded 87.4 bushels per acre, an excellent yield for an early maturing hybrid. Although elite early maturing but susceptible hybrids average about 90 bushels per acre, OhioK24 performed better in moderate and severe infestations.[34]

The use of Oh51A in hybrids left unresolved the problem of short husks in an inbred. To overcome this problem, Stringfield derived Oh43 around 1941 and released it in 1949. He had derived Oh43 by inbreeding the progeny of a cross between an experimental line with good drought tolerance and a strong stalk and a Wisconsin line with good resistance to the first brood in its first two instars. Oh43 possessed good drought tolerance and lodging resistance, fuller husks than Oh51A, and better resistance to the first two instars than Oh51A, though its range of resistance was less broad. Oh43 also possessed better stalk rot resistance than Oh51A.[35]

So valuable was its strong stalk, drought resistance, and borer resistance that Stringfield also released Oh45, a sibling of Oh43, in 1949. Despite the fact that Oh45 suffered from short and tight husks like Oh51A, its similarity to Oh43 made it a valuable inbred.

The most important hybrid that Stringfield derived from Oh43 and Oh45 was OhioK62, which he released in 1949. Oh43 and Oh45 contributed good first brood resistance, and the other two inbreds contributed strong root anchorage, excellent stand, fast drying kernels, and early maturation. Like OhioK24, the early maturation of OhioK62 lent itself to cultivation in northern Ohio. The highest yielder in the K group, OhioK62 had averaged 90 bushels per acre at Toledo in trials between 1943 and 1949. This yield was competitive with elite early hybrids.[36]

These inbreds and hybrids established Stringfield's reputation as a leading corn breeder. As late as 1958, farmers in northern Indiana widely grew OhioK62, and as late as 1961 commercial hybrids throughout the world contained Oh51A, Oh43, or Oh45 in their pedigrees. For example, DeKalb's XL 45, a single cross hybrid, contained Oh43 in its pedigree. XL 45 had a broad range of adaptation, and during the 1950s farmers probably planted it more than any other hybrid in the Corn Belt. During this decade, it comprised one-sixth of DeKalb's sales. Unfortunately, one cannot determine with precision how widely the large seed companies used Stringfield's inbreds because they concealed their pedigrees. However, periodic surveys by the American Seed Trade Association, a national organization of private seed dealers, indicated that in 1956, more than 5% of hybrids sold in the United States had Oh43 in their pedigrees. Although this may not seem impressive, only one other public inbred, at 7.4%, surpassed Oh43. In 1964, Oh43 was the most extensively used public inbred, comprising 15.7% of hybrids sold in the United States. In contrast, the second most extensively used public inbred that year mustered just 11.9%. As late as 1974, the Iowa Agricultural Experiment Station used Oh43 and Oh45 in its breeding program, and as late as 1988 it still used Oh43. No other public inbred can claim such longevity.[37]

SPREAD OF HYBRID CORN IN THE CORN BELT

The intense interest of scientists and farmers to improve corn spawned the new seed corn industry around the turn of the twentieth century. The beginnings of this industry were modest. In 1901, Eugene D. Funk, an Illinois farmer who had attended Yale University, began to sell ear-to-row seeds in Illinois. By the late 1910s, however, Funk and everyone else interested in corn breeding discarded ear-to-row seeds.[38]

This was a blow to farmer-entrepreneurs like Funk who hoped to develop and market their own corn varieties because the ear-to-row method had not demanded knowledge of the new science of genetics. The future of corn improvement lay in the techniques of hybridization, which breeders at the experiment stations and USDA were developing. The new knowledge of genetics made the farmer-entrepreneur dependent on the expertise of professional breeders.

The new relationship between farmer-entrepreneurs and professional breeders is evident in the agreement to breed hybrids that Funk sought and obtained in late 1917 with the USDA. Four years earlier, Henry A. Wallace, the third and most distinguished member of the Wallace clan and a gifted breeder, had begun hybridizing corn. By 1926, he had formed Hi-Bred Seed, which he later prefaced with the moniker Pioneer. The next year, Eugene Funk founded Funk Brothers Seed Company. In 1933, Lester Pfister, an Illinois farmer who had dropped out of high school, founded Pfister. Finally by 1934, DeKalb Agricultural Association had coalesced from a loose confederation of scientifically minded farmers in DeKalb County, Illinois.[39]

The formation of these companies preceded significant demand from farmers for hybrid seeds. The agricultural depression of the 1920s persisted into the 1930s and dampened farmers' willingness to risk money on hybrids, which was a new commodity. The agricultural novelty of the 1920s was the tractor, and the farmer who had stretched his resources to buy one was not ready to spend additional money on hybrids.[40]

Also, the inertia of tradition initially worked against hybrid corn. Farmers had traditionally grown open pollinated varieties, which involved no expense for seed because they simply saved a portion of their harvest every year. In particular, they saved seed from "show corn": plump ears with even rows of perfect kernels, and often missed the point that no correlation existed between esthetic appeal and yield. Hybrid corn, in contrast, required corn growers to buy new seed every year because second-generation seeds yield poorly. Many farmers initially refused to believe that they could not save seed from hybrids as they had with open pollinated varieties. The poor harvests from second-generation seed reinforced farmers' perception that hybrids were less reliable than traditional varieties. Equally alarming, the first hybrids were often less well adapted to local conditions than the standard varieties of a region. Finally, hybrids were too long and slender to impress those accustomed to show corn.[41]

To surmount this skepticism, seed companies enticed farmers to try hybrids for a year or two on a few acres. Once hooked, these farmers became regular customers. In 1929, for example, Funk launched a massive campaign in newspapers and farm journals throughout Illinois. Farmers who bought seeds received both open pollinated and hybrid in equal amounts to compare yields. They received half a bushel of free hybrid seeds for every 5-bushel order. Those who supplied Funk with the names and addresses of other farmers also received free seeds. To market seeds at the grass roots level, companies recruited farmers to sell seeds to neighbors and friends. Such farmers grew company hybrids in their own cornfields, which illustrated the high yield of hybrids.[42]

Perversely, the most fortuitous agents in the adoption of hybrids were the droughts of 1934 and 1936 in the Corn Belt. The dense root systems of hybrids enabled them to tolerate drought better than open pollinated varieties and so to yield better under subnormal rainfall. In the predrought year of 1933, corn growers planted hybrids on only 1% of acreage in the Corn Belt. Only 10 years later, farmers in the Corn Belt planted hybrids on 90% of corn acreage.[43]

Iowa growers accomplished this transition most rapidly. In 1927 and 1928, they planted hybrids on only 0.3% and 0.7% of corn acreage, respectively, and in 1929 they planted hybrids on just 2% of acreage. By 1934, hybrids reached 15.4% of corn acreage in Iowa. In 1937, the figure rose to 61% and by 1940 hybrids had conquered nearly all acreage in the state.[44]

One sociologist has accounted for this expansion by arguing that until 1934 farmers bought only small amounts of seeds to test hybrids. He interpreted the years between 1927 and 1934 as an experimental period. The drought of 1934 demonstrated the superiority of hybrids and ended the trial period. Thereafter, adoption of hybrids was rapid.[45]

This explanation, though interesting, does not account for Iowa's leadership in the hybrid corn revolution. Farmers everywhere had the same opportunity to buy small quantities of seed for trials on a small portion of their land, yet only in Iowa did they have 15% of acreage to hybrids by 1934. This explanation overlooks that seed corn companies had established their headquarters in Illinois and Iowa and concentrated their marketing efforts on these states. The contest for leadership in this revolution pitted Illinois against Iowa. Because the first hybrids were not widely adaptable, seed companies found greater initial success in Iowa, which possessed greater uniformity of soils and climate than Illinois.[46]

Other Corn Belt states were only slightly less spectacular than Iowa in their transition to hybrids. By 1940, farmers planted hybrids on 77% of acreage in Illinois, 66% in Indiana, 54% in Minnesota, and 51% in Wisconsin. By 1950, Illinois had joined Iowa with 100% of acreage in hybrids. Even Nebraska, which had trailed other Corn Belt states in 1940 with 23% of acreage in hybrids, reached 94% in 1950. By that year, farmers planted hybrids on nearly 95% of acreage in the Corn Belt, as Graph 1 indicates.[47]

The large seed companies benefited from the rapid spread of hybrids. DeKalb alone increased production of hybrid seeds from 90,000 bushels in 1936 to 500,000 bushels in 1939. By 1940, DeKalb hybrids accounted for 10% of all corn seeds planted in the United States, including more than 4 million acres throughout the Corn Belt.[48]

The ultimate result of the hybrid corn revolution has been an impressive increase in corn production in the United States. Between 1933 and 1965, corn acreage in the United States decreased from 106 to 55 million acres, yet between these years corn production doubled from 2 to 4 billion bushels. During these years, yields per acre increased steadily as Graph 2 illustrates.[49]

The rapid spread of hybrid corn has had important implications for agricultural science and for agriculture. The experiment stations of the 1890s that offered a mix of services—analyses of fertilizers, water, and soils, inspection of crops and livestock for disease, seed germination tests, and weather forecasts—ceased to exist during the era of hybrid corn. Although the generalist at the turn of the twentieth century had made occasional house calls to help farmers combat a problem, the breeders of the 1930s confined their activities to the experimental plot. This resulted partly from the Smith Lever Act of 1914, which freed the experiment stations from extension work as well as from the desire of breeders to emulate their peers at the seed companies. They attended the same conferences as company men and even presented or published technical papers of no value to farmers.

Perhaps, most importantly, the hybridization of corn accentuated the quest for greater production in an era of food surpluses. In this era, service to farmers increasingly meant only farmers with enough land to squeeze profit from low crop prices. Only large farmers could afford heavy applications of fertilizers, pesticides, and herbicides that hybrids required for maximum yield.[50]

Unfortunately, the transition to hybrids increased the vulnerability of corn to epidemics. Because hybrids possessed greater genetic homogeneity than open pollinated varieties, they exposed farmers to greater risk of enormous losses from diseases or insects, as the Southern Corn Leaf Blight revealed in 1970.

In 1931, the discovery in Texas of a male sterile variety of corn quickened the pace to homogeneity. This discovery raised the possibility that seed companies could eliminate the tedious task of detasseling the rows of corn plants designated as female to permit hybrid crosses. More than merely tedious, by 1950, seed companies spent roughly $20 per acre, $10 million annually, to detassel female rows.[51]

Yet, the use of male sterile lines would only be feasible if crosses with it yielded fully fertile offspring. By 1950, Paul C. Mangelsdorf, then an agronomist at the Texas Agricultural Experiment Station, and Donald Jones had solved the riddle by locating the genes for male sterility in the cytoplasm of a corn cell rather than in the nucleus. Shifting their attention from the nucleic to the cytoplasmic genome, they identified lines that when crossed with the male sterile line yielded fertile offspring.

In 1951, corn growers began to obtain hybrids by planting male sterile lines as the female rather than detasseling these rows. By 1970, seedmen in the United States produced 85% of hybrids by this method. Because few male sterile varieties existed and only a few varieties combined with them to produce fertile offspring, this practice made the U.S. corn crop especially uniform.[52]

Unfortunately, the male sterile lines possessed receptors at the T sites in the cytoplasm for leaf blight, which made these lines vulnerable to damage in the leaf, stalk, husk, shank, and ear. Normal male lines are susceptible only to leaf damage. Although a few scientists in the Philippines had expressed concern about the susceptibility of corn to leaf blight during the 1960s, the fungus had never manifested virulent strains. Most scientists saw susceptibility to blight as an annoyance rather than cause for consternation. This changed in the summer of 1970 when South Corn Leaf Blight, a virulent race of the fungus *Helminthosporium maydis*, emerged. The humid summer that year facilitated its spread, and by harvest it had damaged between 15% and 30% of corn from Florida to Ontario and from Maine to Kansas and had cost farmers $1 billion. Blight diminished yields in the South by 50%–75%. Some growers reported total losses.[53]

In Ohio, nearly 90% of farmers reported damage from blight. Damage was particularly severe along the Ohio River, where humidity aided the fungus. There, with few exceptions, farmers lost their entire crop. Overall, Ohioans lost $50 million of the state's $200 million corn crop in 1970.[54]

The catastrophe did not recur in 1971 because seed companies abandoned the T-cytoplasmic lines, and farmers plowed under corn stalks to deprive fungi of a source of nourishment. Farmers planted blight-resistant hybrids. Nevertheless, the catastrophe of 1970 exposed the vulnerability of a crop with little genetic diversity as perhaps no disease had since Late Blight of Potato struck Ireland in the 1840s.

Despite spectacular damage from the Southern Corn Leaf Blight in 1970, one can hardly fault scientists for developing hybrid corn. By nature, scientists strive to gain as complete an understanding of nature as possible, though the use of such knowledge, whether to build a more destructive bomb or derive higher yielding corn, depends on the agenda of policymakers. In the case of hybrid corn, state and federal legislators financed agricultural research in anticipation of bountiful harvests. Agricultural scientists were realistic enough to appreciate this agenda and catered to it by touting the contribution of hybrid corn to feeding the Allies during World War II. More recently, Mangelsdorf asserted that the development of hybrid corn postponed a Malthusian crisis by a century. Large increases in yields from hybrids not only provided a concrete measure of returns from public investment in the agricultural sciences, but also satisfied public demand for low food prices.[55]

The percentage of acreage planted to Stringfield's hybrids is, however, a clinical and insufficient measure of his career. He derived the first highly resistant inbreds and hybrids. In this regard, he was pioneer. That the borer has never caused catastrophic losses to Ohio's corn crop is due at least partly to Stringfield's work. One participant in the plant-breeding symposium at Iowa State University in 1966 remarked that "the success of the breeding program [to derive borer resistant hybrids] is indicated by the fact that in Ohio the borer is no longer a major problem." In 1966, Governor James A. Rhodes named Stringfield one of the first three Ohioans to be inducted into the Ohio Hall of Fame. Significantly, Stringfield was the only one of the three who was not a native Ohioan, suggesting that Ohio farmers had adopted him as one of their own. Most importantly, for institution building, Stringfield's emphasis on breeding a crop resistant to an environmental hazard set the agenda at the OAES. "Likely no other man," OAES director Leo Rummell had remarked about Stringfield after his death, "was responsible to such high degree to increasing food production."[56]

HYBRID CORN IN THE DEVELOPING WORLD

Economists and agronomists have been alert to the gulf between practice and potential in parts of the developing world. In 1962, the Food and Agriculture Organization of the United Nations in Rome, Italy, began, 20 years after the success of hybrid corn in the United States, to encourage farmers in the developing world to adopt hybrids. Africa provides a case study of the successes and failures of hybrid corn in the developing world. During the twentieth century, corn became important in Southern Rhodesia (now Zimbabwe), Northern Rhodesia (now Zambia), and Kenya, as well as other countries.[57]

Because Southern and Northern Rhodesia were, before independence, part of the British empire, the whites who settled these regions took the best land for their own use, all with the encouragement of Britain. The white planter elite adopted hybrid corn at much higher rate than did the black masses. In the 1960s, the Salisbury Agricultural Research Station, the equivalent to an agricultural experiment station, developed an elite hybrid, SR-52, a single cross whose yield stood 50% higher than the double crosses of the time. Yet, akin to the Green Revolution crops, it needed irrigation and had too long a growing season for more temperate locales. Like other hybrids, the farmer could not save seeds of SR-52 and so had to buy new seeds every year. The expense of these seeds priced many African farmers out of the market. SR-52 was thus a crop of the white elites rather than the black smallholders. In the late 1960s, matters changed dramatically. A new hybrid, R201, was a good match for African farmers because it matured early and tolerated dry conditions and poor soils, just the type of land that so many blacks farmed.

Oddities likewise buffeted Zambia, where SR-52 and hybrids, in general, were not successful in the early years. As in Zimbabwe, the cost of hybrid seed was an obstacle that challenged scientists to develop something better than SR-52. They made several outcrosses with other lines of corn but the results were disappointing. Only in the 1980s did the Mazabuka Corn Research Institute produce much better hybrids, which African farmers adopted. Hybrid corn was a success after all.

Yet, one must be Pollyannaish to see much success in Kenya. Between 1996 and 2006, crop yields of several staples decreased in Kenya, a disappointment that should have pushed farmers to adopt hybrids. Kenya was clearly out of step with India and Mexico, which had benefited from the Green Revolution. Experimental trials have shown that hybrid corn plus fertilizers may double yields in Kenya. Yet, adoption rates vary from year to year and from farm to farm. MIT economist Tavneet Suri conjectures that many Kenyan farmers have poor lands on which hybrid corn does not excel. What incentive do they have to adopt hybrids? He has demonstrated that nearly one-third of surveyed farmers switched into and out of hybrid corn production between 1996 and 2006. Even from its inception, hybrid corn had trouble making strides in Kenya, where farmers first planted hybrid corn in the late 1960s and early 1970s. Even then, however, they preferred to plant millet and other crops with greater drought tolerance despite the fact that corn hybrids have excellent drought tolerance. Part of the problem may be that three-quarters of Kenya's corn crop comes from small farmers, the least likely to adopt hybrid corn.[58]

NOTES

1. J. R. Holbert, "A Great Industry in One Decade," *Finance*, 25 September 1944, 8; Frank J. Welch, "Hybrid Corn: A Symbol of American Agriculture," *Proceedings of Sixteenth Annual Hybrid Corn Industry—Research Conference, 1961* (Washington: American Seed Trade Association, 1962), 95–97; Paul C. Mangelsdorf, "Hybrid Corn," *Plant Agriculture: Readings from Scientific American*, ed. Jules Janick, Robert W. Schery, Frank W. Woods, and Vernon W. Ruttan (San Francisco: W. H. Freeman, 1970), 133.
2. U.S. Department of Agriculture, *Agricultural Statistics*, 1946 (Washington: GPO, 1946), 39; Robert D. Wych, "Production of Hybrid Seed Corn," *Corn and Corn Improvement*, ed. George F. Sprague and John W. Dudley, 3d ed. (Madison: American Society of Agronomy, 1988), 566; A. R. Hallauer, Wilbert A. Russell, and K. R. Lamkey, "Corn Breeding," Corn and Corn Improvement, 464. Glen H. Stringfield and D. H. Bowman, *Annual Report of Ohio Cooperative Corn Investigations* (Washington: GPO, 1942), 104. USDA, *Agricultural Statistics*, 1949 (Washington: GPO, 1949), 47; Glenn S. Ray, Eldon E. Houghton, and James R. Kendall, "Ohio Agricultural Statistics, 1942–1946," *Bulletin 691 of the Ohio Agricultural Experiment Station* (April 1950): 14; David H Boyne, Francis B. McCormick, Dan C. Tucker, Eldon E. Houghton, and Robert L. Griffith, "Ohio Agricultural Statistics, 1960–1965," *Research Bulletin 1019 of the Ohio Agricultural Research and Development Center* (October 1968): 20.
3. George H. Shull, "Beginnings of the Heterosis Concept," *Heterosis: A Record of Researches Directed toward Explaining and Utilizing the Vigor of Hybrids*, ed. John H. Gowen (Ames: Iowa State College, 1952), 14–48.

4. Cyril G. Hopkins, "Improvement in the Chemical Composition of the Corn Kernel," *Bulletin 55 of the Illinois Agricultural Experiment Station* (1899), 205–40; Cyril G. Hopkins, "Improvement in the Chemical Composition of the Corn Kernel," *Journal of the American Chemical Society* 21 (1899): 1039–1057. Mangelsdorf, "Hybrid Corn," 133–141; T. A. Kisselbach, "A Half Century of Corn Research, 1900–1950," *American Scientist* 39 (1951): 647–54. Henry A. Wallace and William L. Brown, *Corn and Its Early Fathers*, rev. ed. (Ames: Iowa State University, 1988), 60–110; Henry A. Wallace, "Public and Private Contributions to Hybrid Corn—Past and Future," *Proceedings of Tenth Annual Hybrid Corn Industry—Research Conference, 1955* (Chicago: American Seed Trade Association, 1956), 109–111; A. Richard Crab, *The Hybrid Corn Makes: Prophets of Plenty* (New Brunswick: Rutgers University Press, 1947), 15–20; Deborah Kay Fitzgerald, *The Business of Breeding: Hybrid Corn in Illinois, 1890–1940)*, 18–22; Singleton, "Early Researches in Maize Genetics," 124.

5. Mangelsdorf, "Hybrid Corn," 135–36; Mangelsdorf, *Corn: Its Origin, Evolution, and Improvement*, 213; Janick, Schery, Woods, and Ruttan, *Plant Science*, 389–90.

6. Fitzgerald, *Business of Breeding*, 23–30; Wallace and Brown, *Corn and Its Early Fathers*, 91–93.

7. George H. Shull, "The Composition of a Field of Maize," *Report of the American Breeder's Association* 4 (1908): 296–301; George H. Shull, "A Pure Line Method of Corn Breeding," *Report of the American Breeder's Association* 5 (1909): 51–59.

8. Donald F. Jones, "Dominance of Linked Factors as a Means of Accounting for Heterosis," *Genetics* 2 (1917): 466–79.

9. Retired entomologist Wilbur D. Guthrie at U.S. Department of Agriculture, interview by author, 1 April 1994, transcript, Ohio Agricultural Research and Development Center Archives, Wooster, Ohio, 1; B. E. Hodgson, "The Host Plants of the European Corn Borer in New England," *U.S. Department of Agriculture, Technical Bulletin 77* (1928): 63; F. F. Dicke, "Studies of the Host Plants of the European Corn Borer, Pyrausta nubilalis Hubner, in Southeastern Michigan," *Journal of Economic Entomology* 25 (August 1932): 869. W. A. Baker and William G. Bradley, "The European Corn Borer: Its Present Status and Methods of Control," *U.S. Department of Agriculture, Farmers' Bulletin 1548* (September 1948): 12–17; William G. Bradley, "The European Corn Borer," *Yearbook of Agriculture, 1952* (Washington: GPO, 1952), 614–16; F. F. Dicke and Wilbur D. Guthrie, "The Most Important Corn Insects," *Corn and Corn Improvement*, 788–91; H. B. Petty and J. W. Apple, "Insects," *Advances in Corn Production: Principles and Practices*, ed. W. H. Pierre, Samuel R. Aldrich, and W. P. Martin, 2d printing. (Ames: Iowa State University, 1967), 396–98.

10. Bradley, "The European Corn Borer," 614; Dicke and Guthrie, "The Most Important Corn Insects," 788; A. J. Ullstrup, "Evolution and Dynamics of Corn Diseases and Insect Problems Since the Advent of Hybrid Corn," *Maize Breeding and Genetics*, ed. David B. Walden (New York: John Wiley, 1978), 284–85.

11. D. J. Caffrey and L. H. Worthley, "A Progress Report on the Investigations of the European Corn Borer," *U.S. Department of Agriculture, Bulletin 1476* (February 1927), 90–103; F. F. Dicke and Wilbur D. Guthrie, "Genetics of Insect Resistance in Maize," *Maize Breeding and Genetics*, 300–01; Claud R. Neiswander and Edwin T. Hibbs, "Corn Borer First Discovered on Middle Bass Island in 1921," *Ohio Farm and Home Research* (July-August 1954): 52; Guthrie, 1994, 1–2; Wilbur D. Guthrie, "Techniques, Accomplishments and Future Potential of Breeding for Resistance to European Corn Borers in Corn," *Proceedings of the Summer Institute on Biological Control of Plant Insects and Diseases*, ed. Fowden G. Maxwell and P. A. Harris (Jackson: University of Mississippi, 1974), 360; Wilbur D. Guthrie and F. F. Dicke "Resistance of Inbred Lines of Dent Corn to Leaf Feeding by 1st-Brood European Corn Borers," *Iowa State Journal of Science* 46 (February 1972): 339.

12. D. J. Caffrey, "Recent Developments in Entomological Research on the Corn Borer," *Journal of the American Society of Agronomy* 20 (October 1928): 1005; Caffrey and Worthley, "Progress Report on the Investigations of the European Corn Borer," 147–48; Bradley, "The European Corn Borer," 614.; Claud R. Neiswander, "An Adventure in Adaptation: The European Corn Borer, *Ostrinia nubilalis* (Hubn.)," *Research Bulletin 116 of the Ohio Agricultural Experiment Station* (June 1962): 4.

13. *Forty-Fifth Annual Report of the Ohio Agricultural Experiment Station for the Year Ended June 30, 1926* (Wooster: Experiment Station Press, 1927, 49–51; Caffrey and Worthley, "Progress Report on the Investigations of the European Corn Borer," 148; W. R. Walton, "Corn Borer," *Yearbook of Agriculture*, 1927 (Washington: GPO, 1927), 202–03.

14. Guthrie, 1994, 1; Hodgson, "Host Plants of the European Corn Borer," 63.

15. Guthrie, 1994; Ullstrup, "Evolution and Dynamics of Corn Diseases and Insect Problems Since the Advent of Hybrid Corn," 284; Reginald H. Painter, *Insect Resistance in Crop Plants* (New York: Macmillan, 1951), 242.

16. R. H. Biffen, "Mendel's Laws of Inheritance and Wheat Breeding," *Journal of Agricultural Science* 1 (1905): 4–48; Ralph O. Snelling, "Resistance of Plants to Insect Attack," *Botanical Review* 7 (October 1941): 544; E. E. Ortman and D. C. Peters, "Introduction," *Breeding Plants Resistant to Insects*, ed. Fowden G. Maxwell and Peter R. Jennings (New York: John Wiley, 1980), 3–4; J. C. Walker, "The Role of Pest Resistance in New Varieties," *Plant Breeding: A Symposium Held at Iowa State University*, ed. Kenneth J. Frey (Ames: Iowa State University, 1966): 220; John F. Schafer "Host Plant Resistance to Plant Pathogens and Insects," *Proceedings of the Summer Institute on Biological Control of Plant Insects and Diseases*, 239.

17. Dicke and Penny, "Built-In Resistance Helps Corn Fight Borers," 10.

18. Wilbert A. Russell and Wilbur D. Guthrie, "Registration of B85 and B86 Germplasm Lines of Maize," *Crop Science* 19 (1979): 565; F. F. Dicke and Wilbur D. Guthrie, "Genetics of Insect Resistance in Maize," *Maize Breeding and Genetics*, 302–07; Guthrie, "Techniques, Accomplishments and Future Potential of Breeding for Resistance to European Corn Borers," 359–80; J. R. Klenks, W. A. Russell, and Wilbur D. Guthrie, "Recurrent Selection for Resistance to European Corn Borer in a Corn Synthetic and Correlated Effects on Agronomic Traits," *Crop Science* 26 (1986): 864; Wilbert A. Russell, Wilbur D. Guthrie, and R. L. Grindeland, "Breeding for Resistance in Maize to First and Second Broods of the European Corn Borer," *Crop Science* 14 (1974): 725; Retired corn breeder William R. Findlay of the U.S. Department of Agriculture, 3 March 1994, transcript, Ohio Agricultural Research and Development Center Archives, Wooster, Ohio, 3–4; Gene E. Scott, F. F. Dicke, and L. H. Penny, "Location of Genes Conditioning Resistance in Corn to Leaf Feeding of the European Corn Borer," *Crop Science* 6 (1966): 444–46; Wilbur D. Guthrie and Glen H. Stringfield, "The Recovery of Genes Controlling Corn Borer Resistance in a Backcrossing Program" *Journal of Economic Entomology* (1961): 784–87.

19. L. H. Patch and L. L. Peirce, "Laboratory Production of Clusters of European Corn Borer Eggs for Use in Hand Infestation of Corn," *Journal of Economic Entomology* 26 (1933): 196–204; Painter, *Insect Resistance in Crop Plants*, 243; Dicke and Guthrie, "The Most Important Corn Insects," 830; Dicke and Guthrie, "Genetics of Insect Resistance in Maize," 301; Wilbur D. Guthrie, E. S. Raun, F. F. Dicke, G. R. Pesho, and S. W. Carter, "Laboratory Production of European Corn Borer Egg Masses," *Iowa State Journal of Science* 40 (1965): 65–83; Guthrie, "Techniques, Accomplishments and Future Potential of Breeding for Resistance to European Corn Borers," 361; Guthrie, 1994, 2.

20. Findlay, 1994, 3; Guthrie, "Techniques, Accomplishments and Future Potential of Breeding for Resistance to European Corn Borers," 360.

21. Dicke and Penny, "Built-In Resistance Helps Corn Fight Borers," 10.

22. Painter, *Insect Resistance in Corn Plants*, 241.
23. Guthrie, 1994, 3; Guthrie, "Techniques, Accomplishments and Future Potential of Breeding for Resistance to European Corn Borers," 359.
24. Retired plant pathologist Lansing E. Williams at Ohio Agricultural Research and Development Center, 2 and 7 March 1994, transcript, OARDC Archives, Wooster, Ohio, 2; E. C. Cotton, "The European Corn Borer," BiMonthly Bulletin 11 and 12 of the Ohio Agricultural Experiment Station 6 (November-December 1921): 180–81.
25. *Forty-First Annual Report of the Ohio Agricultural Experiment Station*, 1922, 31; *Forty-Second Annual Report of the Ohio Agricultural Experiment Station for the Year Ended June 30, 1923*, Wooster: Experiment Station Press, 1923), 39; "Report of the North Central States Entomologists, March 5–6, 1923," *Reports of the North Central States Entomologists, 1923–1935* [No publication information], 6; L. L. Huber and Claud R. Neiswander, "The European Corn Borer in Ohio," *BiMonthly Bulletin of the Ohio Agricultural Experiment Station* 12 (January-February 1927): 5; Neiswander and Hibbs, "Corn Borer First Discovered on Middle Bass Island in 1921," 52.
26. Walter H. Loyd, "Massing for Corn Borer Offensive," *Ohio Farmer*, 19 March 1927, 3.
27. "Report of the North Central States Entomologists, March 5–6, 1923," 6; "Report of the Entomologists, March 2, 1928," *Reports of the North Central States Entomologists, 1923–1935*, 1–2; "Minutes of the Meeting of the North Central States Entomologists, March 1 and 2, 1928," *Reports of the North Central States Entomologists, 1923–1935*, 12.
28. Findlay, 1994, 1; Guthrie, 1994, 2–3; Robert M Salter, L. E. Thatcher, J. T. McClure, "Agronomic Research on the European Corn Borer in Ohio," *Journal of the American Society of Agronomy* 20 (October 1928): 1024; Glen H. Stringfield, "Your Yields Going Up?" *Ohio Farmer*, 1 February 1958, 5; "Three Named to Ohio Hall of Fame," *Ohio Farmer*, 20 August 1966, 15; "Glen H. Stringfield Was Agronomist at OARDC," *Daily Record*, 1 July 1975, 2.
29. That Stringfield largely eschewed extension is evident from the fact that he published only one article in *Ohio Farmer* during his 34 years at the OAES. Stringfield, "Your Yields Going Up?" 5, 44–45.
30. Glen H. Stringfield, "Heterozygosis and Hybrid Vigor in Maize," *Agronomy Journal* 42 (1950): 45–52.
31. "Pedigreed Seed Corn," *Circular 42 of the Ohio Agricultural Experiment Station* (14 September 1905): 96–103; Carlos G. Williams, "Improving the Corn Crop by Selection and Breeding," *Proceedings of the State Farmers' Institute Held in Columbus, Ohio, January 14–15, 1908* (Springfield: Springfield Publishing, 1908), 159–74; William K. Greenbank, *Guidebook of the Ohio Agricultural Experiment Station* (Wooster: Experiment Station Press, 1917), 23.
32. Findlay, 1994, 3–4; Meyers, Huber, Neiswander, Richey, and Stringfield, "Experiments on Breeding Corn Resistant to the European Corn Borer," 6–7. Glen H. Stringfield, R. D. Lewis, and H. L. Pfaff, "The Ohio Cooperative Corn Performance Tests," *Special Circular 59 of the Ohio Agricultural Experiment Station* (January 1940), 1–27; Glen H. Stringfield, R. D. Lewis, and H. L. Pfaff, "The Ohio Cooperative Corn Performance Tests," *Special Circular 64 of the Ohio Agricultural Experiment Station* (February 1942), 1–38; Claud R. Neiswander and Edwin T. Hibbs, "Corn Borer," *Ohio Farm and Home Research* (July-August 1954): 53; Pesho, Dicke, and Russell, "Resistance of Inbred Lines of Corn," 85–98; Russell and Guthrie, "Registration of B85 and B86," 565.
33. Glen H. Stringfield, "Maize Inbred Lines of Ohio," *Research Bulletin 831 of the Ohio Agricultural Experiment Station* (April 1959): 28; "New Hybrid Corn Shows Promise of Reducing Bird Damage," *Ohio Farm and Home Research* (November-December 1952): 82, 96; Maurice L. Giltz, "Blackbirds Have Been a Menace to Corn Production Since 1860," *Ohio Farm and Home Research* (January-February 1959): 3.

34. "Crop Varieties Recommended for Ohio in 1955," *Ohio Farm and Home Research* (November-December 1954): 87; Glen H. Stringfield, R. D. Lewis, and H. L. Pfaff, "The Ohio Cooperative Corn Performance Tests," *Special Circular 64 of the Ohio Agricultural Experiment Station* (February 1942): 5; Stringfield, "Maize Inbred Lines of Ohio," 43, 49, 51.

35. Stringfield, Lewis, and Pfaff, "Ohio Cooperative Corn Performance Tests," 7.

36. Stringfield, "Maize Inbred Lines of Ohio," 31.

37. Guthrie, "Techniques, Accomplishments and Future Potential of Breeding for Resistance to European Corn Borer," 360; Findlay, 1994, 2–3; "Crop Varieties Recommended for Ohio in 1955," 87; Glen H. Stringfield and H. L. Pfaff, "A Summary of Corn Performance Experiments in Ohio: 1943 to 1949," *Research Circular 14 of the Ohio Agricultural Experiment Station* (August 1951): 2; Glen H. Stringfield and H. L. Pfaff, "The Ohio Corn Performance Tests: 1948," *Agronomy Department Series Mimeograph Report 116 of the Ohio Agricultural Experiment Station* (21 February 1949): 5; Edwin T. Hibbs, Wilbur D. Guthrie, and Claud R. Neiswander, "The Yield Performance of a Resistant and a Susceptible Field Corn Hybrid under Different Intensities of European Corn Borer Infestation," *Research Bulletin 818 of the Ohio Agricultural Experiment Station* (October 1958):4.

38. "Miracle Men of the Corn Belt," *Popular Mechanics*, August 1940, 226–229; Hayes, *Professor's Story of Hybrid Corn,* 168–87; Leon Steele, "The Hybrid Corn Industry in the United States," *Maize Breeding and Genetics*, 29–40; Wych, "Production of Hybrid Seed Corn," 565–605; Fitzgerald, *Business of Breeding,* 20–21, 137–38, 150–54; Helen M. Cavanagh, *Seed, Soil, and Science: The Story of Eugene D. Funk* (Chicago: Lakeside Press, 1959), 85–91.

39. Wallace and Brown, *Corn and Its Early Fathers,* 11–20; William L. Brown, "H. A. Wallace and the Development of Hybrid Corn," *Annals of Iowa* 47 (1983): 167–79; Edward L. Schapsmeier and Frederick H. Schapsmeier, *Henry A. Wallace of Iowa: The Agrarian Years, 1910–1940* (Ames: Iowa State University, 1968), 16–29. Fitzgerald, *Business of Breeding*, 150–54, 161, 178–85; 185–89; Hayes, *Professor's Story of Hybrid Corn*, 170–71; Cavanagh, *Seed, Soil and Science*, 80; "Miracle Men of the Corn Belt," 228.

40. Fitzgerald, *Business of Breeding*, 129.

41. Stringfield, "Your Yields Going Up?" 5; Henry A. Wallace, "The Revolution in Corn Breeding," *Prairie Farmer*, 21 March 1925, 1; "Hybrid Produces Poor Seed," *The Missouri Farmer*, 15 November 1938, 7; "Miracle Men of the Corn Belt," 228.

42. Bryce Ryan, "A Study in Technological Diffusion," *Rural Sociology* 13 (1948): 273–85; Cavanagh, *Seed, Soil and Science*, 413–18.

43. R. D. Lewis, "Present and Future Trends in the Breeding of Corn Hybrids," *Proceedings of the Nebraska Crop Growers' Association* 32 (1940): 24–25; "Miracle Men of the Corn Belt," 120A; "The Coming Revolution in Corn Production," *The Missouri Farmer*, 15 November 1938, 5; U.S. Department of Agriculture, *Technology on the Farm* (Washington: GPO, 1940), 21; Hallauer, Russell, and Lamkey, "Corn Breeding," 464.

44. Ryan, "A Study in Technological Diffusion," 273–85; Zvi Griliches, "Hybrid Corn and the Economics of Innovation," *Science* 132 (29 July 1960): 276–77; Lewis, "Present and Future Trends in the Breeding of Corn Hybrids," 23.

45. Ryan, "A Study in Technological Diffusion," 283–85.

46. Kiesselbach, "A Half Century of Corn Research, 1900–1950," 651; Lewis, "Present and Future Trends in the Breeding of Corn Hybrids," 23.

47. U.S. Department of Agriculture, *Agricultural Statistics, 1944* (Washington: GPO, 1945), 39; U.S. Department of Agriculture, *Agricultural Statistics, 1948* (Washington: GPO, 1949), 48; U.S. Department of Agriculture, *Agricultural Statistics, 1952* (Washington: GPO, 1945), 41.

48. Kiesselbach, "A Half Century of Corn Research, 1900–1950," 653.

49. U.S. Department of Agriculture, *Agricultural Statistics, 1937* (Washington: GPO, 1938), 41; U.S. Department of Agriculture, *Agricultural Statistics, 1940* (Washington: GPO, 1941), 47; U.S. Department of Agriculture, *Agricultural Statistics, 1943* (Washington: GPO, 1944), 37; U.S. Department of Agriculture, *Agricultural Statistics, 1945* (Washington: GPO, 1946), 39; U.S. Department of Agriculture, *Agricultural Statistics, 1947* (Washington: GPO, 1948), 39; U.S. Department of Agriculture, *Agricultural Statistics, 1949* (Washington, GPO, 1950), 43; U.S. Department of Agriculture, *Agricultural Statistics, 1951* (Washington: GPO, 1952), 38; U.S. Department of Agriculture, *Agricultural Statistics, 1953* (Washington: GPO, 1954), 32; U.S. Department of Agriculture, *Agricultural Statistics, 1955* (Washington: GPO, 1956), 28; U.S. Department of Agriculture, *Agricultural Statistics, 1957* (Washington: GPO, 1958), 36; U.S. Department of Agriculture, *Agricultural Statistics, 1959* (Washington: GPO, 1960), 30; U.S. Department of Agriculture, *Agricultural Statistics, 1961* (Washington: GPO, 1962), 30; U.S. Department of Agriculture, *Agricultural Statistics, 1963* (Washington: GPO, 1964), 30; U.S. Department of Agriculture, *Agricultural Statistics, 1965* (Washington: GPO, 1966), 31; U.S. Department of Agriculture, *Agricultural Statistics, 1967* (Washington: GPO, 1968), 36.

50. James T. Jardine, "Aiding the War Effort through Experiment Station Research," *Report of the Agricultural Experiment Stations, 1944* (Washington: GPO, 1945), 11; "Corn Crop Enters Fight," *Ohio Farmer*, 6 February 1943, 19; "Research Doesn't Cost, It Pays," *Ohio Farmer*, 20 December 1952, 3.

51. Mangelsdorf, "Hybrid Corn," 138–139; "Genetic Vulnerability of Major Crops," *Genetic Vulnerability of Major Crops*, 10–11.

52. Jugenheimer, *Hybrid Maize Breeding and Seed Production*, 227.

53. Kenaga, Principles of Phytopathology, 276; Heather Johnston Nicholson, "Agricultural Scientists and Agricultural Research: The Case of Southern Corn Leaf Blight," The New Politics of Food, ed. Don F. Hadwiger and William P. Browne (Lexington: Lexington Books, 1978), 94.

54. George F. Sprague, "A Situation Statement of the Southern Corn Leaf Blight," Record Group 22/a/51/24, "Corn Virus Diseases, 1965–1977," The Ohio State University Archives, Columbus, Ohio, 3; C. Wayne Ellett, "Southern Corn Leaf Blight and Ear Rot in Ohio," Record Group 22/a/51/24, "Corn Virus Diseases, 1965–1977," The OSU Archives; Retired director Roy M. Kottman of Ohio Agricultural Research and Development Center, interview by author, 22 March 1993, transcript, OARDC Archives, 7.

55. Jardine, 11; Mangelsdorf, "Hybrid Corn," 134, 140.

56. Walker, "Role of Pest Resistance in New Varieties," 232; "Three Named to Ohio Hall of Fame," *Ohio Farmer*, 20 August 1966, 15; Leo L. Rummell, *"For Land's Sake": A Century of Progress from Farming to Agri-Business* (Tucson: Balkow Printing, 1979), 81.

57. Noel Kingsbury, *Hybrid: The History and Science of Plant Breeding* (Chicago and London: University of Chicago Press, 2009), 248–249.

58. Tavneet Suri, "Selection and Comparative Advantage in Technology," http://www.yale.edu/~egcenter

9 The Green Revolution

GEOPOLITICS AND THE GREEN REVOLUTION

One might trace the roots of the Green Revolution to 1941, when the Rockefeller Foundation and Ford Foundation established the Centro International de Agriculture Tropical (CIAT) in Columbia and the International Institute for Tropical Agriculture (IITA) in Nigeria, and in 1971 the Consultative Group on International Agricultural Research (CGIAR). As early as 1954, Rockefeller Foundation scientists and policymakers had envisioned what would become CGIAR. This is the moment when scientists R. E. Evenson and D. Gotlin imagine the origins of the Green Revolution. In the early 1940s, despite the ravages of World War II, these organizations had the lofty goal of ending hunger and poverty in the developing world. In 1968, William S. Gaud, an administrator in the U.S. Agency for International Development (USAID), coined the phrase Green Revolution. This language was carefully chosen. The revolution was green in order to highlight the role of plants and to contrast with red, the color of Soviet communism. In fact, the Green Revolution was born in a geopolitical context.[1] As early as 1949, the then U.S. President Harry Truman stated the foreign policy goal of wooing unaligned nations to the United States by helping them produce more food through the application of science to farming. As early as 1969, the U.S. House of Representatives acknowledged that the success of the American-led Green Revolution would head off a Malthusian crisis in the developing world that, if allowed to happen, would destabilize nations and perhaps bring them into the orbit of Soviet or Chinese communism. By helping the developing world feed itself, the United States might win converts and allies among the nonaligned nations of the developing world, strengthening America and potentially weakening communism. These hopes for developments were important at a time when the United States was fighting a war in Vietnam to prevent Southeast Asia from following China into communism. In this context, Ford Foundation policymaker Wolf Ladejinsky, in 1969, dismissed the notion that science could be socially, economically, and politically neutral.[2]

OVERVIEW OF SCIENCE AND THE GREEN REVOLUTION

The Green Revolution was not even biologically neutral. In India, scientists favored wheat over rice even though rice was the more important crop. The India of the Green Revolution planted 125 million hectares to grains, an amount that is three-quarters of the nation's arable land. Of these 125 million hectares, about one-third is planted to wheat and legumes, one-third to rice, and one-third to sorghum and millet. Most of India's rice, as is true of much of South, East, and Southeast Asian rice, is paddy, not the dry land varieties that have had some success in parts of Africa and the United States. In 1970, Indian farmers planted 15% of their rice land to high-yielding varieties (HYVs). By 1980, the figure had more than doubled to one-third. Yield gains have been dramatic, ranging between 1000 and 1300 kg of rice per hectare during this period. Yet, closer analysis reveals problems. Between 1961 and 1976, rice yields in India grew only 1.1% per year because 60% of rice lands were rainfed, suffering from nonuniform distribution with floods alternating with dry spells.

The Green Revolution took advantage of three breeding techniques. The first and the oldest technique involved crossbreeding the best specimens of a plant, yielding the first filial (F_1) generation of offspring.[3] The breeder discarded all but the most promising offspring. The remaining plants, crossed again, yielded F_2. The scientist repeated the procedure as long as necessary to obtain an elite variety with several agronomic traits. One must note, however, that this technique produces genetically uniform plants, a danger when hazards arise. The second technique uses wide crosses between varieties to yield hybrids, a program that has worked well with corn. The crossing of plants with widely different genotypes yields offspring with hybrid vigor and other desirable traits. The first corn hybrids, for example, not only yielded more grain, but they were also drought-tolerant and did not lodge. One of the Centro International de Mejora miento de Maiz y Trigo (The International Center for Corn and Wheat Improvement)'s hybrid corn varieties recorded the second highest yield in Mexican history. When the crosses are too disparate, however, offspring are often sterile, producing few if any viable seeds. The third technique, discussed more fully below, is the shuttle-breeding program that Norman Borlaug (Figure 9.1) conceived.

Historian John H. Perkins conceived plant breeding as the core of the Green Revolution.[4] By about 1950, many scientists were persuaded that the addition of water, fertilizers, and the breeding of disease-resistant wheat were the best ways

FIGURE 9.1 A Nobel laureate, Norman Borlaug, perhaps more than anyone else, was the architect of the Green Revolution.

to increase yields. Also important was a cultivar's ability to compete with weeds and tolerate high insect populations. The HYVs turned out to compete well against weeds. Winter wheat was the focus where winters were moderate enough to permit autumn planting and summer harvest. Where winters were too cold, spring wheat was the only alternative to breeders. Scientists favored durum wheat where noodles or pasta was a staple of the diet. The dwarfing genes that scientists acquired from Japan were first used in the United States, then in Mexico, and from Mexico to India. The Green Revolution coincided with a fertilizer revolution in which fertilizers with nitrogen, potassium, and phosphorus—the big three—became cheap enough for farmers in the United States and Europe to afford, though the poor in the developing world were at a disadvantage.

In some ways, Japan, with a growing population and limited land, was a precursor to the Green Revolution. In the twentieth century, Japan labored to breed HYVs of wheat and rice.[5] The use of fertilizers in Japan had only limited success because the extra nitrogen caused a plant to elongate its stalk, which then lodged under the weight of heavy grains. Japanese scientists were among the first to understand that dwarf wheat and rice were necessary in any breeding program that wished to prevent lodging. The United States took notice of these dwarfs, acquiring them for its seed banks. As early as 1911, Japan shared these dwarfs with France and Italy.[6] Japan experimented with crossing these dwarfs with U.S. and Korean wheat and rice varieties. A few of the offspring were important to the launching of the Green Revolution.

In the United States, University of Minnesota agronomist H. K. Hayes worked to breed rust-resistant spring wheat in 1934, obtaining Thatcher, the first rust-resistant spring wheat and a cultivar that the Green Revolution would use as a source of rust-resistant genes.[7] After World War II, General Douglas MacArthur invited Hayes' protégé and USDA wheat breeder Samuel Cecil Salmon to Japan to oversee the efforts to feed the masses. Salmon immediately focused his attention on Japanese dwarf wheat Norin 10, sending seeds to breeders throughout the United States. In 1946, scientists in Arizona planted four precious seeds on an experimental plot.[8] The next year, the arrival of more Norin 10 seeds allowed the USDA to direct the planting of them at Washington State University. Crosses were also made in Oregon and Idaho. In 1961, Washington State released Gaines, which had Norin 10 as the female parent. Gaines was a winter wheat that yielded 5%–50% more grain than its competitors. In Washington, breeder Orville Vogel followed these successes by breeding Orfed in 1943 and Marfed in 1947. These semidwarfs got their shortness not from Norin 10 but from varieties of club wheat that Vogel had screened for yield and lodging resistance. Vogel finally received Norin 10 seeds in 1949, crossed the plants with Brever, a local variety. Crossing the offspring in turn yielded the F_2 generation, some of which were ideally short. Because Norin 10 tends to be male sterile, it was used as the maternal line in crosses.

In 1953, Vogel sent the offspring of Norin 10 and Brever to Borlaug at Centro Internacional de Mejoramiento de Maiz y Trigo (CIMMYT), better known in the English-speaking world as the International Maize and Wheat Improvement Center.[9] The iconic figure of the Green Revolution, Borlaug was born in 1914 in Iowa. His frequent references to scripture lead one to suppose that he was a religious man.[10] From early manhood, he was acutely aware of the danger of a Malthusian crisis in the developing world. In addition to his scientific work, Borlaug was a popular lecturer who spread the gospel of higher yields to anyone who would listen. Initially a student of forestry at the University of Minnesota, Borlaug heeded the advice of Elvin Charles Stakman, a plant pathologist at the University of Minnesota with an interest in Mexican agriculture, and he switched to plant pathology and crop breeding. Although some faculty were skeptical that Borlaug had the intellect for advanced studies, he received a BS, MS, and PhD from the University of Minnesota, where he benefited from a faculty that had dual interests in plant pathology and crop breeding. In 1942, Stakman recruited Borlaug as the first director of the Division of Wheat Cultivation at the CIMMYT. One might have thought Borlaug an odd choice because he was not familiar with the breeding techniques for wheat, corn, or beans and he knew no Spanish.[11]

Borlaug's colleague at CIMMYT, J. George Harrar screened 277 varieties of local wheat, selecting 42 for test crosses that yielded 5 varieties with superior yields to the traditional Mexican varieties.[12] In 1945, Harrar moved toward a series of administrative posts, leaving Borlaug to take over this breeding program. He decided to focus his efforts on the highlands near Mexico City, where smallholders predominated, and in northwestern Mexico, where large affluent farmers were already irrigating wheat. Borlaug focused on deriving rust-resistant wheat variety and then on yield. In his choice of locales, Borlaug pioneered the shuttle breeding method. The method was unorthodox enough to lead Borlaug's superiors to end the program, but the indignation of affluent farmers in the Northwest, people that CIMMYT did not dare offend, led to Borlaug's immediate reinstatement.

By 1955, Borlaug released eight varieties of rust-resistant wheat (Figure 9.2).[13] As Borlaug may have foreseen, crosses between highland and northeastern wheat varieties yielded progeny adapted to both locales. This technique also allowed Borlaug to breed varieties that were less sensitive to differences in the length of day at different

FIGURE 9.2   The semidwarf wheat varieties were a success of the Green Revolution.

latitudes. On his experimental plots, Borlaug raised wheat yields from 750 kg/ha in 1945 to 1370 kg/ha in 1956. Where total output in Mexico had been 365,000 metric tons in 1945, it leapt to 1.2 million metric tons in 1956.

Borlaug's initial work with Norin 10 was complicated by the variety's susceptibility to rust. Yet, by 1955, several crosses between Norin 10 and Borlaug's earlier rust-resistant cultivars yielded rust-resistant semidwarfs. In 1962, Borlaug released Penjamo 62 and Pitin 62, both of which had Norin 10 as the maternal line. Sonora 64 and Lema Rojo 64 followed in 1964. Later released included Jaral 66, Tobar 66, INIA 66, Noreste 66, Norteno 67, and CIANO 67. These new varieties yielded 6000–6500 kg/ha compared with 4000–4500 kg/ha for Borlaug's earlier varieties. The benefit to Mexico was enormous. Importing as much as 278,000 metric tons of grain per year in the 1940s, Mexico by 1963 exported 72,000 metric tons of grain.[14]

By the early 1960s, success in Mexico attracted India's attention. Scientists wanted Norin 10 seeds for their own research. Borlaug teamed with Indian plant breeder Monkonba S. Swaminathan (Figure 9.3).[15] Born in 1925 in Kumbakonam, India, Swaminathan was the son of a physician. The father imbued his son with nationalistic fervor and a commitment to self-reliance. In 1944, at the age of 19, Swaminathan earned a BS from Travanoon University, receiving a second BS in agriculture from Coimbatore Agricultural College in 1947. A diploma from the Indian Agricultural Research Institute followed in 1949. That year, the United Nations Educational, Scientific, and Cultural Organization (UNESCO) offered Swaminathan a fellowship

FIGURE 9.3   Indian scientist Monkonba Swaminathan led a team of scientists to bring the Green Revolution to India. (From J. H. Perkins, *Geopolitics and the Green Revolution: Wheat, Genes, and the Cold War* (New York and Oxford: Oxford University Press, 1997). With permission.)

to study genetics at the Netherlands Agricultural University. The next year found Swaminathan studying potatoes at the University of Cambridge, from which he received a PhD in 1952. Between 1952 and 1953, he joined the University of Wisconsin to conduct research on potatoes. Although the university wished to hire Swaminathan on the tenure track, he returned to India in 1954.

Despite his expertise on potatoes, the Central Rice Research Institute offered him work as a rice breeder. In this capacity, he crossed the japonica varieties from Japan with the indica varieties of India to yield progeny with a yield between 5 and 6 metric tons per hectare, whereas indica varieties yielded no more than 2 metric tons per hectare. Some of the offspring of these crosses were grown in Malaysia between the 1960s and 1990s. In 1954, Swaminathan joined the Indian Agricultural Research Institute (IARI) as an assistant geneticist, where he stayed until 1972. By 1970, he and his colleagues had amassed more than 200 publications and trained undergraduate students. At IARI, Swaminathan switched his focus to wheat, irradiating seeds to induce mutations as Nobel laureate Hermann Mueller had done to fruit flies. Along the way, Swaminathan received a large number of awards from international societies, including election to the Royal Society of London, of which Isaac Newton had been the president. Even the Soviet Union took interest in Swaminathan's work.

In 1959, Swaminathan learned of Vogel's work with semidwarf wheat. Swaminathan contacted Vogel, only to learn that the latter's work was with winter wheat and could not be transferred to India. Vogel instead suggested that Swaminathan contact Borlaug. In the meantime, agronomist S. P. Kohli bred semidwarfs from Indian dwarf varieties. The semidwarfs were well adapted to India's climate, though their yield was not especially noteworthy. This work nonetheless convinced Swaminathan that one could undertake a successful breeding program in India. By 1962, the IARI was convinced that Borlaug's semidwarfs were suitable for India, and the next year Swaminathan invited Borlaug to tour India. While in India, Borlaug became convinced that Swaminathan and his colleagues could breed semidwarfs, using Norin 10 as the maternal line, suitable for India. Borlaug understood, however, that some of India's soils were unfertile and would need liberal applications of fertilizer and water as the wheat matured. Borlaug believed that the Indian government or some national agency should coordinate breeding efforts. He recommended, as he had done in Mexico, the adoption of shuttle breeding.

When he returned to Mexico, Borlaug sent Swaminathan 100 kg of seeds each of Lerma Rojo 64A, Sonora 63, Sonora 64, and Mayo 64 along with seeds from 600 additional varieties that Borlaug judged not ready for release but which would be fine breeding stock. Lerma Rojo 64A and the other cultivars yielded spectacularly at IARI's four experimental plots. All yielded 2900–3700 kg/ha, well above the yields of what had been thought to be elite lines in India. Some of Borlaug's varieties yielded 4600 kg/ha. In other words, Borlaug's wheats yielded roughly three times more grain than traditional Indian varieties. Borlaug nudged the IARI to increase the use of fertilizers on the experimental plots in the belief that they would boost yields to some 6000 kg/ha. He believed that Senora 63 and Senora 64 would yield particularly well with more fertilizer.

In 1964, Borlaug toured India again, this time to robust and favorable media coverage.[16] Yet, Borlaug had critics. Some scientists noted that rice was India's principal grain and wondered why Borlaug focused so much attention on wheat. Others thought India should develop hybrid corn varieties suitable for India. Yet, Swaminathan never wavered in his support for Borlaug. Under his leadership, the Indian Council of Agricultural Research in 1964 asked Borlaug to send it 20 metric tons of Sonora 63 and Sonora 64, a course of action the Ford Foundation applauded. Borlaug complied immediately, though the death that year of Indian Prime Minister Jawaharlal Nehru left Borlaug to wonder whether India would remain committed to agricultural research. Fortunately, the new Prime Minister, Lal Bahadur Shastri, had an intense commitment to ICAR. He echoed Borlaug's call for better varieties, more fertilizers, and irrigation where necessary. He believed that agriculture could only progress with the help of science. Shastri toured Mexico, meeting with Borlaug. In 1965, Indian farmers could at last plant Lerma Rojo 64A and Sonora 64. Borlaug sent India an additional 200 metric tons of Sonora 64 and 50 metric tons of Lerma Rojo 64A.

In 1966, Shastri, asking Borlaug for still more seeds, foresaw the day when India would be self-sufficient in grain. India, which imported both food and weapons from the United States, wished to lessen its dependency. For its part, the United States wished to tie measurable gains in food production in India to U.S. aid. Shastri's death that year led to the rise of Indira Gandhi, Nehru's daughter, as prime minister. She sent three scientists to Mexico to inventory Borlaug's seed stocks. These delegates requested 18,000 metric tons of Lerma Rojo 64A, estimating that these seeds could plant in 1 million acres. Yet, India had 33 million acres to wheat and not enough land under irrigation to save the harvest from drought for the second consecutive year. Nonetheless, India persisted, planting 504,000 ha of HYVs in 1966 and more than 10 million hectares in 1972. During these years, grain imports fell nearly fivefold.

Rice (Figure 9.4) was another success of the Green Revolution. Green Revolution rice breeding began in India, spreading to the Philippines and the rest of Southeast Asia. In 1949, the FAO established a rice-breeding center in Cuttack, India, which began to screen germplasm for dwarfs. Scientists crossed dwarf japonica varieties from Japan with tall indica varieties that were suited to the tropics and subtropics. The results were ADT-27 and Mahsuri, which yielded well and were widely adapted to India. ADT-27 was especially popular in India. In 1960, the Ford and Rockefeller Foundations established the International Rice Research Institute (IRRI) in the Philippines. IRRI scientists screened more than 10,000 rice varieties worldwide in search of novel genes. Collecting dwarfs, they began shuttle breeding as Borlaug had done with wheat. In 1962, IRRI breeder Peter Jennings crossed 38 varieties from seed banks. The eighth cross, between the Chinese dwarf Dee-geo-woo-gen and the tall Indonesian Peta, yielded a single plant with 130 precious seeds, the basis for IR 8. Jennings planted the F1, allowing the progeny to cross. One-quarter of the F2 were dwarfs, the typical Mendelian segregation for a single recessive gene. These offspring were crossed over several additional generations. The F5 contained a single promising plant, IR 8. It matured in 130 days, whereas traditional varieties needed 160–170 days. It yielded 5 metric tons per hectare without fertilizer, five times the average yield of traditional varieties. With fertilizer, IR 8 yielded 10 metric tons per hectare. The president of the Philippines visited India to see IR 8 and presumably took seeds home. In Pakistan, IR 8 yielded 11 metric tons per hectare. The IRRI

FIGURE 9.4   The semidwarf rice varieties were a success of the Green Revolution.

crossed IR 8 with 13 other varieties to derive IR 36, which matured in 105 days and was disease- and insect resistant. In 1990, IR 72 yielded even more grain than IR 8 and IR 36. In the 1970s and 1980s alone, 11 Asian countries, mostly in South and Southeast Asia, increased yields by 63%. By 1982, one-third of rice land in Latin America was planted to HYVs. In 1992, Asia yielded 460 million metric tons of paddy rice on 140 million hectares. Two-thirds of this land was planted to HYVs. Asia planted HYVs on 91% of its rainfed rice lands for consumption by 2.7 billion people. CGIAR admitted, however, that most of these gains accrued to affluent farmers. In the 1990s, scientists bred wheat, corn, and rice that needed less water and so were not as dependent on irrigation and were better suited to the small farmers who could not afford irrigation.

The efforts in the 1950s to breed HYVs of rice and wheat began to expand throughout Mexico and India by the mid-1960s. Asia and the Americas were the initial targets of the Green Revolution. By the late 1950s, breeders had assembled large aggregates of germplasm with which to derive new varieties. The breakthrough was the incorporation of genes for dwarfing into elite lines of wheat and rice, making them short and so less likely to lodge. These varieties also put most biomass into seeds rather than stems. Farmers adapted these varieties more rapidly where rainfall was adequate or irrigation was available.

ACHIEVEMENTS OF THE GREEN REVOLUTION

Many scientists and historians mark the Green Revolution as a phenomenon of the 1960s and 1970s that was largely complete by 1980.[16] The literature of the 1980s evaluates the Green Revolution as an event that occurred in the past. The bias of these decades, as we have seen, favored wheat and rice, with little attention to cassava and other root crops, tubers like the potato, and the many varieties of beans, which were and are important sources of protein. Yet, if one extended the Green Revolution to 2000, the perception is different. By then researchers had broadened their treatment of crops to include those they had neglected earlier and released more than 1000 HYVs to more than 100 countries in the developing world.[18] This procedure was not rapid at first. Only during the 1980s, for example, did breeders concentrate on sorghum, millet, and barley.[19] Root crops, tubers, and legumes were also slow to capture the attention of breeders. Yet, even with an initial focus on wheat, corn, and rice, the Green Revolution was able to expand its geography beyond Mexico and South Asia, spreading to North Africa and western Asia. As early as the 1960s, breeders attempted to target the HYVs to sub-Saharan Africa, though only wheat was adopted rapidly and then only in temperate regions of the continent. The attempt to introduce HYVs from Asia and Latin America into sub-Saharan Africa yielded poor results because these cultivars were not adapted to African conditions.

R. E. Evenson and D. Gotlin identified two steps in the Green Revolution. The first step in breeding the HYVs was to seek adaptation to a latitude and climate.[20] Second, breeders sought to adapt these varieties to the disease and pest biota of a given latitude and climate. The success of this approach grew over time. The first HYVs of rice were planted on about 35% of rice lands in India.[21] By 2000, the percentage leapt to more than 80, principally because these new HYVs had better resistance to several brown planthopper populations. International agencies deserve credit for these successes. The IRRI and CIMMYT bred more than 35% of HYVs of rice, wheat, and corn by 2000.[22] Another 15% of HYVs had one parent from an international repository of germplasm and the other parent derived from a national breeding program. At this instance, Indian scientists made the final crosses between a local variety and an HYV from the IRRI for adaptation to a particular region of India.

For clarity, Evenson and Gotlin divided the Green Revolution into two parts. The early Green Revolution occurred between 1960 and 1980, whereas the late Green Revolution developed between 1981 and 2000.[23] Using this approach, Evenson and Gotlin found two bursts in yield gains, one in the 1960s and the other in the 1990s, when many scientists thought the Green Revolution had long ended. The early Green Revolution succeeded in Latin America and Asia, where HYVs contributed to 17% of the gain in total output of cereals. The rest came from intensification: the use of irrigation, fertilizers, pesticides, and mechanization. The 1990s burst in yield gains came at last to sub-Saharan Africa. Growth elsewhere slowed. In the 1990s, Latin America scaled back its land to HYVs, whereas sub-Saharan Africa rapidly expanded hectares to HYVs. North Africa and western Asia also mounted yield gains. In Africa, HYVs have accounted for some 80% of gains in total output. The success of HYVs has therefore been greater in the late Green Revolution.[24] Benefiting from these cultivars, farmers in Africa, North and sub-Saharan, and in western Asia have needed fewer fertilizers, pesticides, and less water than Latin America and South Asia had needed in the early Green Revolution. Sub-Saharan Africa has had access to HYVs of cassava, rice, and corn adapted to local conditions. The late success of the Green Revolution in sub-Saharan Africa had much to do with its different environments and poor soils, hurdles over which scientists had to leap before they could develop suitable HYVs. By one estimate, the Green Revolution made possible adequate nourishment for 32–42 million people in the developing world who otherwise would have suffered malnourishment.[25] The Green Revolution rescued children from malnutrition. Without the Green Revolution, food might have been scarce and prices high. Farmers probably would have expanded agriculture to the worst soils, destroying habitat in the process.

It is possible to envision the Green Revolution as a linear process. In 1956, Green Revolution wheat and rice varieties occupied no hectares in the developing world. In 1970, HYVs of wheat and rice accounted for 14 and 15 million hectares, respectively, and 20% of the developing world's hectarage for these crops.[26] In 1980, HYVs of wheat totaled 39 million hectares, 50% of the developing world's hectares to this grain. That year, HYVs of rice tallied 55 million hectares and 43% of the developing world's hectares of rice. In 1990, the figures had risen to 60 million hectares and 70% of hectares for HYVs of wheat and 85 million hectares and 65% of hectares to HYVs of rice. Finally, in 2000, HYVs of wheat totaled 70 million hectares and 84% of wheat hectares in the developing world. HYVs of rice totaled 100 million hectares and 74% of rice hectares in the developing world. Total grain production in the developing world rose from 309 million tons in 1956 to 468 million tons in 1970, 618 million tons in 1980, 858 million tons in 1990, and 962 million tons in 2000. Inputs of water, fertilizers, and the number of tractors increased along with these gains. One key to the Green Revolution has been the training of local scientists across Asia, Latin America, and Africa to breed crops, chiefly grains.[27] Both the FAO and the Rockefeller Foundations gave universities in the developing world money to train a new generation of agricultural scientists. These agencies gave these new scientists seeds of HYVs to use as breeding stock.

The results have been marked. As early as 1956, Mexico was self-sufficient in food production.[28] Between 1961 and 1991, corn yields in Latin America more than doubled and wheat yields doubled.[29] Between 1961 and 1991, rice yields more than doubled in South Asia and wheat yields increased fourfold. In East and Southeast Asia, rice output was 52 million metric tons in 1961 and 125 metric tons in 1991.

Between 1965 and 2000, the price of grain, adjusted for inflation, fell by 40%. The Green Revolution surely must be responsible for rising longevity in the developing world. Critics of the Green Revolution often miss the point that there does not appear to have been an alternative way to avert the Malthusian Crisis. Yet, the future of agricultural research is difficult to forecast. International and national agencies have lost funding in real dollars since the 1990s. This decline may be due to the public's fear of biotechnology, which has raised alarm about the uses of agricultural research. There is also a debate over whether private agricultural research, performed by Monsanto and other behemoths, can and should replace or at least compete with the public sector.

The contribution of Norman Borlaug to the Green Revolution is difficult to overstate. His immediate goal was to breed HYVs of wheat resistant to stem rust, an ancient disease and one that troubled Mexican farmers. He established two experimental plots, one near Mexico City and the other in northwestern Mexico. At these, he perfected the technique of shuttle breeding. Growing wheat in the southern experimental plot, he took the harvest to the north, making new crosses along the way so that he could derive two generations of a new variety in a single season, shortening the time necessary to derive a new cultivar. Although traditional breeders needed about 10 years to develop a new cultivar, Borlaug accomplished the task in 5 or fewer years. By insisting that his HYVs be adapted both to the south and north, Borlaug derived wheat cultivars with an unusual breadth of adaptability. His initial wheat varieties were so fecund that they lodged too frequently. This problem caused Borlaug to seek dwarfs to cross with his new varieties to yield short wheat plants with sturdy, stout stalks to resist lodging. Dwarfs from Japan were particularly promising for this approach, yet Borlaug needed 7 years of backcrossing because the Japanese dwarfs were susceptible to rust. That is, Borlaug crossed the progeny of a Japanese dwarf with a rust-resistant wheat, backcrossing the dwarf offspring to the rust-resistant parent over 14 generations to achieve semidwarfs with a high level of rust resistance. The resultant semidwarfs yielded 8 tons more wheat per hectares than had Borlaug's original tall HYVs of wheat. In the 1960s, Borlaug concentrated on adapting these wheats to lands south of Mexico and to Asia, especially Pakistan and India. In these two countries, rice and wheat yields increased by 2.1% per year between 1950 and 1990. During these years, aggregate output of these grains nearly tripled, though since 1990 yields in world rose only 0.5% per year. Yet, one might attribute this apparent slowdown to the disastrous harvests in the final years of the Soviet Union and in the new but fragile Russia and the former republics.

Often taken as the beginning of the Green Revolution, in 1960 the Rockefeller and Ford Foundations, the USAID and the Filipino government created the IRRI in Los Babos, Philippines to transfer wheat's gains to rice but using the same breeding methodology.[30] Again the goals were disease and pest resistance, high yields, and short stature. By 2000, the IRRI focused on deriving short HYVs of rice with even thicker, stouter stalks to enable the production of even more massive seeds without lodging. At the same time, scientists at Cornell University and their colleagues in China and Malaysia screened thousands of wide and exotic rice varieties in a Malaysian gene bank, finding a single wild plant that when crossed with elite lines yielded 15%–50% more rice than extant HYVs.[31] The Cornell scientists mapped the genome of these new varieties, finding two genes from the wild variety that may have contributed to the yield gains.

Funded by a Japanese philanthropist and urged by the former U.S. president Jimmy Carter, Borlaug in 1986 came out of retirement to coordinate a program to bring the Green Revolution to sub-Saharan Africa.[32] By 2000, Borlaug had established some 400,000 breeding centers throughout sub-Saharan Africa to derive high-yielding cultivars of sorghum and millet to a variety of local conditions. As early as 1991, Borlaug and his colleagues had bred the first HYV of sorghum suitable for parts of Africa. Thanks to this and other varieties, Ethiopia's grain harvest totaled 6 million tons in 1994 and 11.7 million tons in 1996. This gain came at only a fraction of the fertilizers that the United States adds to its grain land. Japan, however, supported Borlaug's work only in the countries of sub-Saharan Africa that had a stable government. In the last decade of his life, Borlaug focused on breeding more nutritious varieties of corn, soybeans, rice, wheat, and pasture grasses and on breeding HYVs adapted to soils with too much aluminum, often a toxin to plant growth.

Wheat is planted at the end of the rainy season in Mexico and India and harvested in the following summer. The soils on which wheat was traditionally planted do not hold moisture well, especially because long use has deprived them of organic matter. Accordingly, irrigation was important as the crop matured. The irrigation of wheat in Mexico and India began in earnest in the late 1960s, and by the 1980s Indian farmers irrigated 60% of their wheat.[33] Because of irrigation and the application of agrochemicals, wheat yields in India between 1960 and 1976 increased by 3.4% per yield and total output increased by 4.1% per year, leading Pierre Spitz to label India's Green Revolution as a "wheat revolution."[34] By the 1980s, wheat had become so dominate that it threatened traditional practices of mixed crops, for example, the simultaneous cropping of wheat and chickpeas, an important food in Indian cuisine. Yet, small farmers remained true to their heritage, growing traditional, low-yielding corn or wheat, *Phaseolus* beans from the Americas, the mung bean, peas, and cowpeas in rotation.

In sub-Saharan Africa, sorghum suffers from witchweed, a parasitic plant, in addition to diseases, insects, and birds. Witchweed destroys crops, sorghum among them, by feeding on their roots. Legumes and grains are both susceptible. Also a problem in South Asia, witchweed may have caused 15%–40% yield losses in sub-Saharan Africa.[35] All important to Africa, sorghum, corn, and millet are vulnerable to witchweed. In 1995, Purdue University scientists bred the first witchweed-resistant sorghum. Farmers in Ethiopia, Chad, Mali, Rwanda, and Sudan embraced the new cultivar. An understanding of the biology and chemistry of witchweed was important in this success. A single witchweed plant may yield as many as 500,000 seeds. These remain viable in the soil for 20 years, apparently innocuous until the roots of a host plant release a chemical signaling witchweed to germinate. Activated by two chemicals at two different stages, witchweed produces a haustorium, the underground organ that penetrates the host plant's roots to steal water and nutrients. Witchweed does most of its damage underground before the plant breaks the soil's surface, by which time it is too late to eradicate. The first HYV of sorghum, Hageen Dura-1, was drought-tolerant but vulnerable to witchweed. In 1992, a graduate student at Purdue discovered the chemical gorgolactose that causes witchweed to germinate. Another graduate student began screening sorghum germplasm for a plant that did not produce this chemical. He grew a single sorghum plant per Petrie dish filled with witchweed seeds, discovering that one sorghum plant, SRN-39, had a recessive gene that minimized sorgolactome production. He backcrossed this plant to eight popular sorghum cultivars over several generations, later mapping the gene on its chromosome for cloning. In 1999, 200,000 farmers in 12 sub-Saharan African countries grew witchweed-resistant sorghum varieties.

Borlaug's "miracle wheat" doubled and even tripled yields per hectare in regions of the developing world capable of sustaining wheat cultivation.[36] Scientists achieved similar successes with corn and rice, though one must remember that the breeding program for corn differed from those of wheat and rice. Between 1965 and 1970, the HYVs that Borlaug and other achieved spread rapidly throughout many regions of Latin America and Asia, though sub-Saharan Africa did not initially benefit from the early stages of the Green Revolution.[37] By 1970, these HYVs occupied 10 million hectares in the developing world, with large concentrations in Mexico, Pakistan, and

India, the earliest beneficiaries of the Green Revolution. By 1970, Pakistan produced enough wheat and rice to feed its population, a remarkable turnabout for a nation that had been so dependent on U.S. imports of grain. By then, Sri Lanka and the Philippines recorded record harvest of rice and North Africa, if not sub-Saharan Africa, and South America were making strides in adapting HYVs.

On the verge of famine in 1967, India by 1972 produced enough wheat and rice to feed its people without the need for U.S. imports. India even managed to record surpluses in the 1970s, which it wisely kept to avert famine during the drought of 1979. Between 1961 and 1980, India increased its wheat harvest more than threefold.

Because of the success, at least partial, of the Green Revolution, international funding of agricultural research increased more than $200 million between 1979 and 1984. The effect of the Green Revolution on labor is much debated. One thesis holds that the Green Revolution impoverished small holders, who sometimes lost their land and became day laborers.[38] In northern India, however, the dense stands of wheat matured at the same time and put a premium on labor because even in the 1980s few landowners had a mechanical harvester. The dearth of labor at this critical time put workers in a better bargaining position.

Famine in 1984 in Ethiopia underscored the lack of progress in the Green Revolution in sub-Saharan Africa.[39] In the late Green Revolution, however, Ethiopia nearly doubled food production between 1991 and 2000. By 2000, Ethiopia had sufficient surpluses to export 200,000 tons of grain to Kenya. Borlaug equated this new success as a miracle of science.

As early as the 1960s and 1970s, much of Asia was self-sufficient in food. Yet, the Green Revolution may have postponed rather than ended the Malthusian crisis. The United Nations has predicted that the global population may rise to 7.1 billion by 2030 (see Chapter 11). At the same time, the affluent West consumes meat, which requires that corn and other food grains and legumes be fed to livestock rather than humans.

Yet, because of the Green Revolution, by 1994 only 10% of the world's people subsisted on fewer than 2200 calories per day.[40] The CGIAR has challenged the notion that the wealthy benefited disproportionately. Between 1972 and 1983, small farmers saw their income rise by 90%.[41] Landless laborers saw their wages increase by 125% during these years. The poor have access to more calories and protein. The CGIAR emphasizes that the pace of gains quickened during the Green Revolution. In the 1970s, rice yields increased by 2.7% per year and during the 1980s by 3% per year. Although these gains have leveled off in some parts of Asia, rice yields in Thailand and Indonesia rose in the 1990s. That decade, Java recorded 4.5 metric tons of rice per hectare. The HYV IR 8 still holds the record yield in Asia. As early as 1966, it yielded 10 metric tons per hectare on an experimental plot. In the 1990s, the CGIAR concentrated on breeding rice with more biomass as seeds and less as foliage, believing that it could increase the number of seeds per panicle from about 100 to 250. These new varieties might increase yields by 25%–40%. China has taken a different route to the Green Revolution, developing hybrid rices that boast the yield gains of hybrid corn. According to Chinese scientists, these hybrids yielded 20% more grain than semidwarf rice. Accordingly, the IRRI funded the development of hybrid rices in India, Indonesia, Korea, Malaysia, and Thailand.

The adoption of hybrid corn in Africa had uneven results. Large farmers adopted it, reaping 14 metric tons per hectare, whereas the rest of Africa averaged just 1 metric ton per hectare. In his 1970 Nobel Address, Borlaug cited India, Pakistan, the Philippines, Afghanistan, Sri Lanka, Indonesia, Iran, Kenya, Malaysia, Mexico, Thailand, Tunisia, and Turkey as beneficiaries of the Green Revolution.

RESPONSE TO CRITICISM

Although the Green Revolution had critics, to be fair, Borlaug, in his Nobel Address in 1970 and in other writings, emerged as a humanitarian who wished to benefit all farmers, not just the rich. He envisioned the Green Revolution as a kind of anti-poverty program. Moreover, the CGIAR, which receives funding from the World Bank, the United Nations Development Program, and the United Nations Food and Agriculture Organization (FAO), was perhaps the leading coordinator of Green Revolution research and promoted the use of the agricultural sciences to ameliorate poverty. It has promoted integrated pest management to lessen small farmers' reliance on pesticides. It focused research on deriving disease and pest-resistant crops, again with the intent of diminishing farmers' reliance on pesticides. One may recall the breeding of European Corn Borer-resistant hybrid corn in the previous chapter. The CGIAR has aimed to breed crops that need fewer fertilizers to save poor farmers from this expense. The CGIAR founded the International Crops Research Institute for the Semi-Arid Tropics (ICRISAT) in Hyderabad, India, to pursue these goals not only in India, but also in neglected sub-Saharan Africa. The ICRISAT was to devote particular attention to developing drought-tolerant crops. Hybrid corn is a good example of this model because hybrids are more drought-tolerant than the traditional cultivars that hybrids supplanted. These efforts have made clear the desire to create a more flexible, resilient Green Revolution less dependent on lavish and costly inputs of water and agrochemicals.

Green Revolution plant breeders have become more sophisticated in their approach, breeding multiple genotypes of rice, each with a unique genotype in hopes that if one variety falls prey to the planthopper, the other genotypes will contain resistant genes, thereby slowing the evolution of a new type of planthopper. If success has been slow to achieve and brief in duration, breeders have nonetheless had some success with the chickpea and pigeon peas, important legumes in the developing world, the chickpea being an important crop in India. Breeders have derived chickpeas and pigeon peas that tolerate insects well enough to yield without the need for insecticides.

The sharp rise in energy prices since the Organization of Petroleum Exporting Countries' embargo in 1973 and 1974 has forced scientists to concentrate their efforts on breeding crops that used less fertilizer with greater efficiency, and so yield well with fewer inputs. This course of action has been necessary because the Haber-Bosch process, for example, uses natural gas, to make nitrogenous fertilizers. The overuse of fertilizers worried scientists and farmers, who understood that rice plant roots absorb as little as 30% of the fertilizer supplied to it, a waste of money and a danger to the environment. This realization has led scientists to attempt to breed crops that, like legumes, fix nitrogen in the soil. They have attempted to derive soil bacteria and algae capable to fixing nitrogen in the soil. To avert the Malthusian crisis, scientists have called on public and private agencies to redouble their commitment to agricultural research, investing millions more over the coming years.[42] There is a need to focus money on the agricultural sciences as the United States once poured money into nuclear weapons and the Apollo program.

TOWARD THE FUTURE: THE AGENCIES THAT MADE THE GREEN REVOLUTION POSSIBLE

A principle that has united the Green Revolution agencies focus on the importance of increasing yields to keep abreast of population growth and to keep food prices low. Yet, since the 1990s, Green Revolution research has turned increasingly to the difficult problem of alleviating rural poverty. In the United States, only the historically black land grant universities appear to be vigorous in pursuing this aim. One must remember that not all scientists are in this camp. Some eschew applied research because it lacks status. Others follow some "trendy" topic. As in the United States, the Green Revolution agencies, for lack of public funding, have turned to competitive grants. Worldwide between 1990 and 2007, funding for agricultural research remained constant, with no prospect of increase.[43]

The CGIAR has turned from an emphasis on yields to one geared to combat poverty, improve water and soils, enhance health and nutrition, and promote sustainability.[44] The group funds research in biotechnology, branding it the "New Agriculture."[45] CGIAR promotes the cultivation of vegetables and fruits, which fetch higher prices than grains. Diverse to the core, CGIAR funds research to conserve genetic resources, use genetic resources in conventional breeding and biotechnology, conserve water, manage forests, improve soils, increase the yield of high value foods, and focus on issues that affect smallholders.

In 1973, CGIAR founded the International Potato Center (CIP) in Peru, the cradle of the potato.[46] The aim has been to increase potato yields throughout the developing world. In 1985, CGIAR expanded CIP's mission to include the sweet potato. Taking on this mission, the CIP established potato and sweet potato gene banks and expanded research to include Andean tuber and root crops that the indigenes had grown for millennia but that never became world staples. As funding began to stagnate in the 1990s, the CIP reassessed its research. By 1995, it, like the CGIAR, focused increasingly on rural poverty.[47] In 2005, the CIP judged that antipoverty research should receive even higher priority. The CIP also increased its research on the sweet potato. It has focused anew on new races of *Phytophthora infestans*, a dangerous pathogen of potatoes (see Chapter 6). Identifying new genes for resistance, the CIP has backcrossed them into local cultivars, which one presumes are not elite. The CIP aims to identify other barriers to yield and to find new markets for sweet potatoes. The center has bred drought- and virus-tolerant potatoes and HYVs of sweet potatoes. The CIP aims to find new ways to combat the potato tuber moth, the sweet potato weevil, the sweet potato Andean weevil, and several sweet potato viruses. The CIP has undertaken research in Rwanda, Burundi, Zaire, China, Peru, Tunisia, the Dominican Republic, Cuba, Vietnam, India, and Egypt. The CIP has also emphasized the planting of seeds, perhaps to avoid genetic uniformity, though one wonders whether this practice is widespread.

The International Institute of Tropical Agriculture focuses on several crops, yams perhaps being the most important. Between 2001 and 2010, yams commanded $7.5 billion in research.[48] Each $1 of investment yielded $1.31 in benefits to farmers. After yams, research has focused on cassava, corn, rice, cowpea, citrus fruits, sorghum, plantain, millet, and peanut. Vegetables and fruits, though not citrus, remain secondary. Like the CGIAR and CIP, the IITA focuses research on alleviating poverty. Corn may have the greatest potential to decrease poverty and, we have seen, is an important item of research. IITA estimates that farmers who adopt elite hybrids may increase income by 5%. Yet research in Nigeria fails to demonstrate that corn culture helps the poorest. It may be better to plant cowpea, millet, sorghum, peanut, and soybean. Cassava, yam, and cacao appear to have increased income inequality in Nigeria.

The CGIAR founded the ICRISAT in 1972 with the aim of increasing the yields of sorghum, pearl millet, peanut, chickpea, and pigeon pea.[49] Since then, research has expanded to focus on sustainability and environmental concerns, equity, gender issues, and health and nutrition. Between 1999 and 2007, the ICRISAR took aim against poverty, noting it to be the primary cause of hunger, diseases, and environmental decay.[50] Because the farmers are poor, the center must target research on HYVs and efficient cultural methods. The center encourages farmers to diversify crops so that the failure of one will not endanger them. The ICRISAR identifies environmental decay as a cause of poverty. In this context, research on the environment is antipoverty research. The center emphasizes the value of extension.

The Africa Rice Center (WARDA) serves 22 countries in sub-Saharan Africa, working in concert with the West Africa Rice Research Network, the Inland Valley Consortium, the East and Central African Rice Research Network, and the Africa Rice Initiative.[51] Research emphasizes productivity and marketing with attention to upland, lowland, and irrigated rice. In Gambia, Guinea Conakry, Guinea Bissau, Nigeria, Sierra Leone, and Senegal, WARDA focuses on intensive rice cultivation in mangrove swamps. Like the CIP, WARDA is building a germplasm reservoir by including seeds of numerous varieties of traditional African rices. WARDA created New Rice for Africa to accomplish this aim. Like its kindred agencies, WARDA seeks to alleviate poverty and improve health and nutrition. The center pursues biotechnology not merely to increase yields but to engineer more nourishing varieties of rice. Maintaining seed stocks for distribution to farmers, WARDA emphasizes sustainability, soil improvement, the combating of pests and pathogens, the improvement of water quality, reforestation, the adoption of technology to save labor, and a diminution in the use of pesticides, fertilizers, and herbicides, protect soils from erosion, treat saline and acidic soils, increase incomes, create jobs, and produce more abundant food.

During the 1990s, CIMMYT followed its sister organizations in shifting research from yields above all else to the alleviation of poverty and sustainability.[52] This was a departure for an agency that had built its reputation on HYV and the technologies of production. Now CIMMYT tailors wheat varieties to the soils and climate of Kazakhstan and to breeding wheat and corn varieties for marginal lands. Since 2003, CIMMYT has devoted more resources to corn than wheat.[53] The organization has also scaled back research in Latin America, pivoting to East, South, and Southeast Asia, a surprise given the importance of rice to these lands. CIMMYT has intensified efforts to derive corn hybrids more drought-tolerant and with more lysine and tryptophan.

The International Center for Agricultural Research in the Dry Areas (ICARDA), like kindred agencies, focuses on poverty and sustainability and pegs agricultural production with the trend in population. The center aims to help farmers become self-sufficient. It promotes efficiency and equity. By mandate, the ICARDA focuses on arid regions of the temperate developing world, for example, North Africa and parts of sub-Saharan Africa, Central, West, and South Asia, China, and Latin America. These areas total roughly 3 billion hectares and 1.7 billion people, 41% of who are farmers.[54] High population growth spurs research on increasing yields. The climate spurs research on developing drought-tolerant crops. Wheat and barley are the primary crops with sorghum in Sudan and cotton in Egypt and Syria. Syria irrigates cotton. In many areas, faba beans, chickpea, and lentil are important. Where irrigation is available, farmers grow potatoes, oilseed crops, and sugar beets. In other regions, olives, almonds, figs, pistachios, apples, apricots, peaches, hazelnuts, grapes, quinces, dates, cucumbers, and melons dominate. The demand for food is outstripping production. Many countries must import food. Research focuses on raising soil fertility and the breeding of crops tolerant of drought, heat, salinity, frost, insects, pathogens, and weeds.

NOTES

1. Felicia Wu and William P. Butz, *The Future of Genetically Modified Crops: Lessons from the Green Revolution* (Santa Monica, CA: Rand Corporation, 2004), 28.
2. Pierre Spitz, "The Green Revolution Re-Examined in India," in *The Green Revolution Revisited: Critique and Alternative*, ed. Bernhard Glaeser (London: Allen & Unwin, 1987), 56.
3. Wu and Butz, *The Future of Genetically Modified Crops*, 13.
4. John H. Perkins, *Geopolitics and the Green Revolution: Wheat, Genes, and the Cold War* (New York and Oxford: Oxford University Press, 1997), 20.
5. Ibid, 216–217.
6. Ibid, 217.
7. Ibid.
8. Ibid, 218.
9. Ibid, 222.
10. Norman Borlaug, "Nobel Lecture," www.nobelprize.org/nobel_prizes/peace/laureates/1970/borlaug-lecture.html
11. "Norman Borlaug—Biographical," www.nobelprize.org/nobel_prizes/peace/laureates/1970/borlaug-bio.html
12. Perkins, *Geopolitics and the Green Revolution*, 225.
13. Ibid, 230.
14. Ibid, 231.
15. Ibid, 232–235.
16. Ibid, 236–237.
17. Bernhard Glaeser, "Agriculture between the Green Revolution and Ecodevelopment: Which Way to Go?" in *The Green Revolution Revisited: Critique and Alternative*, ed. Bernhard Glaeser (London: Allen & Unwin, 1987), 2.

18. R. E. Evenson and D. Gotlin, "Assessing the Impact of the Green Revolution, 1960 to 2000," *Science* 100 (2003): 758–759.
19. Ibid, 760–761.
20. Ibid, 758–759.
21. Ibid, 759.
22. Ibid, 759.
23. Ibid, 758.
24. Ibid, 761.
25. Ibid, 761.
26. Wu and Butz, *The Future of Genetically Modified Crops*, 14.
27. Ibid, 15.
28. Glaeser, "Green Revolution," 1.
29. Wu and Butz, *The Future of Genetically Modified Crops*, 15.
30. Evenson and Gallin, "Assessing the Impact of the Green Revolution," 758; Charles Mann, "Reseeding the Green Revolution," *Science* 277 (1997): 1039.
31. Mann, "Reseeding the Green Revolution," 1042.
32. Ibid, 1041.
33. Spitz, "Green Revolution Re-Examined in India," 61.
34. Ibid.
35. Mann, "Reseeding the Green Revolution," 1040.
36. Gleaser, "Green Revolution," 1.
37. Evenson and Gallin, "Assessing the Impact of the Green Revolution," 759.
38. Spitz, "Green Revolution Re-Examined in India," 69–71.
39. Mann, "Reseeding the Green Revolution," 1041.
40. Ibid, 1039.
41. Michael Collinson, *Towards a New Green Revolution: A Perspective from the CGIAR Secretariat* (Washington, DC: World Bank, 1992), 6.
42. Mann, "Reseeding the Green Revolution," 1043.
43. David A. Raitzer and George W. Norton, "Introduction to Prioritizing Agricultural Research for Development," in *Prioritizing Agricultural Research for Development: Experiences and Lessons*, eds. David A. Raitzer and George W. Norton (Cambridge, MA: CABI, 2009), 3.
44. Peter Gardiner, "Method and Approach to Identify the Consultative Group on International Agricultural Research (CGIAR) System Priorities for Research," in *Prioritizing Agricultural Research for Development: Experiences and Lessons*, eds. David A. Raitzer and George W. Norton (Cambridge, MA: CABI, 2009), 195.
45. Ibid, 194–195.
46. Keith O. Fuglie and Graham Thiele, "Research Priority Assessment at the International Potato Center (CIP)," in *Prioritizing Agricultural Research for Development: Experiences and Lessons*, eds. David A. Raitzer and George W. Norton (Cambridge, MA: CABI, 2009), 26.
47. Ibid, 27.
48. Victor M. Manyong, Diekalia Sanago, and Arega D. Alene, "The International Institute of Tropical Agriculture's (IITA) Experience in Priority Assessment of Agricultural Research," in *Prioritizing Agricultural Research for Development: Experiences and Lessons*, eds. David A. Raitzer and George W. Norton (Cambridge, MA: CABI, 2009), 52.
49. Jupiter Najeunga and Cynthia Bantilan, "Research Evaluation and Priority Assessment at the International Crops Research Institute for the Semi-Arid Tropics (ICRISAT): Continuing Cycles of Learning to Improve Impacts," in *Prioritizing Agricultural Research for Development: Experiences and Lessons*, eds. David A. Raitzer and George W. Norton (Cambridge, MA: CABI, 2009), 83.
50. Ibid, 91.

51. Aliou Diagne, Patrick Kormawa, Ousmane Youm, Shellemich Kaya, and Simon Nioho, "Priority Assessment for Rice Research in Sub-Saharan Africa," in *Prioritizing Agricultural Research for Development: Experiences and Lessons*, eds. David A. Raitzer and George W. Norton (Cambridge, MA: CABI, 2009), 117.

52. John Dixon and Roberto La Rovere, "Highlights of the Evolution of Priority Assessment and Targeting at the International Center for Maize and Wheat Improvement (CIMMYT)," in *Prioritizing Agricultural Research for Development: Experiences and Lessons*, eds. David A. Raitzer and George W. Norton (Cambridge, MA: CABI, 2009), 137.

53. Ibid, 150.

54. Kamil Shideed, Mahmoud Solh, Ahmed Mazid, and Mezen El-Salh, "The International Center for Agricultural Research in the Dry Areas' (ICARDA) Experience in Agricultural Research Priority Assessment," in *Prioritizing Agricultural Research for Development: Experiences and Lessons*, eds. David A. Raitzer and George W. Norton (Cambridge, MA: CABI, 2009), 158.

10 Plant Biotechnology

OVERVIEW

Before the rise of plant biotechnology, the only means of transferring a desirable gene from an ordinary variety to an elite variety of the same species was to cross-breed the two, backcrossing the progeny that manifest this trait with the elite cultivar over a number of generations. Typically, a breeder needed 10 years to derive an elite cultivar with the desirable gene by this laborious process. With his shuttle breeding method, U.S. agronomist and plant pathologist Norman Borlaug could scarcely reduce the time much below 5 years.

Biotechnology has erased much of the time and effort that went to deriving suitable varieties of plants for a number of environments by making possible the transfer of a single gene, or at times a number of genes, from one plant or organism of another type quickly into a plant. No longer was it necessary to confine one's efforts to plants of a single species because it was now possible to transfer genes from microbes to plants, a technique impossible with traditional plant breeding. The products of biotechnology have significantly affected agriculture worldwide. Although there have been problems along the way, biotechnology holds great promise.

BASICS

In plants and animals, deoxyribose nucleic acid (DNA) (Figure 10.1) stores a cell's genetic information. This DNA contains the instructions for making the chemistry of life. It is difficult to overstate the importance of this molecule. Segments of nucleotide bases in DNA represent genes. Each gene codes for the expression of a single enzyme or protein. (Enzymes are proteins.) As a rule, plants and animals contain large numbers of genes, not all of which function in a given organism. The chromosome is a linear arrangement of genes, though U.S. geneticist Barbara McClintock demonstrated that genes can move on chromosomes, an insight that startled the scientific community but eventually led to her crowning as a Nobel laureate.

Plants, like animals, have genes in their nucleus and mitochondria. Unlike animals, plants also have genes in their chloroplasts, a structure unique to plants. Genes in mitochondria and chloroplasts form loops, unlike the linear sequence of genes in the nucleus. Like animals, plants contain most of their genes in the nucleus. If one opens the nucleus of a plant cell, one can extract a sequence of nucleotide bases several meters long, giving one a sense of the astonishingly large number of genes in the nucleus.

The active parts of a genome, transcription units, synthesize ribonucleic acid (RNA) (Figure 10.2). Unlike nuclear DNA, RNA operates outside the nucleus to direct the synthesis of an enzyme or protein. The ribosomes contain ribosomal

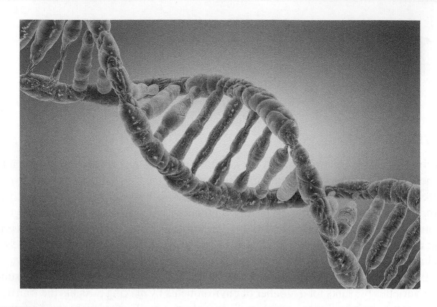

FIGURE 10.1 DNA is the macromolecule of heredity. (From Shutterstock.)

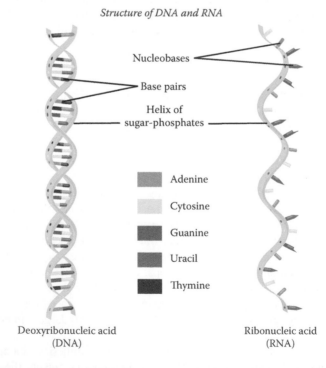

Structure of DNA and RNA

Nucleobases

Base pairs

Helix of
sugar-phosphates

Adenine

Cytosine

Guanine

Uracil

Thymine

Deoxyribonucleic acid
(DNA)

Ribonucleic acid
(RNA)

FIGURE 10.2 DNA translates its information, nucleotide base by nucleotide base, into a molecule of RNA. (From Shutterstock.)

RNA, rRNA. Transcription RNA (tRNA) assembles proteins outside the nucleus. Messenger RNA (mRNA) carries the genetic code from the nucleus to the cytoplasm of a cell. During translation, mRNA acts as a template upon which amino acids, the building blocks of proteins, arrange in a definite order.

A codon—a sequence of three nucleotide bases—directs the synthesis of a single amino acid. In some cases, different codons may direct the synthesis of the same amino acid. Several factors regulate the behavior of genes. As scientists sequenced the human genome, they have done the same with a few plants, though they began with *Arabidopsis thaliana*, an agronomically inconsequential relative of mustard, cabbage, broccoli, Brussels sprouts, and cauliflower: the cole crops. Although it is not a food plant, *A. thaliana* has a small genome, which quickened its sequence, and it has served as a model for biotechnology. Thereafter, the focus has been on the grains, notably rice, given its worldwide importance and its status as a staple of humankind. Between 2002 and 2005, the International Rice Genome Sequencing Project completed this task for both subspecies of rice, one native to Asia and the other to Africa.[1] The inclusion of African rice is important because it was likely the building block of rice culture in the Americas in the eighteenth century. The sequencing of rice demonstrated that rice shares a number of proteins with corn, wheat, and barley—not surprising given the fact that all are grasses.

A transgene is a foreign gene introduced into another organism, in this case a plant. The basics of plant biotechnology require the regeneration of a plant from a few cells. Typically these cells contain one or more transgenes. This regeneration is known as tissue culture. Tissue culture requires the introduction of nutrients, water, sunlight, heat, and growth regulators to stimulate the cell or cells to generate a new plant. An explant, parts of roots or some other structure, may be used in tissue culture. The younger the explant, the greater is the success of tissue culture. In some cases, a protoplant—a cell without the cell wall and often derived from shredded leaves—may be used in tissue culture. In some cases, the anther or a grain of pollen may be used as an explant without the need for sexual reproduction. Because gametes, sex cells, have only half the genome of a complete plant, this is haploid tissue culture. Alternatively, one may transform a gamete into a somatic cell by exposing it to colchicines, which doubles the haploid chromosomes to yield a full complement of chromosomes in a diploid cell. These nascent plants are sometimes called di-haploids because they have two pairs of the same chromosomes rather than the diverse genome that one derives from sexual reproduction. The grains are traditionally cultured from embryo cells. Curiously, elite cultivars of many crops are much more difficult than weeds to derive a new plant through tissue culture.

TECHNIQUES FOR PLANT TRANSFORMATION

The bacterial genus *Agrobacterium*, which naturally infects plants, may be used to insert transgenes into a plant.[2] *Agrobacterium* works much better on dicots than monocots, leading scientists to seek different vectors for corn, wheat, rice, barley, and other agronomically important monocots. Much of this work has centered on *Agrobacterium tumefaciens*, a soil-borne species of bacteria. When *A. tumefaciens* infects a plant—grapes, walnuts, apples, or rosebush, for example—it transfers its own genes into the plant's cells. In the early stages of plant biotechnology, scientists wondered whether they could modify *A. tumefaciens* to insert agronomic genes into crops. *A. tumefaciens* typically enters a plant at the site of a wound. A wound releases chemicals that attract *A. tumefaciens*, which attaches directly to a plant cell or cells. *A. tumefaciens* introduces its own transfer DNA (T-DNA) into plant cells.[3] Early efforts at transferring agronomic genes via *A. tumefaciens* focused on the dicot tobacco, a worldwide plant that nonetheless does irreparable harm to humans. Ground tobacco leaves have been placed in a medium with *A. tumefaciens*, which contains an agronomic transgene. The medium is allowed to stand 30 min during which *A. tumefaciens* binds with tobacco cells from the shredded leaves. The cells and *A. tumefaciens* are then put into a Mureschigo and Skoog medium for 2 days, during which the T-DNA conveys the transgene to the nucleus of tobacco cells. A new plant emerges with the transgene.

Alternatively, eschewing *A. tumefaciens*, plant biotechnologists may bombard, for lack of a better word, plant cells with the desirable DNA hoping that the cells take up the new material.[4] In this method, particles of tungsten or gold are coated with the desirable DNA and shot at high speed into plant cells. Unfortunately, the success rate of this method is low, though there have been notable successes, for example, the transfer of genes from the bacterium *Bacillus thuringiensis* (Bt) to corn, enabling it to express a protein toxic to the European Corn Borer, an important pest of corn. These techniques have been used extensively to create herbicide-resistant plants, including Monsanto's Roundup Ready (RR) sugar beets, corn, soybeans, and cotton.[5] This success has given these crops the ability to withstand doses of the herbicide glyphosate, marketed by Monsanto as Roundup. The weeds succumb, leaving only RR crops to thrive in fields barren of weeds. These crops have only grown in importance over time. To date, herbicide- and insect-resistant crops have been among the major achievements of plant biotechnology.

Herbicide resistance was a natural progression of biotechnology because even before the introduction of plant biotechnology, scientists knew much about the ways herbicides worked on plants and the biochemical pathways they affected. Biologists were already studying the phenomenon of herbicide resistance. Often a single gene confers resistance to a herbicide, and this gene has been a good candidate as a transgene. Monsanto had an economic incentive to patent the herbicide Roundup and the RR crops to lock in farmers. Monsanto produced these RR crops by inserting a transgene into the T-DNA of *A. tumefaciens* for insertion into an explant. The bioengineering of a herbicide-resistant crop usually, we have seen, involves the transfer of only a single gene to confer resistance and so is attractive to scientists.

BIOTECHNOLOGY OF HERBICIDE RESISTANCE

Although the land grant complex bred crops in the early- and mid-twentieth century, by the 1980s and 1990s, private companies had taken up much of this work. Biotechnology has arisen as a largely private affair. Monsanto understood that RR crops would be popular among farmers because the latter knew that weeds competed with crops and so diminished yields. They were ready for a new input to save them from these losses. Farmers also knew that weeds could harbor insects and pathogens, the classic case being Johnsongrass, which harbored the aphids that spread maize dwarf mosaic virus (MDMV) and maize chlorotic dwarf virus (MCDV) to corn, engulfing cornfields along the Ohio and Mississippi rivers during the early 1960s.[6]

The development of broad-spectrum herbicides posed a dilemma. They were useful against lots of weeds, but could only be used at a time when they would not kill the crop. Herbicide-resistant crops have allowed farmers to spray herbicides whenever they are most effective without fear of injuring a crop.[7] Herbicides may target pathways of photosynthesis or nutrient synthesis, processes that occur in the chloroplasts. Often herbicides will kill weeds by interfering with the production of a single enzyme or protein. Herbicides are classified by mode of action, that is, by the pathway they disrupt. Herbicides sulfomylureas and imidazolinomes target the enzyme acetolactate sythase. Roots or leaves often absorb herbicides. Herbicides also differ in timing. Some are sprayed preplanting, whereas others pre-emergence, postemergence, or preharvest. Some are biodegradable, whereas others persist in the environment. One difficulty is that a weed may produce a super abundance of an enzyme or protein and so is less susceptible to a herbicide that targets this enzyme or protein. It is desirable to derive an enzyme or protein that is a mutant of the target enzyme or protein and that can be incorporated in a crop, displaying resistance to that class of herbicide. Another biotechnological approach is to insert a transgene into a crop that detoxifies the herbicide, thereby imparting resistance to that crop.

Killing 76 of the world's 78 worst weeds, glyphosate works by inhibiting the production of the enzyme 5-enolpyruvylistikimate-3-phosphate synthase (EPSPS) in weeds.[8] By binding to this enzyme, glyphosate prevents it from catalyzing important chemical reactions. In short, glyphosate, by inhibiting the production of EPSPS, denies weeds the ability to manufacture proteins. Consequently, a weed yellows and dies. Weeds are especially vulnerable to glyphosate when they are growing vigorously because this is the time of maximum protein synthesis.

Monsanto has taken two approaches to bioengineering glyphosate-resistant crops. First, it has engineered glyphosate resistance by causing a crop to overexpress the target enzyme so that glyphosate will not bind to all of it. This approach was first successful with the petunia, involving the growing of transgenic petunia in ever increasing doses of glyphosate, selecting only those plants that survived for use as an ornamental. Monsanto derived a petunia with 20 extra copies of the gene that expressed this enzyme. Monsanto used cauliflower mosaic virus to insert the extra copies of this gene into petunia. The gene codes for the production of the 35 S promoter, which can be inserted into the T-DNA of petunia.

Second, Monsanto inserted mutant copies of the enzyme to which glyphosate could not bind, leaving the crop free to synthesize proteins. Copies of the gene to produce the mutant were added to the plasmid DNA of *A. tumefaciens* for insertion into a crop. Yet, this strategy produced crops with only modest tolerance to glyphosate. It was better to insert a transgene to direct the production of the chimeric enzyme to catalyze the production of proteins in a crop. The bacterium *Escherichia coli* has also been used to insert transgenes into crops that coded for the production of enzymes to which glyphosate did not bind. RR soybeans and cotton are products of this process. Curiously, RR corn is derived in a different way, containing a transgene that expresses the EPSPS enzyme to which glyphosate does not bind. RR oilseed rape also contains this transgene. Monsanto has inserted multiple transgenes into canola to give it glyphosate resistance.

Monsanto and other companies have bioengineered a number of herbicide-resistant crops: alfalfa, cotton, oilseed rape, soybeans, wheat, sugar beets, corn, rice, potatoes, tomatoes, flax, canola, and sunflower. In 1997, U.S. farmers planted 17% of their soybean hectarage to RR soybeans and 10% of their cotton hectarage to RR cotton. By 2001, these figures had risen to 68% and 56%, respectively. As the use of RR crops has increased, so too has the use of glyphosate. In 1997 and 1998 alone, farmers registered an 81% increase in glyphosate use on fields planted to RR soybeans. This change may not be as perilous as it appears. Much of the increase in glyphosate use came from the fact that farmers abandoned the use of other herbicides. The reduction in the use of other herbicides has led some scientists to claim a net benefit to the environment. Because it is cheaper to produce soybeans using only glyphosate and RR seeds, farmers have saved about $25 per hectare. Overall, farmers who plant RR soybeans appear to use less total herbicides, though they do use more glyphosate. By one account, U.S. soybean growers saved $770 million in 2000 alone by switching to RR soybeans. This appears to contradict the assumption that herbicide-resistant crops will goad farmers to use more herbicides. Monsanto believes that glyphosate poses fewer environmental risks than other herbicides.

Some environmentalists, however, fear the loss of biodiversity in the loss of species of weeds.[9] More serious from farmers' perspective is the rise of glyphosate-resistant weeds. This was bound to be a problem. A glyphosate-resistant corn plant in an RR sugar beet field is by definition a weed. In 2013, scientists reported the emergence of glyphosate-resistant weeds in Oregon, Nebraska, Wyoming, Minnesota, and North Dakota. Nevertheless, the rise of RR sugar beets has been astonishing. In 2008, the first year Monsanto released large seed stocks, U.S. farmers planted RR sugar beets on 59% of hectarage. The next year, farmers planted RR sugar beets on 95% of hectarage, and the figure approaches 100%, though RR sugar beets have not been a success in California.[10] Moreover, the emergence of glyphosate-resistant weeds has led scientists at Michigan State University to believe that Monsanto will soon need to bioengineer a new variety of herbicide-resistant sugar beets to combat the problem of resistant weeds. Through the process of gene flow and mutation, herbicide-resistant weeds have arisen, and as is clear with sugar beets, has quickly become a significant problem. Moreover, pollen from an RR crop may fertilize related weeds, conferring glyphosate resistance to them. The problem is particularly notable with RR sugar beets and in theory should be a serious problem with corn because it outcrosses, that is crossbreeds with other plants, usually of the same species. Otherwise the offspring, if there are any, are likely sterile. Not all crops outcross. Soybeans and peas, and several grasses, are examples of crops that self-fertilize. Viruses may be vectors, transferring glyphosate-resistant genes to weeds.

BIOTECHNOLOGY AND INSECT-RESISTANT CROPS

The threat of pathogens, insects, and other scourges is difficult to overstate. By one calculation, only 63% of a potential crop is available for harvest, whereas the rest having succumbed to weeds, insects, and diseases.[11] By one estimate, insects and other pests consume some 13% of crops worldwide. Insects are the worst problem, but not to be neglected are nematodes, birds, and some mammals, which can do damage, as any home gardener knows. The search for higher yields typically produces a plant with even lower tolerance to insects. Elite cultivars are the most vulnerable to insects. The Green Revolution highlighted this problem (see Chapter 9). Green Revolution rice, planted in narrow rows, provided an environment, moist and shady, for brown leafhopper, the chief pest of rice, whose population exploded in Southeast and South Asia.

As with herbicides, farmers turned to the chemistry of control, using pesticides as defense against insects and other pests. One might recall the mania that surrounded dichlorodiphenyltrichloroethane (DDT) in the immediate postwar years. But DDT and other insecticides hastened the evolution of resistant insects. Insect-resistant crops held the promise of less reliance on insecticides and reduced a crop's vulnerability to insects. This model of control predated biotechnology. In 1905, British scientist Rowland H. Biffen discovered the first instance of an insect-resistant crop when he chanced upon a variety of wheat resistant to the Hessian fly.[12] In the 1940s, USDA breeder Glen Herbert Stringfield derived inbreds of corn resistant to the European Corn Borer for use in hybrid crosses (see Chapter 8). Stringfield also derived his own European Corn Borer-resistant hybrids of corn.

When one thinks of the plague of locusts in the Old Testament, one infers that adult insects are the culprits when, in fact, larvae do the most damage. The larvae of the European Corn Borer are a classic example. So too are the larvae of other moths, butterflies, and flies. In addition to these, mosquitoes, grasshoppers, crickets, aphids, and beetles also damage crops. Some insects magnify the problem by transferring pathogens to crops. A notable example concerned the aphids that fed on corn, transferring MDMV and MCDV to corn plants, causing an epidemic in the early 1960s.

Biotechnology has used the bacterium *B. thuringiensis* (Bt) to combat insects.[13] Scientists discovered Bt in the gut of dying silkworms and so understood that it produced a toxin to insects nearly a century before biotechnology took advantage of this trait. Bt produces an insecticide, the insecticidal crystal protein, often known as ICP or Cry proteins.[14] Bt carries Cry genes that code for these proteins in its DNA loops in plasmids. When Bt infects an insect, it transfers its Cry genes into the gut of insect larvae, where the proteins expressed by the genes destroy the cells in the gut by causing an imbalance of cations in these cells. Organic gardeners may purchase Bt spores as a natural pesticide so it cannot be as dangerous as critics once charged. These spores may be placed on the leaves of plants every 40 days or in the soil every 2 years.

Biotechnology's first attempt to use Bt was the insertion of Cry 1A and Cry 3A genes with cauliflower mosaic virus as the vector. Alternatively, the Cry genes were inserted into the T-DNA in *Agrobacterium*. These early attempts produced modest but not spectacular results in tobacco, tomato, and potato, all members of the Nightshade family. Using Bt as the vector produced much better results, causing a plant to express 100 times more Cry proteins than had been possible with the early method.

In the mid-1990s, the U.S. Environmental Protection Agency (EPA) approved the first Bt crops. Since then, the list has grown. Bt potatoes possessing the Cry 3A gene are resistant to the Colorado beetle. Bt cotton possessing the Cry 1Aa and Cry 2Ab or Cry 1Fa genes is resistant to the tobacco budworm, cotton bollworm, and pink bollworm. Bt corn with the Cry 1Ab, Cry 9C, and Cry 1Fa genes, the infamous Star Link variety, was resistant to the omnipresent European Corn Borer. Bt corn with the Cry 34Ab and Cry 35Ab genes was resistant to the corn rootworm. Bt corn, resistant to both the European Corn Boer and the corn rootworm, had the Cry 1Fa, Cry 34Ab, and Cry 35Ab genes. A second Bt corn resistant to these larvae has the Cry 1Ab and Cry 3Bb genes. Yet another Bt corn, resistant to the corn rootworm, has the mCry 3Aa gene. Among these transgenic crops, Bt potato is no longer grown. Monsanto, Dow, Syngenta, Aventis, Pioneer Hi Bred, and DuPont are the leaders in bioengineering Bt crops. By 2004, Bt corn resistant to the European Corn Borer was planted on 15% of corn hectarage worldwide.[15]

Problematically, some Bt lines are effective against only the first brood of the European Corn Borer. Farmers in states like Iowa, where first and second broods are both omnipresent, did not grow these Bt lines. To their credit, Bt lines give greater protection against insects than insecticides had. Yet, this fact did not prevent controversy. In the United States and Europe, the infamous Star Link controversy wrecked one Bt line of corn. The cautious EPA refused to approve Star Link for human consumption because of fears that it might irritate the stomach, approving Star Link only to feed livestock. Because the vast majority of corn feeds livestock rather than humans, this ruling did not disconcert corn growers. In 2000, however, reports surfaced that Kraft Foods had used Star Link, probably inadvertently, to make taco shells, prompting the EPA to order a recall.[16]

A second controversy concerned a Cornell University study that Bt pollen is toxic to the larvae of Monarch butterflies.[17] Larvae (caterpillars) that feed on leaves dusted with Bt pollen grew slowly and ate less. Environmentalists, conscious of the Monarch butterfly's widespread appeal, amplified this concern. Television news programs, even the Public Broadcasting System (PBS), focused on this story. Scientists in the United States and Canada determined to get to the truth, publishing their own study in 2001 in the *Proceedings of the National Academy of Sciences*. The researchers found, not surprisingly, that different Bt lines had different levels of Cry proteins in their pollen. Levels were highest in Bt 176, but even its pollen posed little risk to Monarch butterflies.

Perhaps more significant has been the danger of insects evolving resistance to Bt crops. Usually just one susceptible gene mutates in an insect to confer resistance. The larger the number of genes, however, the smaller is the probability of an insect's evolving resistance. Biotech companies have incorporated multigenetic resistance by crossing different Bt lines that express different Cry proteins. For example, Monsanto's Bollgard II cotton combines both Cry 2Ab and Cry 1Ac genes to try to thwart the cotton bollworm, beet armyworm, and fall armyworm from evolving resistance to Bollgard II. The approach has been similar with corn. Monsanto's Yield Guard Plus and Dow's Hercules Xtra contain multiple Bt genes. Yield Guard Plus has Cry 1Ab and Cry 3Bb, whereas Hercules Xtra contains Cry 1Fa, Cry 34Ab, and Cry 35Ab genes. Syngenta's Agrissure CB, resistant to both European Corn Borer and corn rootworm, has the Cry 1Ab and mCry 3Aa genes. Biotechnology firms are seeking other agronomic genes and proteins, one candidate being the Vip protein, which appears to be toxic to a range of insects but not humans. Again Bt is the vector. Bt when cultivated in a Petrie dish with Vip will absorb the protein. Already Sygenta has developed COT120 cotton with Vip 3A gene to kill a range of butterfly and moth larvae. Bacterium *Photorhabdis luminescens* produces toxin A (Tod A) to kill nematodes. Tod A has been expressed in plants resistant to the tobacco hornworm and southern corn rootworm.

Adherents of integrated pest management advocate the rotation of Bt and non-Bt crops and advise against the cultivation of different Bt plants that, though they may be resistant to a number of insects, are susceptible to the same insect. Even as the hectarage of Bt cotton has risen from 12% in 1996 to 35% in 2000, heavy infestations of insects still require insecticide use.[18] Farmers who grow Bt cotton on 100 ha must plant 25 ha to non-Bt cotton on which herbicide is applied to control the tobacco budworm, the cotton bollworm, and the pink bollworm.

As with herbicide-resistant crops, biotechnologists have engineered a number of insect-resistant crops and trees favored for their wood: oilseed rape, poplar, potato, tobacco, apple, lettuce, rice, corn, strawberry, sunflower, sweet potato, tomato, wheat, birch, alfalfa, pea, some beans, grapes, and sugarcane. In 2004, the International Rice Research Institute in the Philippines planted the first Bt line of rice in Iran. Thereafter, it spread to China, but evidence is sparse of its cultivation in Southeast Asia, India, the United States, and nations in Africa. If this is paddy rice, it will not find a home in the United States.

New research is focusing on a gene from *Galanthus rivalis*, an ornamental commonly known as Snowdrop, for its potential to confer resistance to the brown leafhopper, the world's most serious pest of rice. The leafhopper causes $250 million in rice losses per year worldwide. It also has the ability to infect rice plants with tungro virus adding another $90 million in damages.

With only a single exception, all transgenic pest-resistant plants have Bt genes for insect resistance.[19] Worldwide nearly half of cotton is either Bt, RR, or both in hopes of reducing the use of pesticides and perhaps herbicides. By one estimate, 25% of the world's insecticides are sprayed on cotton.[20] In West Africa, 80% of insecticides are sprayed on cotton.

BIOTECHNOLOGY AND THE SEARCH FOR PLANTS RESISTANT TO PLANT DISEASES

In the United States alone, plant diseases may cost farmers $33 billion per year.[21] These diseases have been a scourge of humanity for millennia. Ever since humans developed the written word, they have recorded the destruction by wheat rust. The pathogen that caused Late Blight of Potato caused the last subsistence crisis in Europe, destroying the potato crop between 1845 and 1849 (see Chapter 6). Because the masses depended on the potato for sustenance, 1 million starved and another 1.5 million fled Ireland. The Southern Corn Leaf Blight, a fungal disease, cost U.S. corn growers between 15% and 30% of their crop in 1970. The disease recurred the next year, but the damages were not as severe.

As was true of both the potato and corn, monoculture encouraged the aggregation of high densities of pathogens. Scientists divide pathogens into two types.

Necrotrophs kill plants, continuing to feed on them after their death. Biotrophs need a living plant to complete their life cycle. Even though biotrophs may not kill plants in every instance, they still do considerable damage. Fungi and bacteria may be either necrotrophs or biotrophs. Viruses are biotrophs. Obviously viruses can kill, as was evident in the epidemic of MDMV and MCDV in the 1960s.

Resistance to plant diseases is also of two types. Nonhost resistance means that an entire species is resistant to a pathogen. This resistance is multigenetic and is conserved across the entire species. This is a goal of biotechnology. Host resistance, on the other hand, means that a single variety is resistant to a single pathogen. Other varieties in the same species may be vulnerable. This is the gene-to-gene interaction to which U.S. plant pathologist Howard Flor referred. For a single gene of resistance in a plant, there exists a single gene for susceptibility in a pathogen. Resistance means that a pathogen may enter a plant but will not spread throughout it. Because, in Flor's case, only a single gene confers resistance in a plant, a pathogen that mutates its susceptible gene may be able to overcome the plant's resistance.

Among plant pathogens, fungi are the most troublesome. *Bipolaris maydis* causes Southern Corn Leaf Blight. Various species of *Fusarium* bedevil wheat, cotton, tomato, and other crops, costing growers billions of dollars per year. *Phytophthora infestans*, though no longer considered a fungus, causes Late Blight of Potato. Other *Phytophthora* species attack soybeans and cacao. *Phytophthora* causes $430 million in losses to cacao in Africa per year. U.S. soybean growers suffer $120 million in losses from *Phytophthora*, especially Phytophthora root rot. Worldwide losses from all species of *Phytophthora* total more than $10 billion per year.[22] The genus *Puccinia* attacks wheat and is the cause of infamous rust. *Rhizoactoria solari* attacks several crops.

Among viruses, MDMV and MCDV attack corn. Barley yellow dwarf virus focuses on small grains, causing more than $300 million in losses in the United States per year.[23] Cacao swollen shoot virus attacks cacao, costing African farmers more than $50 million per year. Cassava African mosaic virus targets cassava, causing more than $2 billion in losses worldwide. Rice tungro virus totals more than $1.5 million per year in losses to rice grown in Southeast Asia. Tomato spotted wilt virus costs tomato growers in Georgia $20–40 million per year. Tomato mosaic virus was once a serious disease in the greenhouse tomato industry of the Great Lakes and has long been a scourge of tobacco worldwide.

Among bacteria, *Pseudamonus syringae* attacks many varieties of New World beans. *Xanthomonus campestonia* targets tomatoes, peppers, and beans, all New World crops. These diseases cost Florida growers several million dollars per year.

We have seen that fungi are the worst class of pathogen. Some 8000 fungi attack agronomic plants. They enter a plant through a wound, stomata, or degradation of cell walls from another source. Interestingly, *P. infestans*, the pathogen that caused the Irish Potato Famine is, we have seen, no longer classed a fungus but rather a water mold. *P. infestans* is a diploid organism, whereas fungi are haploids. New pathogens evolve all the time, for example, *Phytophthora ramenum*, which kills oak trees. In the 1990s alone, this pathogen killed millions of oaks in the United States and has since spread to Europe.[24] By 2004, it had broadened its hosts to ash, beech, rhododendron, and camellia in addition to oak. The bacterium *Pantoes ananatis* has jumped from humans to plants, infecting eucalyptus and onion. Onions grown in Georgia suffer acutely. The pathogen also infects cantaloupe, honeydew, muskmelon, pineapple, and tomato. Most viral pathogens are single strands of RNA, though they may also exist as loops of RNA or even the classic double-stranded DNA. In this context, one must distinguish between rice tungro bacilliform virus (double-stranded DNA) and rice tungro spherical virus (single-stranded RNA). These viruses may be present together, and the brown leafhopper spreads both as a species of aphid spreads both MDMV and MCDV. Rice tungro spherical virus stunts plants, as do MDMC and MCDV. This seems surprising given that rice cultivars are already short because of the Green Revolution breeding program.

The United States spends $700 million per year on fungicides, though concern has arisen that these fungicides harm humans and the environment.[25] For example, methyl bromide is used against a range of pathogens and insects. In the United States, farmers apply some 27,200 tons of the chemical per year on soils before planting. Methyl bromide depletes the ozone layer, and in this context may be partly responsible for the rise in the rate of skin cancer in humans. Because methyl bromide is capable of killing all forms of life, it is a dangerous chemical. In 2013, California banned its use, requiring farmers to switch to methyl iodide, which in itself is hardly benign.

Those who look beyond the chemistry of control tout biological control. The presence of ladybugs in the garden is desirable because they eat other insects, yet this approach does not appear feasible for commercial farms. To date, despite aggressive efforts by biotechnology, the only commercial successes have come against viruses. The initial approach involved pathogenesis-related (PR) proteins, a large group of proteins that can be safely inserted into a plant's genome, but which may cause allergies in humans. PR proteins contain the chemical benn, which may be the active ingredient in resistance. Biotechnologists have inserted genes into plants that code for the expression of the PR enzymes chitinase and glucanase, which broach the cell walls in fungi, causing their death. Biotechnologists have inserted the genes that code for these enzymes into rice and barley. These biotech rice and barley do not suffer a loss in yield, which may occur in other biotechnological techniques. Yet, these bioengineered rice and barley have not won converts, perhaps for fear that they are unsuitable for human consumption given the possibility that they might cause allergies. The gene coding for chitinase, taken from the bacterium *Serratia marcescens* and inserted into tobacco, enhanced its resistance to the pathogen *R. solari*. The gene that codes for chitinase has also been inserted into cucumbers. Using *Agrobacterium* as the vector, the gene has been inserted into apple trees, making them resistant to the fungus *Venturia inequalis*, which causes apple scab. Yet these apple trees grow more slowly and so are not good candidates for commercial orchards. This gene is also limited by its ineffectiveness against the many *Phytophthora* pathogens. Glucanase enzymes, however, show promise in enhacing the expression of chitinase in plants. The 35 S promoter gene, isolated from cauliflower mosaic virus and inserted into tobacco, causes the latter to express the enzyme B-1, 3 glucanase, which shows promise against plant pathogens. Some success has come to tobacco by inserting strands of rRNA, which kill the cells of invading fungi. Bioengineered wheat has expressed more than one of these rRNA strands in addition to an enzyme that confers resistance to *Fusarium graminearum*. Puzzlingly, this wheat only expresses resistance in the greenhouse and is a failure in the field.

A gene from alfalfa, alfAFP, introduced into potatoes, confers resistance to the fungus *Verticillium dahlia*. Portions of this gene may also be extracted from bees and silkworms and then fused into a single gene. In this instance, the gene is known as CEMA and may, as is the case with alfAFP, be inserted into potatoes with the bacterium *Erwinia carotovora* as the vector. These potatoes are resistant to soft rot. CEMA has been introduced into tobacco to fortify its resistance to the fungus *Fusarium solari. Agrobacterium,* proving its worth again, may be used to insert the gene that expresses the enzyme T4 lysozyme into apple and pear to confer moderate resistance, perhaps better thought of as tolerance, to fireblight, a serious bacterial disease in Europe caused by *Erwinia amyloyora* and capable of killing a tree in a single growing season. Even in a moderately resistant tree, some 5% of shoots may still be damaged. All of these efforts have led to experimental trials but not to commercial releases, at least not yet.

The story has been different with the attempt to achieve viral resistance. In 1986, the first bioengineered tobacco plant was resistant to tobacco mosaic virus. Since then, biotechnologists have developed virus-resistant alfalfa, citrus, papaya, peanut, potato, rice, squash, sugar beet, and wheat. In some cases, ceruloplasmin (CP) genes, which mimic the etiology of a virus, have been introduced into plants, a process that may be somewhat akin to vaccination in humans. For example, nematodes infect grapes and raspberries with arabis mosaic virus. The virus is a single strand of RNA. Finding the gene that expresses this RNA in shredded leaves, scientists sequenced the nucleotide bases of this gene. From this, scientists were able to derive a CP protein in the laboratory for insertion into grapes and raspberries to confer resistance to arabis mosaic virus.

Another strategy is to use antisense RNA. Ribozymes introduced into plants break apart the invading virus and so prevent replication. This approach, though promising, is not yet commercial. The use of CP genes may lead to the phenomenon of gene silencing by expressing a stop codon that the genes insert into the midst of an invading virus so that it cannot complete transcription. This phenomenon is also known as RNA interference. It involves the use of double-stranded RNA to inhibit the expression of genes in an invader, a pathogen, for example. This strategy also works against a variety of insects and arachnids: ticks, the light brown apple moth, the trintomine bug, termites, the tsetse fly, the western corn rootworm, and the beet armyworm. As these organisms feed on transgenic crops, they ingest the double-stranded RNA, which will not allow transcription in insect and arachnid cells. Monsanto has used this approach to bioengineer transgenic squash, including zucchini, resistant to the viruses ZYMV, watermelon mosaic virus, and CMV. Transgenic papaya is resistant to PRSV, a serious disease in Hawaii. Researchers in Hawaii and at Cornell University have engineered two varieties of transgenic papaya, SunUp and Rainbow. Rainbow has been outcrossed with susceptible lines using conventional techniques. Aphids that transmit PRSV, when they feed on resistant papaya, may be cleansed of the virus so that they cannot transmit it to susceptible papaya in a neighboring field. This approach therefore benefits more than just transgenic papaya.

Monsanto sold NewLeaf, a transgenic potato with Bt genes for insect resistance and genes for resistance to several viruses such as PLRV and PVY. Curiously given its benefits, NewLeaf did not sell well, forcing Monsanto to abandon it in the way that automakers jettison unpopular models.

One goal of biotechnology is to identify and insert transgenes into agronomic plants that express antibiotic chemicals: oxygen compounds, metabolites, hydrolytic enzymes, peptides, and other proteins. This approach may hold promise against fungi and bacteria, an area where biotechnology has had mixed success. The insertion of transgenes that express proteins with protease inhibitors confers disease resistance in tobacco. Single-gene insertions, we have seen, are rarely durable, because the pathogen need only mutate the single gene of susceptibility to become immune to the plant's defenses. It is better to "stack" several genes, all of which a pathogen must defeat to gain immunity. It is also desirable to insert genes, which target different diseases or both diseases and pests, into an agronomic plant. Stacked rice, for example, has several genes that confer resistance to the pathogen *Xanthomonas oryzae*. This rice has the R genes Xa-4, xa-5, xa-13, and Xa-21.

Climate change may hasten the spread of pathogens, especially viruses, many of which have origins in the tropics and subtropics. As the climate warms, these pathogens should be able to spread farther from the equator. Most new plant pathogens are viruses. Globalism makes it easy to transmit pathogens quickly across continents. This problem has grown since the Columbian Exchange. About 80% of viruses depend on vectors: insects, fungi, and mites.

BIOTECHNOLOGY AND ABIOTIC STRESS

In addition to biotic agents, abiotic stresses may cost farmers part or all of their crops. Water deprivation, saline soils, temperatures too high or low, and other factors all diminish yields.[26] In field trials, even moderate water stress has diminished wheat yields by 13%. Depletion of the ozone layer and climate change will surely worsen these stresses. As population rises, the temptation will increase to plant on marginal land in marginal climes, worsening abiotic stresses. Biotechnologists have attempted to meet these challenges by, in one case, transferring a gene that codes for the expression of a protein in tomato. Grown in saline soils, this transgenic tomato takes up sodium and chlorine ions without any ill effect on growth and yield. Moreover, the fruit do not retain this sodium and chlorine and so are safe for human consumption, meeting all the requirements of a low-salt diet. The same effect has been found in transgenic canola. The plant will take up the sodium and chlorine ions to no ill effect and the seeds and oil will not possess excess salt. Other saline-tolerant successes include transgenic wheat, corn, rice, tobacco, and cotton. As a bonus, these transgenic plants are more drought-tolerant than conventional cultivars because saline and water stress involve the same biochemical pathways.

Another goal of biotechnology has been to lengthen the shelf life of food, an important consideration given the large quantities that spoil. In 1994, in the first success of its kind, Calgene bioengineered the FlavR Saver tomato (Figure 10.3).[27] It had genes that slowed ripening and so extended the shelf life. Because these tomatoes ripened slowly, they could be picked red and ripe without fear of spoilage. The

FIGURE 10.3   The Calgene tomato, known as the Flavr Savr tomato, was an early geneti-cally engineered crop.

product, having attained maturity, had more flavor and nutrients than traditional tomatoes that were picked green, sprayed with ethylene gas, and then allowed to ripen during transit. Anyone who has bought a tomato from the grocer knows that because it was picked immature, it will never reach the peak of flavor. What one has is a firm but insipid mass. The genes in FlavR Saver reduced pectin methyl esterase (PME) enzyme activity, slowing ripening. PME enzyme strengthens the cell wall, slowing the process by which it deteriorates, one of the characteristics of decay. Transgenic mango, peaches, and pears also ripen more slowly.

Scientists are investigating the addition of genes to express antisense ACC syn-thase or ACC oxidase in ornamentals, the rosebush, for example, to delay ethyl-ene gas production with the result that cut flowers will remain fresh longer. The potential of delayed ripening will be important in the developing world, where poor infrastructure delays the transit of food to market. An additional bonus is that some transgenic tomatoes are dwarfs that require less staking and so less labor. Transgenic wheat and corn also contain genes for dwarfing, a characteristic desirable in wheat but presumably of no value in corn because hybrids do not lodge.

Despite 20 years of work, biotechnologists have identified only a few genes that improve yield under abiotic stress. Perhaps additional research in this area is war-ranted. The transgene mtlD, derived from the bacterium *E. coli*, improves drought and saline tolerance in wheat. Tomato, cotton, and rice are the foci of current research. The introduction of the transgene AVP1 aids plants in developing more fibrous root systems with presumably greater efficiency extracting water and nutrients from the soil. The transgene P5CSF129A, inserted in chickpea, directs the expression of proline, leading the legume to improve modestly the efficiency of transpiration so that chickpea loses less water through this process. The transgene PSCS improves drought and saline tolerance in rice.

PUBLIC ACCEPTANCE OF PLANT BIOTECHNOLOGY AND ITS PRESENT STATE

Concerns over transgenic crops focus on four issues: safety to humans and the environment, private ownership of transgenic crops, ethical issues about intellectual property, and the labeling of transgenic foods as such.[28] As of 2009, the United States planted transgenic corn and cotton on 64 million hectares.[29] India tended 8.4 million hectares of transgenic cotton. Brazil planted 21.4 million hectares to transgenic corn and cotton. China planted 3.7 million hectares to transgenic cotton and poplar. Argentina tended transgenic corn and cotton on 21.3 million hectares. South Africa planted these two transgenic crops on 2.1 million hectares. Canadian farmers grew transgenic corn on 8.2 million hectares. The Philippines grew transgenic corn on 490,000 ha. Australia and Burkina Faso planted transgenic cotton on 230,000 and 115,000 ha, respectively. Uruguay and Spain grew transgenic corn on 790,000 and 76,000 ha, respectively. Mexico grew transgenic cotton on 73,000 ha. Chile grew transgenic corn on 32,000 ha. Colombia planted transgenic cotton on 24,000 ha. Honduras, the Czech Republic, Portugal, Romania, Poland, Egypt, and Slovakia all grew transgenic corn on 15,000 or fewer hectares. Costa Rica planted transgenic cotton on about 2000 ha. Worldwide, in 2009, farmers planted transgenic crops on 131 million hectares, though one notes that most of this land goes to transgenic corn and cotton. To be truly successful, biotechnology must produce a large number of transgenic crops.

In 1996, transgenic soybeans occupied 500,000 ha worldwide, in 2001 33.3 million hectares, and in 2006 58.6 million hectares.[30] The last figure represents 68% of land planted to soybeans worldwide. In 1996, transgenic corn occupied 300,000 ha worldwide, rising to 9.8 million hectares in 2001 and to 25.2 million hectares in 2006.[31] The last figure represents 18% of corn land. In 1996, transgenic cotton was planted on 800,000 ha worldwide. Thereafter, the figure rose to 6.8 million hectares in 2001 and to 13.4 million hectares in 2006. The latter represents 39% of land planted to cotton. In 1996, transgenic canola occupied 100,000 ha worldwide, rising to 2.7 million hectares in 2001 and to 4.8 million hectares in 2006. The last figure totals 21% of land planted to canola.

In 2006, transgenic soybeans occupied more than 90% of the area planted to soybeans.[32] Outcomes depend on the crop. RR soybeans led to a diminution in herbicide use and to modest yield gains. RR cotton led to stasis in herbicide use and to notable yield increases. In the developing world, pesticide use decreased by 15% between 1996 and 2005.[33]

Biotechnology must strive for quality, yield, and sustainability. It must help wean farmers off fossil fuels and agrochemicals. It is difficult to foresee how the world will avert a Malthusian crisis without biotechnology. Biotechnology must form the bedrock of a second Green Revolution. Most transgenic crops have been designed to benefit farmers, whether through resistance to insects, pathogens or herbicides, or improved yield in saline soils or during drought. By 2011, 15.4 million farmers in 29 countries grew transgenic crops on 366 million acres.[34] Sixty-three nations have plant biotechnology programs. By one prediction, biotechnology may reduce the need for chemicals, fossil fuels through minimum tillage, which may in turn conserve soil and water.

Between 1996 and 2007, global farm income increased to $33.8 billion, partly due to biotechnology.[35] Pesticide use fell nearly by 16%. During these years, corn and cotton farmers have decreased their use of fossil fuels by 25%. Bt corn and cotton have put significant downward pressure on the use of pesticides.[36] Transgenic corn has increased U.S. farm income by $23 million per year. Bt cotton has allowed some Chinese farmers to reduce insecticide use by as much as 80%. Per unit weight Bt cotton can be produced nearly 30% cheaper than traditional varieties. Bt rice may likewise reduce insecticide use by 80%, though yields have grown only by 8% between 1996 and 2007.[37] In India, Bt cotton has reduced insecticide use by 40%. Yields have grown nearly one-third between 2002 and 2006. Profits per hectare have nearly doubled. One hopeful sign is that poor farmers in China and India are planting Bt cotton and corn, increasing their income and diminishing their use of pesticides. During these years, farmers in India saw their collective income rise to $5.1 billion, and at the same time they were saving money on pesticides.[38] Once a cotton importer, biotechnology has made India an exporter. India is now the world's largest cotton exporter and the second largest producer, trailing only the United States or China, depending on whom one consults. Other transgenic crops, possibly eggplant, tomato, potato, and rice, are poised to enter India.

In 2008, Egypt was the first Arab country and only the third in Africa behind Burkina Faso and South Africa to adapt transgenic crops, planting Bt corn on 700 ha that year.[39] Inexplicably, France has been hostile to this effort, even hinting that its relationship with Egypt might deteriorate over the planting of transgenic crops. Future efforts will continue to focus on biotic stress, but more research is likely to target abiotic stresses. The emphasis on the latter is likely to increase with climate change. Monsanto and Dow are collaborating on the development of a large number of stacked cultivars. It may be possible to bioengineer plants that need less nitrogen in the soil and so decrease fertilizer use. The derivation of grain with nodules in their roots comparable to legume roots or a tomato potato hybrid has stirred the interest of scientists for generations, but there is not yet evidence that biotechnology can achieve these milestones. Another goal is to bioengineer plants that may be sown earlier in spring to lengthen the growing season.

Monsanto's RR 2 Yield soybeans give 7%–11% more beans than conventional soybeans, suggesting that yield increases are within the compass of biotechnology.[40] The bioengineering of rice that tolerates flooding will be ever more important as sea levels rise and the distribution of rainfall varies. Another goal of biotechnology is to engineer plants that tolerate aluminum in the soil. Perhaps even more important, at least to consumers, will be the bioengineering of plants with higher quantities and quality of nutrients, for example, potatoes with a better balance of amino acids, rice with beta carotene, high protein corn and sweet potatoes, mineral dense rice and corn, and crops that are more efficient in their uptake of minerals like iron and zinc from the soil. Public research agencies have introduced two transgenes—daffodil phytoene synthase and rice endosperm specific promoter—to make rice express beta-carotene, which the body converts into vitamin A. This achievement has the potential to save people who consume a diet too rich in polished rice from becoming blind. In the continuing struggle against drought, Monsanto has introduced genes into corn that express the protein dehydration-responsive element binding. This corn yields 30% more grain than conventional varieties during dry spells. Transgenic canola has yielded similar gains.

In 2010, half of the Americans supported transgenic crops as necessary to feed the world's population.[41] Americans have responded positively to the notion that biotechnology has the capacity to increase the density of nutrients in food plants. Seventy-seven percent of Americans favor the bioengineering of insect-resistant crops if the goal is to reduce the use of pesticides. Seventy-three percent of Americans expressed a willingness to buy bread, pasta, and other items made from transgenic wheat. These results are remarkable given that many Americans know of biotechnology only through sensational headlines in tabloids, which distort reality. These same Americans tend to distrust government and so are not eager to read balanced, careful reports about biotechnology from government agencies such as the USDA. Much data about attitudes toward biotechnology come from polls by Texas A&M University, Harris, Gallup, and the Pew Institute. Curiously, given unfavorable attitudes toward government, Americans want government to regulate transgenic plants. Some Americans dislike biotechnology because they associate it with profit hungry Monsanto and other multinational corporations. Monsanto and others have stirred discontent by forcing farmers to buy new seeds every year or to introduce terminator genes that make the F2 infertile. With corn this is not an issue because F2 hybrids yield poorly, but, where it permissible, farmers could easily save a portion of the soybean harvest for planting next year. Worldwide some 1.4 billion farmers rely on saved seeds for next year's crop. The perception remains, therefore, that biotechnology benefits corporations not people.

Surveys in the United Kingdom have shown little public interest in plant biotechnology, perhaps because the United Kingdom is not at the forefront of research. Of those who have an opinion, however, only 14% is favorable, 40% is unfavorable, and the rest are less strident.[42] In the 1990s, the public in the United Kingdom expressed support of bioengineered foods. Supermarkets sold bioengineered tomatoes and foods that contained genetically modified (GM) soybean oil or meal. Yet, the public became frightened in 1998 when a report claimed that bioengineered potatoes irritated the intestines of rats. In the ensuing uproar, the lead scientist of this report lost his job and peer-reviewed journals refused to publish his paper. The Royal Society of London reported that the findings were incorrect. British supermarkets nonetheless removed bioengineered foods from their shelves.

NOTES

1. Adrian Slater, Nigel W. Scott, and Mark R. Fowler, *Plant Biotechnology: The Genetic Manipulation of Plants*, 2d. ed. (Oxford: Oxford University Press, 2008), 32.
2. Mark D. Curtis, "Recombinant DNA, Vector Design, and Construction," in *Plant Biotechnology and Genetics: Principles, Techniques, and Applications*, ed. C. Neal Stewart, Jr. (New York: Wiley, 2008), 166.

3. John Finer and Tanya Dhillon, "Transgenic Plant Production," in *Plant Biotechnology and Genetics: Principles, Techniques, and Applications*, ed. C. Neal Stewart, Jr. (New York: Wiley, 2008), 251–252.
4. Finer and Dhillon, "Transgenic Plant Production," 256–257.
5. Peter B. Kaufman, Soo Chul Chang and Ara Kirakosyan, "Risks and Benefits Associated with Genetically Modified (GM) Plants," in *Recent Advances in Plant Biotechnology*, eds. Ara Kirakosyn and Peter B. Kaufman (New York: Springer, 2009), 334.
6. Christopher Cumo, "Public Science and Private Enterprise: Battling Ohio Valley Corn Viruses," *Timeline* 16 (May-June 1999): 34–43.
7. G Brooks, "Plant agriculture: The impact of biotechnology," in *Plant Biotechnology and Genetics: Principles, Techniques, and Applications*, ed. C. Neal Stewart, Jr. (New York: Wiley, 2008), 7.
8. Kenneth L. Korth, "Genes and Traits of Interest for Transgenic Plants," in *Plant Biotechnology and Genetics: Principles, Techniques, and Applications*, ed. C. Neal Stewart, Jr. (New York: Wiley, 2008), 197.
9. Kaufman, Chang and Kirakosyan, "Risks and Benefits," 342.
10. Christopher Cumo, "Roundup Ready Sugar Beets" [unpublished manuscript]
11. Graham Brooks, "Plant Agriculture," 7–8.
12. R. H. Biffen, "Mendel's Laws of Inheritance and Wheat Breeding," *Journal of Agricultural Science* 1 (1905): 4–48.
13. Korth, "Genes and Traits of Interest," 200.
14. Ibid, 201–202.
15. Slater, Scott, and Fowler, *Plant Biotechnology*, 146.
16. Andrew Pollack, "Kraft Recalls Taco Shells with Bioengineered Corn," New York Times, 23 September 2000 http://www.nytimes.com/2000/09/23/business/kraft-recalls-taco-shells-with-bioengineered-corn.html
17. Detlef Bartsch and Achim Gathmann, "Field Testing of Transgenic Plants," in *Plant Biotechnology and Genetics: Principles, Techniques, and Applications*, ed. C. Neal Stewart, Jr. (New York: Wiley, 2008), 314–315; Douglas Powell, "Why Transgenic Plants Are So Controversial," in *Plant Biotechnology and Genetics: Principles, Techniques, and Applications*, ed. C. Neal Stewart, Jr. (New York: Wiley, 2008), 349–350.
18. Slater, Scott, and Fowler, *Plant Biotechnology*, 144–145.
19. Kaufman, Chang and Kirakosyan, "Risks and Benefits," 339.
20. Slater, Scott, and Fowler, *Plant Biotechnology*, 144–145.
21. Ibid, 156.
22. Ibid, 159.
23. Ibid.
24. Ibid, 160.
25. Ibid, 160.
26. Kaufman, Chang and Kirakosyan, "Risks and Benefits," 344.
27. Ara Kirakosyan, Peter B. Kaufman and Leland J. Cseke, "Plant Biotechnology from Inception to the Present," in *Recent Advances in Plant Biotechnology*, eds. Ara Kirakosyn and Peter B. Kaufman (New York: Springer, 2009), 11.
28. Kaufman, Chang and Kirakosyan, "Risks and Benefits," 344; Alan B. Bennett, Cecilia Chi-Ham, Gregory Graff, and Sara Boettiger, "Intellectual Property in Agricultural Biotechnology: Strategies for Open Access," in *Plant Biotechnology and Genetics: Principles, Techniques, and Applications*, ed. C. Neal Stewart, Jr. (New York: Wiley, 2008), 329–331.
29. H. S. Chawla, *Introduction to Plant Biotechnology*, 3d ed. (Enfield, NH: Science Publishers, 2009), 499–500.
30. Slater, Scott, and Fowler, *Plant Biotechnology*, 128, 318–319.
31. Ibid, 138–139.

32. Ibid, 318.
33. Ibid, 128.
34. Martina Newell McGloughlin, "Prospects for Increased Food Production and Poverty Alleviation: What Plant Biotechnology Can Practically Deliver and What It Cannot," in *Plant Biotechnology and Agriculture: Prospects for the 21st Century*, eds. Arie Altman and Paul Michael Hasegawa (Amsterdam: Elsevier, 2012), 552.
35. Ibid.
36. Matias D. Zurbriggen, Nestor Carrillo and Mohammad-Reza Hajirezaei, "Use of Cynobacterial Proteins to Engineer New Crops," in *Recent Advances in Plant Biotechnology*, eds. Ara Kirakosyn and Peter B. Kaufman (New York: Springer, 2009), 66.
37. McGloughlin, "Prospects for Improved Food Production," 553.
38. Ibid, 554.
39. Ibid.
40. Ibid, 555.
41. Ibid, 558.
42. Slater, Scott, and Fowler, *Plant Biotechnology*, 317.

11 The Malthusian Crisis

AN ESSAY ON THE PRINCIPLE OF POPULATION

In 1798, British cleric Thomas Robert Malthus (Figure 11.1) published his influential *An Essay on the Principle of Population*. In it, he was perhaps the first to propose that the human population has limits and cannot grow unabated. This idea appears to square with the awareness of the ancients. *The Book of Revelation*, for example, proposes that war, famine, and disease limit population.[1] Malthus uses these unpleasantries as checks on human population growth. Malthus understood that "the means of subsistence" (the food supply) determines the limit above which population cannot grow.[2] The level of subsistence checks every facet of human growth, be it the economy, society, or politics. The Enlightenment faith in progress through scientific rationalism did not impress Malthus, who ventured two postulates. First, food is essential to human survival. Second, the libido is powerful, causing humans to reproduce at a rate that is unlikely to abate in the future. Human sexuality is akin to an impulse that few resist. In this context, the "power of population is indefinitely greater than the power in the earth to produce subsistence for man."[3] Population, Malthus held, increases geometrically unless catastrophe befalls it, whereas food supply increases only arithmetically. That is, population cannot outstrip the food supply without causing high mortality to match the birth rate, creating a dismal equilibrium. Humankind, and here Malthus counters the optimism of the Enlightenment, cannot reason its way out of these laws of nature. Under these conditions, humans cannot perfect themselves or society, in general. Famine is perhaps the most obvious limiting factor correlated to food supply, for when food is scarce people starve. The food supply is not as elastic as the human capacity for reproduction so that population tends to outrun its food supply, causing a check on population. The checks to population fall heavily on the poor, who cannot afford sufficient food for their children so that infant mortality is high. As population increases, food becomes too expensive for the poor. Again high mortality ensues. Agricultural improvement is possible but cannot continue unabated. Moreover, the conversion of land from crops to pasture exacerbates the population crisis.

Malthus' ideas are remarkable when one considers that humans, by their biology, tend not to produce too many offspring. Although flies and mosquitoes and other insects lay millions of eggs to contribute to the next generation, only a fraction of which survive, a woman gives birth to only a few offspring over the course of several years and then invests her energy raising these children in hopes they will survive into adulthood. True, humans tend to have more children when the death rate is high in hope of seeing a few through to maturity. Eighteenth-century German composer Johan Sebastian Bach appears to be a classic case (see Chapter 3). He and his wives had 20 children. Only half survived to adulthood. Moreover, in low-tech

FIGURE 11.1 Thomas Malthus warned readers that human population tends to outrun its food supply. (From Shutterstock.)

rural settings, humans tend to have large families so that the children can help on the farm. Because population outruns the food supply, mortality rates remain high. In this instance, Bach provides a curious example. Well paid, he had no problem feeding his family. He worked for several cities in the course of his life, however, where contagion took its toll on his children. Malthus understood the importance of disease in curbing population. Possibly through sexual restraint, humans might limit population, but often the external checks act as the correctives. Over the long-term, population cannot evade these checks and so, through high mortality, people must live within the limits of the food supply. One must note that Malthus wrote at a comparatively early time in the history of the agricultural sciences, even if one traces these developments to the Greeks, Romans, and Chinese (see Chapter 5). For this reason, he appears to have underestimated the ability of humankind to increase the food supply in such dramatic fashion after his death. Of course, he cannot have foreseen the revolutions in the plant sciences: hybrid corn, the Green Revolution, and biotechnology. He cannot have anticipated the use of fertilizers, the intensification of irrigation, and the growth of other agrochemicals.

POPULATION PROBLEM

About every hour, the world produces more than 10,000 new mouths to feed.[4] Every tick of the second hand witnesses the birth of six people and the death of three for a net growth of three people per second or 180 people per minute. By 2050, earth is likely to have between nine and 10 billion people to feed. In this context, the big question is whether agriculture can provide enough food to feed so many people. The answer, obviously, must be yes if population grows so large by 2050, but there are no certainties. Today, demographers tend to use the phrase "carrying capacity" to denote the number of people that earth can truly sustain at a moment in time. Obviously, this number has not been constant over time. Periods of demographic stress, when, for example, diseases almost extirpated the Amerindians between the sixteenth and eighteenth centuries, have often been followed by population resurgence. This trend does not appear to apply to the Amerindians during the Columbian Exchange, but Europeans appear to have rebounded after the Black Death of the fourteenth century. Throughout the twentieth century, human population grew more rapidly than in the past even though the two world wars massacred millions.

Fundamental to this discussion, we have seen, is earth's carrying capacity, sometimes rendered the "Malthusian limit" or the "Malthusian cap." When food production stagnates, so does the rate of human increase. This is the moment of equilibrium as each death negates the potential increase of each birth. The death rate is thus capable of canceling the potential for growth and is thus nature's corrective to fecundity. This Malthusian equilibrium must have prevailed over at least parts of the vast millennia of human prehistory and history. Only in the twentieth century did humans witness rapid, one might almost say reckless, gains. In some societies, humans have tried to live within the Malthusian constraints by using birth control or infanticide. From the inception of anatomically modern humans about 200,000 years ago, humans only reached about 250 million people by the time of Jesus.[5] By the seventeenth century, the population approached 500 million. Even this number appears open to question given that the sixteenth and seventeenth centuries witnessed the virtual extirpation of the Amerindians. Other figures illustrate the trend. From the inception of anatomically modern humans about 200,000 years ago to 1800 CE, human population totaled about 1 billion, doubling in 1930 to about 2 billion, tripling in 1960 to 3 billion and in 2011 reaching 7.1 billion.[6] Note the huge growth in the twentieth century. The world must feed an additional 84 or so million new people each year. By 2025, the population may swell to 8 billion and perhaps more. It seems possible that earth has already reached its carrying capacity. If so, the Malthusian crisis must be upon us. Estimates of this carrying capacity are all over the place from the seventeenth century to the present, the low being 1 billion people and the high being 15 billion people, a staggering total. Most estimates range between 4 and 6 billion, both inadequate to express the current dilemma.[7] By modifying the environment, humans have increased earth's carrying capacity. Global warming and climate change, deforestation, and what appears to be a mass extinction all point to the reality that humans have reached the limit of environmental modification. Figure 11.2 demonstrates the dimensions of famine during a Malthusian crisis.

These numbers owed much to scientific and technological advancement. Humans went from using a digging stick to extract roots as hunter-gatherers to using a hoe as they became farmers. Subsequent developments included the plow, irrigation, the terracing of hills, crop rotation, the use of manures, and, in the nineteenth century, the use of synthetic fertilizers. These innovations may have led to bursts in populations followed by relative stasis so that human populations grew during brief periods, leveling off as the full potential of a scientific or technological adoption reached maturity. Human population has not and is not growing linearly. Science and technology have had the effect of raising the Malthusian limit. Periods of decreasing productivity and other evils led to population collapse as may have happened to the Maya and as did happen to the Irish in the 1840s. Malthus wrote his *Essay* in perhaps that last interval between a comparatively compact global population and the beginning of a rise in population, especially significant in the twentieth century. During this period, North America, parts of Europe, and Australia emerged as food exporters, a factor that must have driven the population increases of the twentieth century.

FIGURE 11.2   Famines make clear the correctness of Malthus' prediction.

As food production has increased, advances in public health, particularly the provision of reliably potable water, diminished the death rate. In the New World, European settlers populated the land rapidly, multiplying eightfold every 70 years, a rate that is not sustainable.[8] The enslaved population in British North America must also have swelled. American colonists needed few slave imports, relying instead on natural increase. In fact, the U.S. Constitution, written by slaveowners, ended the U.S. participation in the slave trade in 1808.[9]

After 1600, China's population began to grow, probably because of the adoption of sweet potatoes, corn, and other American crops. Europe did not initially press against the Malthusian ceiling because as the death rate fell, Europeans, especially the elites, began having fewer children. That is, the birth rate began to level off. Even in the United States, the birth rate began to fall in the late nineteenth century. The more stable a society, the more educated its people, and the lower the birth rate was. Two points deserve mention. The education of women was particularly important because the more education a woman has, the more likely she is to seek a career, to delay the decision to have children, and to have few children over the course of her reproductive life. The movement from countryside to city was also an important factor. On the farm, parents welcomed the birth of many children to help on the farm. Once in the city, however, this stimulus disappeared so that urbanites tended to have few children. This has not meant the end to the Malthusian crisis but a reorientation toward it. Cities continue to grow because people leave the countryside for the hope of opportunity in the cities. As cities continue to grow, they continue to need even more food so that urbanization appears to be one manifestation of the Malthusian crisis.

The Malthusian crisis is also not going away because the death rate appears to be declining. In 1921, India reported a decline in the mortality rate, a curious announcement given that the Spanish flu had devastated the planet in 1918 and 1919.[10] Egypt, Sri Lanka, Java, the Philippines, Taiwan, Korea, and Manchuria all reported the same phenomenon. By the early 1950s, populations were rising rapidly in what are today Zimbabwe and Uganda, though longevity remained low. The Caribbean Islands, Central America, Fiji, Malaysia, Iraq, Turkey, Colombia, Nicaragua, Panama, Ecuador, Paraguay, and Peru all total higher birth rates, a circumstance that gave urgency to the Green Revolution (see Chapter 9). Only about 1965 did the growth in birth rates begin to taper in parts of the developing world.[11] China took the extreme measure of coercing its women not to have more than one child. Birth rates, however, grew unabated in sub-Saharan Africa. Throughout the world, the Catholic Church opposed the use of birth control, family planning, and other efforts to slow the birth rate. Since 1950, the population in the developed world has increased 44%. The figure in the developing world is 166%. Since 1950, the population of sub-Saharan Africa has more than tripled.[12]

The big question is how many more people can the agricultural sciences and technology feed? No consensus exists. Ultimately land must be a limiting factor because not only it is finite, but also the amount that is arable is smaller still. Science and technology too must have limits. Population growth is so rapid that one must wonder whether science and technology can keep pace, even if revenues for research were much larger than they now are. How much money are governments and private firms willing to spend on agricultural research, and will the findings of the agricultural sciences really increase yields? We have seen that GM crops do not necessarily yield more food than conventional high-yielding varieties (HYVs), except under duress.

Complicating matters is the fact that humans know very little about the frequency and severity of famines throughout history. To try to put flesh on the bones, Ottersdorf and Weingartner have attempted to quantify 22 famines between 1770 and 1996 in an attempt to show that Malthus' ideas might have been new, the reality of his correctives must be much older.[13] In fact, Exodus records predictions of 7 years of famine following 7 years of plentitude. In 1770, about 10 people died in Bengal, India. In 1846 and 1847, famine may have carried away 2–3 million people in Ireland, though as I suggested in Chapter 6 the real mortality rate was not that high. In 1866, famine consumed 1 million people in India, in 1869 another 1.5 million, and in 1876 and 1878 another 5 million. Between 1888 and 1892, one-third of Ethiopia's population succumbed to famine. In 1920 and 1921, "several million" perished in Russia. Doubtless, political discord contributed to famine. In 1929, China lost 2 million people to famine. The next year, Russia lost 3 million to famine. In 1943, deaths from famine in Bengal ranged from 1.5 to 3 million people. Between 1946 and 1948, China lost a staggering 30 million people to famine. Between 1966 and 1970, 2 million Nigerians succumbed to famine. In 1973 and 1974, "millions" starved when drought struck Sahel, Africa. The phenomenon repeated in 1983 and 1984. "Millions" more perished from famine in 1992 and 1993 in East and southern Africa. The same regions suffered "millions" more to famine in 1995 and 1996. One might note that Asia suffered acutely before the Green Revolution. More recently, Africa has borne the brunt of famine. No figures appear to exist for pre-Columbian America.

Since the 1930s, famine, though not at an end, has ceded some ground to malnutrition as the chronic state of affairs. Malnutrition is hardly less serious, truncating lives and leaving others with serious deficiencies. Quality of life cannot be high wherever hunger threatens. We cannot look to ourselves for comfort. Since 1950, more people have joined the world's population than in all the previous 200,000 years during which anatomically modern humans have lived.[14] By 2100, the world may have 12 billion people, if food production can keep pace. Otherwise, one must revise the numbers downward. Between 1990 and 2025, population may increase by 160% in sub-Saharan Africa, 107% in western Asia and North Africa, 59% in Latin America, 56% in Asia, 20% in Eastern Europe, and 14% in western Europe, the United States, and Australia.[15] These figure are bad news to those who expect population to stabilize in the United States and its partners. Yet, in some circles, optimists hold that Africa can double food production by 2025. Between 1970 and 1990, food production worldwide increased by 2.3% per year.[16] The Green Revolution concentrated much of this growth in Latin America and Asia, with slow gains in Africa until much more recently. Yet gains in food production have slowed just as population continues to dash ahead. Lester Brown, founder of the Worldwatch Institute and founder and president of the Earth Policy Institute, has warned repeatedly that dense populations in the developing world still face food shortages. He warns that as humans degrade land and other resources, food will become more, rather than less, difficult to produce.

The Food and Agriculture Organization of the United Nations (FAO) is optimistic enough to forecast more rapid increases in food production in Africa. At the moment, many parts of Africa and some parts of southwestern Asia, notably Saudi Arabia, depend on food imports to sustain populations. One should be cautious about land and irrigation. As scarce resources, they are likely to check increases in food production. To leverage science and technology to try to stave the onset of the Malthusian crisis, the world must intensify investments in education, research, and extension, requiring new commitments of money and other resources. Yet, this is the very money when investments in education, research, and extension have slackened. We have seen that the Green Revolution agencies are putting their money into antipoverty programs rather than in traditional research to increase yields. Where is the money that we will need to applied and basic research? Again governments and private firms must step forward with resources and ideas. The world needs a new Green Revolution of innovation. Much effort must target the developing world, again emphasizing education, research, and extension. The record is mixed. As one measure of education, about 70% of Africans are literate, whereas the percentage is only 30 in Latin America and South Asia.[17] There is also gender imbalance, with women less likely to be literate than men. The developing world lags well behind the West in extension. In the West, one extension agent serves about 400 farmers. In the developing world, the ratio is one agent for 2500 farmers.[18] Such a heavy caseload must hamper all but the most minimal extension in the developing world. As the World Bank moves governments toward austerity, it is difficult to imagine improvements in the funding of research and extension. For that matter, one wonders how closely research and extension coordinate in the developing world. Some extension agents lack a car. How can they reach more distant farmers?

The proponents of increasing food production pin their hopes on two factors. First, they aim to bring more land into cultivation, an exercise in land extensification. We have seen that little suitable land remains to be cultivated. Second, they hope to increase yields per acre, the magic bullet that was the savior during the Green Revolution. This method is agricultural intensification. This method is uncertain and in any case would require a vast commitment of new funding for agricultural research, both basic and applied. Yet funding has stalled for traditional areas of research and is being reshaped to aim at alleviating the problems of rural poverty. Ideally, the developing world should build a land grant complex similar to what exists in the United States. Governments and private firms should emphasize plant breeding, genetics, biotechnology, and mechanization. Will such investments help? Optimists point to the success of the agricultural sciences to keep pace with ever larger increases in population. Skeptics warn that plants may have reached their biological limits at the same time as population marches inexorably ahead. Consensus, however, exists at many levels: increase funding for agricultural research, make contraceptives more available, make agriculture sustainable because progress is impossible in a degraded environment, stabilize nations, and mitigate warfare.

TRENDS IN FOOD PRODUCTION

Since Malthus' day, food production has actually surged ahead of population growth. Doubtless, this occurrence would have surprised Malthus. The amount of food and so calories available per person has actually increased since roughly 1950.[19] This bounty may pose a problem in the West, where abundant food and a robust junk food industry and a lack of vigorous activity promote obesity and other diseases. Sciences like plant breeding, biotechnology, and the chemistry of insect, pathogen, and weed control have made these gains possible. The Green Revolution (Chapter 9) is perhaps the most well-known coalescence of these factors. But how much longer can science and technology perpetuate these gains? This question may hinge partly on problems of distribution and fairness. The United States, Europe, and Australia produce a consistent food surplus, whereas hunger, malnutrition, and death plague people in parts of Africa, Asia, and Latin America. Perhaps, some distinction is necessary about Africa. North Africa appears to be ahead of sub-Saharan Africa in food production per person. Should not the developed world distribute food to the hungry worldwide rather than make more potato chips for people who already struggle with their weight? Unfortunately, poverty and malnutrition appear to be part of the human condition. Corporate chief executive officers and hedge fund managers make millions if not billions of dollars and seem disinclined to share their wealth. The same appears true of food surplus nations. The perpetuation of food aid may not goad developing nations to become self-sufficient in food.

The amount of energy one can derive from eating plants is limited because even though the sun emits enormous amounts of energy, plants convert only about 1% or 2% of the sunlight that reaches earth into the sugar glucose, upon which so many forms of life on earth depend.[20] Plants lose considerable energy because they can absorb only a limited set of wavelengths of light, just as humans can see only a limited spectrum of light. Other wavelengths of light, even those within the zone of capture, may not activate a molecule of chlorophyll. In other words, of the light available for capture—being of the right wavelength and hitting a molecule of chlorophyll—a plant may absorb only 30% of this light, making the total capture about one-third of 1% or 2%. All these factors limit a plant's capacity for photosynthesis (Chapter 1). In addition, plants use some of the light they capture to maintain cellular function. This light does not promote photosynthesis. If one considers this process inefficient, a solution, at least in theory, might be to use science, particularly biotechnology, to derive crops that conduct photosynthesis more efficiently than appears possible at present. Yet, this possibility remains theoretical. As we read in Chapter 2, natural selection has had 430 million years to fine tune land plants' ability to photosynthesis. How much additional photosynthetic capacity can science add in the span of say two generations, when the population is to increase dramatically? Yet, the Gates Foundation proposes to do just that, touting the slogan "Using the sun to end hunger."[21] Much of this research focuses on rice, a good choice given that it feeds more people than any other crop. The aim is ambitious: boost yields by 50% while cutting water and fertilizer use. In this regard, the Gates Foundation aims to go beyond the chemical- and water-intensive days of the Green Revolution. Such achievements appear to rest on engineering new pathways of photosynthesis. The most efficient photosynthesizers use the C4 pathway (see Chapter 1). Rice is not among this group, using instead the less efficient C3 pathway. It should be possible, again in theory, to identify the genes that control the C4 pathway and insert them into rice and other C3 pathway crops. This would be a triumph of basic and applied genetics and of biotechnology, but we noted in Chapter 10 that biotechnology is controversial in some parts of the world. Were such rice varieties developed, how would governments and consumers respond? Another issue concerns a plant's edible biomass. We noted that the Green Revolution made important strides by breeding wheat and rice varieties that converted more of their plants into edible biomass by deriving plants with thick heads of seeds and short, stout stalks. Are further gains in edible biomass possible? If so, how are the gains to be achieved? The use of the phrase "edible plant" or "edible biomass" must be taken in context. Even edible plants are seldom edible in their entirety. Humans cannot eat tomato and potato plants because the foliage is toxic. The appetite must confine itself to the fruits of the tomato plant and the tubers of the potato plant. Grains do not pose a much better alternative. Only the seeds of wheat, corn, rice, and several other grains are edible. The same is true of peas and other seed legumes. By one calculation, an average human consumes only 7% of an edible plant. Plant residues may serve other functions, holding promise as biofuels, for example, but these efforts do not abate hunger.

Rice serves as an example of a crop capable of yield gains. Asia grows 90% of the world's rice.[22] Smallholders grow rice and eat much of the harvest. Only about half Asia's rice enters the global market. Bangladesh and Myanmar derive about 70% of their calories from rice. Asia plants about 150 million hectares to rice (about 10% of the continent's arable land), an increase from 87 million hectares in 1950. Asia irrigates about half its rice. Of the remainder, about half receives sufficient rainfall to obviate the need for irrigation. Some of these lands yield as much rice per hectare as irrigated farms. The remaining land is infertile, providing just 6% of Asia's rice output. These farmers are among the poorest in Asia and never benefited from scientific farming. The HYVs of the Green Revolution transformed Asia. By 2000, China, India, and Japan planted HYVs on all irrigated land.[23] The rate of adoption is between 75% and 90% in Vietnam, Indonesia, the Philippines, Malaysia, and Sri Lanka. Thailand has resisted the HYVs in favor of traditional varieties that fetch a fine price. Even at lower yields, Thailand has no trouble exporting rice. Despite the overall good news, Asia will probably have trouble sustaining growth in rice yields. Irrigation is expensive and unlikely to be expanded to new lands. Research funds are limited and unlikely to go toward irrigation. An expansion in land is also unlikely because what is left is generally infertile, arid, or otherwise deficient. At the same time, cities may swell, taking farmland out of production. The very success of the agricultural sciences may be problematic, leaving the impression that science has conquered the Malthusian crisis. Development organizations and banks have scaled back their commitment to agricultural research at the moment when a redoubling of effort is necessary. Science has made gains in breeding pest-, pathogen-, and herbicide-resistant crops, but their maximum yield is not higher than other elite crops. Research remains wedded to improving agricultural production on the best soils, bypassing the millions of farmers who occupy marginal or infertile lands. As populations rise in the developing world, farmers will find themselves with less land on which to produce more food. Other limits emerge. Phosphorus is one of the Big Three macronutrients for plants, yet agriculture may exhaust the world's stocks by 2150.[24] How will humans adequately fertilize crops then? Perhaps, we will be forced to return to manure, as Varro and Columella had in Rome (see Chapter 5). The United Nations predicts a world population of 10.8 billion by 2050.

The science of plant breeding and biotechnology is only part of the equation of feeding more people. One recalls that nineteenth-century German chemist Justus von Liebig and other scientists initiated a fertilizer revolution that century. Along this line of research, German chemists Fritz Haber and Carl Bosch in the twentieth century developed a method for converting atmospheric nitrogen and hydrogen into ammoniacal fertilizers. By one estimate, this process may have doubled the capacity of earth to feed people, in effect allowing the population to double.[25] In turn, the Green Revolution made possible a doubling in world population between 1960 and 2000. One may find in these series of agricultural revolutions four pillars of increasing food production: plant breeding using fairly traditional methods though the novelty of breeding semidwarfs was crucial to success, irrigation, mechanization, and agrochemicals, including fertilizers, herbicides, and pesticides. What one finds then is a series of agricultural revolutions based on manipulation of a plant's genotype, the rise of agricultural chemistry, and mechanization through technological change. The emphasis, so far, has been on grains, especially rice and wheat, rather than the roots and tubers and legumes so important to the developing world. It seems, then, that a shift in focus may be desirable, moving from the grains to roots, tubers, and legumes. I do not suggest that the sciences abandon the cereals, only that developing world crops like cassava, yams, and sweet potatoes receive more attention.

Other developments have played an important role in helping farmers feed more people. The Columbian Exchange made possible increases in population. One need only consider the adoption of the potato in northern Europe. Plant breeding may have made possible half the yield gains that the Green Revolution witnessed. With abundant food came an increase in purchasing power. About 1950, an average person in the United Kingdom spent about one-third of his or her income on food. By the early twenty-first century, that amount had declined to less than 10% of income.[26]

Throughout history, food has been, with some notable exceptions, scarce: the limiting factor that has checked population growth. Most people had to struggle to find enough food to feed the family. Drought and warfare frequently threatened famine. Inequality has priced food out of reach. Into the eighteenth century, for example, wheat bread was too expensive for the French masses, who made do with rye bread and potatoes. Regional shortages were catastrophic in an era of small surpluses and inadequate transportation. The Malthusian correctives—war, famine, and diseases—have always brought population back to the level permitted by the food supply.

PROBLEMS OF PLENTITUDE

Despite these gains, the modern agriculture that feeds so many people may not be sustainable. Already humans use about two-thirds of fresh water to irrigate plants per year. Can our species really maintain this rate of exploitation? So much water goes to agriculture that today the average person has access to only half the potable water that was his or her due in 1950.[27] A single apple consumes about 70 L of water as it matures. Modern agriculture is also intensive in its use of fossil fuels at a time when most reasonable people understand the finitude of such sources of energy. By one estimate, humans will exhaust the world's petroleum by 2100. Agriculture and the science of it must change well before this event. In natural populations, animals spend about one calorie for every 10 calories of food they consume.[28] This ratio appears to have been true of hunter-gatherers and the first farmers. Modern agriculture has distorted what had been natural. Today, every calorie of food comes at the cost of 10 calories of energy, much of it in the form of fossil fuels and agrochemicals, both of which are energy-intensive to produce and use. This trend, too, is unsustainable. Though some people may not care, modern agriculture has harmed wildlife. We noted in Chapter 10 that Roundup Ready crops tolerate glyphosate. Robust uses of this herbicide leave fields nearly bereft of the weeds that herbivores had once eaten. Simply put, modern agriculture is at odds with the maintenance of biodiversity. One wonders whether this habitat distortion might affect the capacity of agriculture to feed more people. It seems possible, as well, that the sciences have pushed plants near their biological capacity to increase crop yields. The trend is disconcerting. Between 1960 and 1990, worldwide rice yields increased more than 2% per year.[29] During these years, wheat did even better with annual gains of 3%. Between 1990 and 2007, however, rice yields had fallen by more than half, achieving just 1% gains in yield per year. Wheat has fallen even further, reaping just 1/2% yield gains per year. It is clear that the initial gains came comparatively easily, but as the sciences have approached the biological limits of what a plant can produce, gains have become harder to achieve. The fact that global food prices have risen in the twenty-first century suggests that gains in food production have slowed relative to population growth. The current debate focuses on whether plants are near their maximum in producing edible biomass or whether large gains are still possible through some new and unexpected avenue of science. Optimists assert that an increase in funding for agricultural research, which has lagged since the 1990s, would restore the growth rates of the Green Revolution.

Between 2007 and 2030, the world demand for food seems likely to increase 50%.[30] Between 2007 and 2050, food production will need to double just to keep pace with population growth. Two factors appear likely to drive this growth. One is the obvious increase in population. The second depends on growing affluence in parts of the developing world. Affluent people tend to eat more than the poor. They also switch dietary habits from a plant-based diet to one based on meat, beef, and pork, in particular. This trend is troubling because livestock are not efficient converters of the grasses they eat into biomass. The 10-to-1 rule roughly applies. A cow or pig needs about 10 kg of corn to produce 1 kg of beef or pork.[31] Obviously, 10 kg of corn can feed more people than 1 kg of pork. The demand for beef will goad stockmen to set aside more land for pasture and so less for cropland. Affluence will bring humans to the Malthusian crisis much faster and with greater consequence than had people clung to a mostly vegetarian diet.

Between 2007 and 2050, per person consumption of meat may increase from about 28 to 52 kg per year.[32] If this occurs and if stockmen move to put more land to pasture, one assumes that the price of bread, to cite one example, should rise. Not only with more corn be diverted to cattle and pigs, more corn will likely be diverted to biofuels. With these pressures, corn prices should rise. This system seems likely to exacerbate global warming and climate change. Most scientists accept the reality of global warming and climate change, ideas first noted by nineteenth-century Swedish chemist Svante Arrhenius. By 2100, global temperatures may rise 2°C.[33] This increase may seem manageable, but we should veer away from Pollyanna. It is not clear how an increase in global concentrations of carbon dioxide will affect plants. Because plants consume carbon dioxide, the natural impulse is to assume increasing productivity with an increase in carbon dioxide concentrations. One must remember, however, that the grasses—which include sugarcane, rice, wheat, corn, rye, oats, and other plants—evolved at a time when carbon dioxide levels were comparatively low, making uncertain a prediction of how they will fare in a high carbon dioxide era. Moreover, floods and droughts should multiply as temperatures increase. Part of the problem is that the production of food itself causes the production of greenhouse gases. Tractor exhaust contains more than enough carbon dioxide for the planet. Cows emit methane. In 2000, agriculture produced about one-quarter of all greenhouse gases. By 2050, the figure could reach 75%. The removal of trees to make way for crops causes a net increase in carbon dioxide concentrations. Other effects are less certain. A warming climate should lengthen growing seasons in Canada and northern Europe, possibly increasing yields. But pests and pathogens should be more numerous too. One must wonder whether southern Europe's summers will become too hot and arid to sustain agriculture. Africa already suffers from low productivity. Matters seem certain to worsen if rainfall diminishes. North Africa, for example, is already quite arid. In the twenty-first century, food production in several regions of Africa may fall by 5%–25%.[34]

Some scientists believe in a "yield gap." That is, by current methods, agriculture provides x amount of food when the real potential y is much higher. Close the gap and science can feed a much larger number of people. Even with existing science and technology, with the elite varieties we have today, significant gains should be possible. Yet, one must remember that the persistent use of irrigation and chemicals makes this agriculture appear unsustainable. Even in Asia, which has seen the greatest successes, scientists believe yield gains of another 20% should be comparatively easy to achieve. Africa, on the other hand, has benefited least from scientific agriculture, particularly in the case of sub-Saharan Africa. The most optimistic forecast predicts that this region of Africa has the potential to increase yields by 200-fold.[35] This prediction appears Pollyannaish but it is worth noting that Africa irrigates only 6% of its crops, whereas Asia irrigates about 40% of its cropland. Africa uses less than 10 kg of fertilizer per hectare given the global average of 100 kg per hectare. Yet, these additional inputs will be expensive and ultimately unsustainable. Where Africa can improve is in the planting of HYVs. Of the grains, Africa plants only 25% of its grain land to elite cultivars, whereas the percentage is 85 in Asia. Chapter 8 reminded the reader that parts of Africa have been slow to adopt varieties of hybrid corn. Malawi has also had experience growing corn, though probably not hybrids. Corn is the leading crop in Malawi, much of it being produced by farmers with less than 1 ha of land. Eighty-five percent of these tiny plots go to corn. By the 1980s, low yields combined with population growth meant that Malawi could not feed its people and so had to import food. About 2000, the government began to promote an increase in the use of fertilizers, subsidizing their purchase, in hopes of making Malawi self-sufficient in grain.[36] Government also subsidized the cost of the seeds of HYVs of corn. By 2010, 65% of Malawi's farmers were planting and fertilizing these varieties. By then, corn yields had doubled. No longer an importer, Malawi could now reap a surplus for export. Child mortality has halved as food has become more available and affordable. Every $1 of subsidy allowed Malawi farmers to produce thrice the food than that same $1 could buy on the world market. Some worry, however, the Malawi, like other regions of the world, has become too dependent on costly fertilizers. Other scientists hope to leverage applied and basic genetics to increase yields of sorghum, millet, cassava, and banana. Of course, such gains, particularly in cassava and banana, should also benefit Latin America, including the Caribbean. Others offer low-tech solutions like the collection of rainwater to reduce the need for irrigation. Kenyan political scientist Calestous Juma believes that Africa remains stuck in the rut of the Green Revolution, using varieties that had been bred in the 1960s. An aggressive switch to today's HYV might, he believes, make Africa self-sufficient in food in a generation. In this context, the problem may not be science and technology but the reliance on outmoded science and technology. Why drive a 1914 Model T when one might have a 2014 BMW? Africa should seek the latest developments in biotechnology and computer modeling of yields. If Africans have embraced cell phones without ever having used a landline phone, why cannot African farmers leap past the Green Revolution into an environment in which scientists conduct the latest research in molecular biology, genetics, and biotechnology? All that is needed, at the most fundamental level, is a commitment to science, technology, mathematics, and engineering, the so-called STEM subjects. Also important will be satellite imagery. GPS systems could instantly allow farmers to target fertilizers and other chemicals to precise positions on a farm. Perhaps nanotechnology might help scientists and engineers to build better pesticides.

These developments may play a role in social, economic, and gender issues. The Malawi program, for example, largely ignores female farmers. Obviously, one cannot make the best use of science by ignoring half the population. Indeed, throughout Africa, women play important roles in food production and processing. Where they farm, women too must have the best crop varieties and the appropriate inputs if humanity is seriously to face the Malthusian crisis. Everywhere, it is important that farmers own their land because they make the best investments in scientific agriculture. Tenants have much less at stake and little incentive to improve the land and crops through the sciences. If the United States has modeled agribusiness as the recipient of applied sciences, the future of the agricultural science and farm productivity appears to lie with smallholders in Africa and other parts of the developing world. In Brazil, agribusiness and smallholders appear to coexist.

Part of the problem facing the agricultural sciences is that between 1983 and 2006 investments in agricultural research in the developing world have diminished.[37] This trend must reverse if the world is to feed ever more people. At the same time, the United States has reduced its commitment to fund agricultural research in the developing world, a contrast from the Green Revolution.

One ecological solution to the problem of insufficient food production may be to abandon monoculture for a diversity of crops. Monoculture leaves one vulnerable to pests and pathogens. One need only remember the Irish Potato Famine to drive home this point. Diversity minimizes risks as well as the buildup of pest and pathogen populations. If the population of pests remains small, farmers should need fewer pesticides. In line with an ecological movement, interest has risen about organic farming, but it may not be the solution because it is labor-intensive and yields about 60% of the food that conventional agriculture produces. One need only consider the Romans. In the ancient world and well into modernity, all agriculture was organic. Yet, the Romans could produce only about a 10% surplus. This is the limited world that organic farming makes possible. Because organic farming is not yield-intensive, it must be land-extensive, requiring more land to crops and pushing farmers to marginal lands and in conflict with ranchers. Again, the Roman example is relevant. In building the empire, the Romans were ever in quest for more land to subdue with the plow. Organic farming thus does not appear to solve the Malthusian crisis. By taking more land, organic farming should decrease biodiversity, as a study in the United Kingdom appears to have confirmed.

Scientists, policymakers, and the public have seen the tension between those who support biotechnology as a natural extension of the plant selection and breeding work that humans have conducted for centuries. Others fear that biotechnology will poison humans with "Frankenfoods" (see Chapter 10). In this context, biotechnology is dangerous and unnatural. Yet, no one can doubt its importance. Genetically modified food and fiber crops totaled 1.7 million hectares in 1996 and more than 17 million hectares in 2012, a more than 10-fold increase.[38] In 2012, more than 17 million farmers worldwide grew GM crops, 90% of these farmers being smallholders in the developing world. In this sense, biotechnology is deeply ingrained in rural life and perhaps is the solution to the Malthusian crisis. In 2012, GM soybeans were the total soybean crop in Argentina, 93% of the soybean crop in the United States, and 88% of the soybean crop in Brazil.[39] About 85% of the soybeans that the European Union imports as livestock feed are GM soybeans. One aim of biotechnology is to engineer grains whose roots would be similar to those of legumes in absorbing gaseous nitrogen in the soil. The difference is that the roots of legumes use nitrogen-fixing bacteria to deliver nitrogen ions to the roots. The proposal of biotechnology appears to aim for a direct absorption of gaseous nitrogen. Was such a goal to be realized, nitrogenous fertilizers should no longer be necessary and the production of grains like rice, corn, and wheat should increase. Such an achievement would gain traction against the Malthusian crisis. Such a prospect might increase grain yields in the developing world by 30%. Current gains in grain yields in the developing world are just 6%. GM crops have reduced the use of pesticides. One may recall from Chapter 10 that Bt corn contains a toxin to the European Corn Borer and so is more effective than pesticides alone in controlling borer populations. The question is what these new crops mean for the future. A plant like Bt corn, by making corn impervious to the borer, should stabilize yields so that in periods of infestation the farmer need fear no loss in yields. But Bt corn does not necessarily yield more corn than a conventional HYV. That is, Bt corn does not push the yield threshold higher. In this context, it is difficult to know whether such crops will help agriculture conquer the Malthusian crisis. GM crops do well in no till settings, a benefit for farmers and our planet, in lowering the consumption of fossil fuels. Again, however, it is difficult to argue that no till agriculture raises yields.

Yet, modern agriculture diverts potential food from the hungry to biofuels. In 2011, 20% of the world's sugarcane and 9% of oilseed crops went to make biofuels.[40] Importantly, staples like corn, wheat, barley, and sorghum go to make biofuels. Does not this diversion subtract from the ability of agriculture to feed the world? The logic behind biofuels is to stretch the supply of petroleum but the production of all these crops requires an input of gasoline to power machines and the use of chemicals, some of them made from petroleum. What is the true return on biofuels? Corn appears to provide little savings, though sugarcane produces real savings in the use of petroleum plus the bagasse may be used to generate electricity, as is true in Hawaii and Brazil. Biofuel crops will also witness no reduction in water use and take land out of food production, a flaw that should be fatal. Food prices are increasing because of the corn lost to biofuels, increasing the price that stockmen pay for feed. The Roman solution of putting more land to the plow is no longer realistic. Only about 3% of land worldwide not in production today may be suitable for cropland. The rest is infertile or too dry to make a difference.

If agriculture is to feed more people, it must embrace plant agriculture and de-emphasize stock raising. The same area of land can feed many more vegetarians than meat-eaters. Yet, meat seems to have been important in our diet at least since the inception of our genus. The consumption of meat may fulfill a biological impulse. Perhaps, it is unrealistic to expect a conversion to vegetarianism. Indeed, the trend in India and China is toward increasing consumption of meat. Even if one were to promote a plant-based diet, the preparation of some such food, tofu for example, is energy-intensive. In some cases, it may be more efficient to eat chicken than highly prepared plant products.

Per unit of land, roots and tubers are perhaps the most efficient use of land. Potatoes, sweet potatoes, and cassava feed millions in the tropics, and in the case of the potato in the temperate zones. Perhaps agriculture, in the quest to feed more people, might convert at least some land from grain to roots and tubers. The production of more food to feed more people depends partly on a reduction in warfare. We have seen that warfare is a check on population growth in the conventional sense that Malthus made clear, but it is also detrimental in destroying crops and killing the people who work the land. In this sense, warfare reduces food production. Where famine caused food contractions and starvation in an earlier era, war may be the limiting factor in our age. Current levels of violence may put as many as 80 million people at risk of starvation worldwide.[41] How can one feed a growing population amid savagery? Another problem, particularly in the developing world, is distribution. If food is available in one region, a portion must be moved to a region threatened by famine. The production of more food, if it is to improve humanity, must mean the production of more protein, vitamins, and minerals. Vitamin A deficiency has been prevalent in parts of Asia too dependent on conventional varieties of rice. The grain does not normally contain much beta-carotene, the precursor of vitamin A. This deficiency causes blindness. The good news is that scientists have bred Golden Rice, which contains sufficient beta-carotene to stave off nutritional disorders. One may trace other deficiencies to the diet. A diet inadequate in iron causes anemia. Poverty is another important factor that limits food production.

At the dawn of the twenty-first century, some scientists believed that the world could double or even triple food production, even in the short term, to avert or at least postpone the Malthusian crisis.[42] Such an achievement would require a large investment in science and technology. Are the United States and other countries willing to make these investments? Are farmers still willing to adopt modified plants and other products of science and technology? The Malthusian crisis, in a curious way, rests in part on urbanization. We have seen the drawback of urbanization, but there may be a silver lining. To make the production of more food viable, a base of consumers without access to homegrown food must exist to buy more food. These people are by definition urbanites, and their demand for more food should provide farmers fair prices for their food and so enhance the desire to grow more food.

Nonetheless, agricultural productivity remains low in Central Asia and sub-Saharan Africa, exactly the regions that need to invest in the science and technology of producing more food. This goal is essential to achieve because the birth rate remains too high in both regions. Despite the efforts of the Green Revolution and the willingness of countries like Argentina and Brazil to invest in biotechnology, Latin America continues to make poor progress. Niger, Malawi, Rwanda, and Burundi have few cities (centers of consumption). Farmers cannot make a living solely by trying to sell food to other farmers. This lack of consumers may impede efforts to increase food production. Where the return on investments in agriculture is so low, how will governments find the will to invest in plant breeding, genetics, and biotechnology? Populations will continue to grow rapidly worldwide so that the greatest need for producing more food lies in the future. Unnervingly, the Green Revolution appears merely to have postponed the Malthusian crisis. Since 1985, the world has added 88 million new mouths to feed per year.[43] This rate must abate, but when? Even if population growth stalls or even declines in the West, parts of the developing world will remain a challenge. At a minimum and in the very near term, food production must rise in Central Asia and sub-Saharan Africa. Through 2060, by one estimate, sub-Saharan Africa must increase food production sufficient to feed an extra 20–25 million people per year. Stated in these terms the Malthusian crisis, if humans are lucky, may emerge as a regional rather than a global problem. One factor that is difficult to pinpoint is the effect of AIDS. It may reduce populations in some areas of Africa, but it is also likely to reduce food production by depriving land of labor. A reduction in the number of laborers would not be so acute were Africa to invest in mechanization and plant breeding. If both population and food production fall in Africa, the net effect may not bear on the Malthusian crisis.

By one estimate, global population may double between roughly 2000 and 2050.[44] This growth would mean a movement from about 6 or 7 billion people to perhaps as many as 14 billion people, a staggering number. Predictions over a longer term forecast population diminution after 2100. Clearly, sometime in the twenty-first century, humans will reach the threshold of earth's carrying capacity. We will reach the biological limit of plant's capacity to provide edible biomass. The Malthusian crisis appears imminent in our century. These changes are likely to occur in a world of fewer farmers. Worldwide only about 5% of the population farms, though in East and South Asia and sub-Saharan Africa the percentage is closer to 70. As in the case of the birth rate, education will play an important role in who farms. Where children seek additional schooling, they are unlikely to return to the farm. The land grant universities witnessed this phenomenon since their founding in the United States in the nineteenth century (Chapter 8). Who will take their place to produce the food a growing world needs? The answer may be science and technology, which has allowed the United States to shrink to farm population to less than 2% of the workforce.

If biotechnology is to fulfill its promise, it must concentrate, first, on increasing yields by increasing the efficiency of photosynthesis, increasing the efficiency with which plants manufacture glucose, improve the efficiency with which plants absorb water and nutrients, and increase tolerance of high and low temperatures. Second, biotechnology, as already is true, must engineer plants resistant to pests, diseases, abiotic stresses, and herbicides. RR crops already have the ability to grow in the presence of glyphosate. Third, research must increase the quality of plant protein, vitamins, minerals, starch, sugar, and oil. Fourth, research should concentrate on engineering plants that make drugs and vaccines. This is fine but it does not solve the population problem. The emphasis should be on yield gains in the developing world. Because biotechnology is much more expensive than traditional breeding, leadership must come from the developed world. Among the disiderata, yield may be the most difficult to increase because it is a polygenetic attribute. As of 2000, biotechnology did not have the ability to identify and extract these genes and to insert them in an agronomic plant. Biotechnology must lead the way in transitioning from the C3 to the C4 pathway of photosynthesis, especially for tropical crops. By traditional measures, biotechnology appears successful. In 2013, GM corn occupied 90 of corn hectarage in the American Midwest.[45] The rise of intensive plant breeding and even more of biotechnology focused on the prowess of the scientist opening a divide between university-trained scientists and less-educated farmers. No longer was the attempt to change the genotype of a plant a farm activity. It was a scientific endeavor reinforced by the rise of industrial agriculture. Modern plant breeding arose after 1900, coinciding with the evolutionary synthesis among Mendelian genetics, population genetics, and Darwinism. The early focus on plant breeding did involve farmers, some of whom developed their own seed companies. In Switzerland, scientists and farmers cooperated in developing new varieties in spelt and wheat. German farmers and scientists focused on developing new varieties of rye and oats. By the 1930s, however, plant breeding as the preserve of scientists and private seed companies came to the fore. The work of farmers in selecting new varieties of a crop had depended on genetic diversity. The focus of professional breeders has narrowed diversity to a handful of elite cultivars. At the same time, aware of the dangers of uniformity, institutions created seed banks to preserve the genetic diversity of a crop.

In the developing world, the focus is on sustainability and providing food that is sufficiently abundant so that even the poor can afford it and so have access to adequate protein, calories, vitamins, and minerals. Agricultural productivity, however well intentioned, is probably not sustainable in the future without the guidance of science. By adding genes for resistance to pests, pathogens, and herbicides, biotechnology makes possible reliable yields. This reliability is important in an uncertain world. For example, fungal diseases ruin 50 million metric tons of rice per year. Insects destroy 26 million metric tons of rice per year. Virus and bacteria each claim about 10 million metric tons of rice per year. As one can apprehend, biotechnology has more work to do. The focus, of course, should not be solely on rice. Similar investments should be made for cassava and other staples.[46]

Intensive agriculture appears to be the basket into which humans have put their eggs. It appears to be the only possibility for feeding billions more people in the future. In this context, the focus should be on expanding our understanding of the biology of yield, as there is little to gain from extensive agriculture. The sciences must develop new, highly productive, sustainable agriculture worldwide. Governments must find the political will to increase funding for basic and applied research on agronomic plants. If the sciences cannot succeed, the alternative is famine. No economy can grow without a concomitant growth in agriculture. To this end, the CGIAR asserts its total focus on "agricultural research," which it targets to the developing world, which will produce most of the additional population growth in the twenty-first century.[47] The group notes that it is essential to jump start research now because it can take 20 years from research to transition from experimental plot to farm. The crops that CGIAR has developed have provided 75% of the developing world's protein and calories. In the near term, the farm must double yields to keep pace with population and without endangering the environment. The science and technology of farming must change if farming is to feed the world. Efforts must concentrate on plant genetics and biotechnology. This research must occupy the careers of scientists in the developing and developed worlds. The CGIAR believes biotechnology has the capacity to increase yields and to tailor its benefits to poor farmers.

According to one expert, we are living amid a "food crisis."[48] The mainstream solution is to replicate industrial agriculture around the world. This seems to have been the premise of the Green Revolution: That is, look to the United States, Canada, Australia, and western Europe for guidance. Others fret over the environmental and social flaws of this approach. Modern agriculture may no longer be viable in the modern world of limitations and environmental degradation. Scarcity may be the new reality, as it was for so deep in the past. The sciences must help increase food production by about 60% by 2050 if they are to feed the world.[49] Grain yields must leap 30% by 2050. The status of agriculture and the agricultural sciences as tools to allow humans to continue to reproduce unabated seems futile. Farmers, according to an old adage, should perfect humans, not simply produce more food.

NOTES

1. Revelation 6:1–6:17 (New Revised Edition).
2. Thomas Malthus, *An Essay on the Principle of Population* (Oxford: Oxford University Press, 2000), 3.
3. Ibid, 12–13.
4. John Krebs, *Food: A Very Short Introduction* (Oxford: Oxford University Press, 2013), 87.
5. John C. Caldwell, "Population Growth: Its Implications for Feeding the World," in *Food Security and Nutrition: The Global Challenge,* eds. Uwe Kracht and Manfried Schultz (New York: St. Martin's Press, 1999), 75.
6. David A. Cleveland, *Balancing on a Planet: The Future of Food and Agriculture* (Berkeley: University of California Press, 2014), 23.
7. Ibid, 27.
8. Caldwell, "Population Growth," 76.
9. U.S. Constitution, Article 1, Section 9.
10. Caldwell, "Population Growth," 77.
11. Ibid, 79.
12. Ibid, 81.
13. Peter von Blanchenburg, "The Feeding Capacity of the Planet Earth: Developments, Potentials and Restrictions," in *Food Security and Nutrition: The Global Challenge,* eds. Uwe Kracht and Manfried Schultz (New York: St. Martin's Press, 1999), 92.
14. Ibid, 94.
15. Ibid, 95.
16. Ibid, 96.
17. Ibid, 102.
18. Ibid, 103.
19. Krebs, *Food*, 92–94.
20. Ibid, 89.
21. Ibid, 90.
22. Klaus J. Lampe, "The Green Revolution, Not Green Enough for Asia's Rice Bowl," in *Food Security and Nutrition: The Global Challenge,* eds. Uwe Kracht and Manfried Schultz (New York: St. Martin's Press, 1999), 419.
23. Ibid, 422.
24. Ibid, 424.
25. Krebs, *Food*, 91–92.
26. Ibid, 94.
27. Ibid, 94.
28. Ibid, 109–110.
29. Ibid, 95.
30. Ibid.

31. Ibid, 96.
32. Ibid.
33. Ibid.
34. Ibid, 97.
35. Ibid, 98.
36. Ibid, 99.
37. Ibid, 101.
38. Ibid, 103.
39. Ibid, 104.
40. Ibid, 108.
41. Uwe Krecht, "Hunger, Malnutrition and Poverty: Trends and Prospects Towards the 21st Century," in *Food Security and Nutrition: The Global Challenge,* eds. Uwe Kracht and Manfried Schultz (New York: St. Martin's Press, 1999), 57.
42. Caldwell, "Population Growth," 81.
43. Ibid, 82.
44. Ibid, 87.
45. Krebs, *Food*, 104–105.
46. Klaus M. Leisinger, "Biotechnology in Third World Agriculture: Some Socio-economic Considerations," in *Food Security and Nutrition: The Global Challenge,* eds. Uwe Kracht and Manfried Schultz (New York: St. Martin's Press, 1999), 490.
47. Ismail Sarageldin, "Overcoming World Hunger: The CGIAR Prepares for the New Millennium," in *Food Security and Nutrition: The Global Challenge,* eds. Uwe Kracht and Manfried Schultz (New York: St. Martin's Press, 1999), 582.
48. Cleveland, *Balancing on a Planet*, 1.
49. Ibid, 2.

Bibliography

Abbott, E. *Sugar: A Bittersweet History*. New York: Duckworth Overlook, 2008.

Agoda-Tanjawa, G., S. Durand, C. Gaillard, C. Garnier, and J. L. Doublier. Properties of cellulose/pectins composites: Implication for structural and mechanical properties of cell wall. *Carbohydrate Polymers* 90; 2012:1081–1091.

Altman, A. and P. M. Hasegawa. *Plant Biotechnology and Agriculture: Prospects for the 21st Century*. Amsterdam: Academic Press, 2012.

Anemone, R. L. *Race and Human Diversity: A Biocultural Approach*. Upper Saddle River, NJ: Prentice-Hall/Pearson, 2011.

Aubry, S., N. J. Brown, and J. M. Hibbard. The role of C3 proteins prior to their recruitment into the C4 pathway. *Journal of Experimental Botany* 62; 2011:3047–3051.

Baatz, S. *Venerate the Plough: A History of the Philadelphia Society for Promoting Agriculture, 1785–1985*. Philadelphia, PA: Philadelphia Society for Promoting Agriculture, 1985.

Barber, S. *The Prendergast Letters: Correspondence from Famine Era Ireland, 1840–1850*. Amherst and Boston: University of Massachusetts Press, 2006.

Biffen, R. H. Mendel's laws of inheritance and wheat breeding. *Journal of Agricultural Science* 1; 1905:4–48.

Blank, C. E. Origin and early evolution of photosynthetic eukaryotes in freshwater environments: Reinterpreting proterozoic paleobiology and biogeochemical processes in light of trait evolution. *Journal of Phycology* 49; 2013:1040–1055.

Boak, A. E. R. *A History of Rome to 565 AD*. 4th edition. New York: Macmillan, 1955.

Boerma, H. R. and J. E. Specht (eds), *Soybeans: Improvement, Production and Uses*. Madison, WI: American Society of Agronomy, 2004.

du Bois, C. M., C. B. Tan, and S. Mintz (eds), *The World of Soy*. Urbana and Chicago: University of Illinois Press, 2008.

Broughton, J. M. and M. D. Cannon. *Evolutionary Ecology and Archaeology: Applications to Problems in Human Evolution and Prehistory*. Salt Lake City: University of Utah Press, 2010.

Brown, W. L. H. A. Wallace and the development of hybrid corn. *Annals of Iowa* 47; 1983:167–179.

Browne, C. A. Agricultural chemistry. *Journal of the American Chemical Society*, 48; 1926:127–201.

Burbank, L. and W. Hall. *The Harvest of the Years*. Boston and New York: Houghton Mifflin, 1931.

Busch, L. and W. B. Lacy (eds), *The Agricultural Scientific Enterprise: A System in Transition*. Boulder, CO: Westview Press, 1986.

Cato, M. P. *On Agriculture*. Trans. W.D. Hooper. Cambridge, MA: Harvard University Press, 1993.

Chawla, H. S. *Introduction to Plant Biotechnology*. 3rd edition. Enfield, NH: Science Publishers, 2009.

Ciochon, R. L. and J. G. Fleagle. *The Human Evolution Source Book*. Englewood Cliffs, NJ: Prentice-Hall, 1993.

Clason, T., T. Ruiz, H. Schagger, G. Peng, V. Zickermann, U. Brandt, H. Michel, and M. Radermagger. The structure of eukaryotic and prokaryotic complex I. *Journal of Structural Biology* 169; 2010:81–88.

Cleveland, D. A. *Balancing on a Planet: The Future of Food and Agriculture*. Berkeley, CA: University of California Press, 2014.

Collison, M. *Towards a New Green Revolution: A Perspective from the CGIAR Secretariat.* Washington, DC: World Bank, 1992.

Columella, L. J. M. *On Agriculture.* 3 vols. Trans. H.B. Ash. Cambridge, MA: Harvard University Press, 1968

Cook, W. J. How old is the universe? *U.S. News and World Report* 123; 1997:34–37.

Cook, L. M. and I. J. Saccheri. The peppered moth and industrial melanism: Evolution of a natural selection case strategy. *Heredity* 110; 2013:207–212.

Cornford, F. M. *Before and after Socrates.* Cambridge, UK: Cambridge University Press, 1932.

Crabb, A. R. *The Hybrid-Corn Makers: Prophets of Plenty.* New Brunswick: Rutgers University Press, 1947.

Cummings, V., P. Jordan, and M. Zevelebil (eds), *The Oxford Handbook of the Archaeology and Anthropology of Hunter-Gatherers.* Oxford: Oxford University Press, 2014.

Cumo, C. Public science and private enterprise: Battling Ohio Valley corn viruses. *Timeline,* 16; 1999:34–43.

Dahlberg, F. (ed.), *Woman the Gatherer.* New Haven and London: Yale University Press, 1981.

D'Amato, P. *The Savage Garden: Cultivating Carnivorous Plants.* Revised edition. Berkeley, CA: Ten Speed Press, 2013.

Darwin, C. *The Origin of Species by Means of Natural Selection or the Preservation of Favored Races in the Struggle for Life.* New York: Modern Library, 1993.

Darwin, C. *The Descent of Man, and Selection in Relation to Sex.* London: Penguin Books, 2004.

Demaree, A. L. The farm journals, their editors, and their public. *Agricultural History* 15; 1941:182–188.

Deng, G. *Development versus Stagnation: Technological Continuity and Agricultural Progress in Pre-Modern China.* Westport, CT: Greenwood Press, 1993.

Dethloff, H. C. *A History of the American Rice Industry, 1685–1985.* College Station, TX: Texas A&M University Press, 1988.

Donnelly, J. S. Jr. *The Great Irish Potato Famine.* Gloucestershire, UK: Sutton Publishing, 2001.

Eldridge, N. *Life Pulse: Episodes from the Story of the Fossil Record.* New York: Facts on File, 1987.

Evenson, R. E. and D. Gotlin. Assessing the impact of the green revolution, 1960 to 2000. *Science,* 300; 2003:754–762.

Fitzgerald, D. K. *The Business of Breeding: Hybrid Corn in Illinois, 1890–1940.* Ithaca, NY: Cornell University Press, 1990.

Fussell, B. *The Story of Corn.* New York: Knopf, 1992.

Gest, H. Homage to Robert Hooke (1635–1703): New insights from the recently discovered Hooke Folio. *Perspectives in Biology & Medicine* 52; 2009:392–399.

Gibbons, A. Who was homo habilis—And was it really homo? *Science* 332; 2011:1370–1371.

Glaeser, B. (ed.), *The Green Revolution Revisited: Critique and Alternatives.* London: Allen & Unwin, 1987.

Gould, S. J. *Punctuated Equilibrium.* Cambridge, MA: The Belknap Press of Harvard University Press, 2007.

Gowen, J. H. (ed.), *Heterosis: A Record of Researches Directed toward Explaining and Utilizing the Vigor of Hybrids.* Ames, IA: Iowa State College, 1952.

Hamilton, R. W. *The Art of Rice: Spirit and Sustenance in Asia.* Los Angeles: UCLA: 2003.

Hanson, N., B. Allen, R. L. Baumhardt, and D. Lyon. Research achievements and adoption of no-till, dryland cropping in the semi arid U.S. Great Plains. *Field Crops Research* 132; 2012:196–203.

Harlan, J. R. *Crops and Man.* 2nd edition. Madison, WI: American Society of Agronomy, 1992.

Hawkes, J., G. Cochran, H. C. Harpending, and B. T. Lahn. A genetic legacy from archaic homo. *Trends in Genetics* 24; 2008:19–23.

Heatherly, L. G. and H. F. Hodges (eds), *Soybean Production in the Midsouth*. Boca Raton, FL: CRC Press, 1999.

Hopkins, C. G. Improvement in the chemical composition of the corn kernel. *Bulletin 55 of the Illinois Agricultural Experiment Station* 1899:205–240.

Howell, F. C. *Early Man*. New York: Time-Life Books, 1965.

Janick, J., R. W. Schery, F. W. Woods, and V. W. Ruttan. (eds), *Plant Agriculture: Readings from Scientific American*. San Francisco, CA: W. H. Freeman, 1970.

Johnson, S. W. The agricultural experiment stations of Europe. *Annual Report of the Sheffield Scientific School of Yale College* 10; 1874–1875:11–31.

Johnson, S. A. *Tomatoes, Potatoes, Corn, and Beans: How the Foods of the Americas Changed Eating Around the World*. New York: Atheneum Books, 1997.

Johnson, D. W. *The Laws That Shaped America: Fifteen Acts of Congress and Their Lasting Impact*. New York and London: Routledge, 2009.

Jones, D. F. Dominance of linked factors as a means of accounting for heterosis, *Genetics* 2; 1917:466–479.

Kingsbury, N. *Hybrid: The History and Science of Plant Breeding*. Chicago and London: University of Chicago Press, 2009.

Kirakosyan, A. and P. B. Kaufman. *Recent Advances in Plant Biotechnology*. New York: Springer, 2009.

Kisselbach, T. A. A half century of corn research, 1900–1950. *American Scientist* 39; 1951:647–654.

Klenks, J. R., W. A. Russell, and W. D. Guthrie. Recurrent selection for resistance to European corn borer in a corn synthetic and correlated effects on agronomic traits. *Crop Science* 26; 1986:864–870.

Knoblauch, H. C., E. M. Law, W. P. Meyer, B. F. Beacher, R. B. Nestler, and B. S. White, Jr. *State Agricultural Experiment Stations: A History of Research Policy and Procedures*. Washington, DC: U.S. Department of Agriculture Miscellaneous Publication 904, 1962.

Knutson, R. D., S. D. Knutson, and D. P. Ernstes (eds), *Perspectives on 21st Century Agriculture: A Tribute to Walter J. Armbruster*. Oak Brook, IL: Farm Foundation, 2007.

Kracht, U. and M. Schultz (eds), *Food Security and Nutrition: The Global Challenge*. New York: St. Martin's Press, 1999.

Krebs, J. *Food: A Very Short Introduction*. Oxford: Oxford University Press, 2013.

Lemonick, M. D. and A. Dorfman. A long lost relative. *Time* 174, 2009:52–45.

Liebig, J. In Lyon P. and W. Gregory (eds), *Chemistry in Its Applications to Agriculture and Physiology*. Philadelphia: T. B. Peterson, 1847.

Liu, K. *Soybeans: Chemistry, Technology, and Utilization*. New York: Chapman & Hall, 1997.

Loebenstein, G. and G. Thottappilly (eds), *The Sweetpotato*. Dordrecht, the Netherlands: Springer, 2009.

Ma, H., K. Chong, and X. W. Deng. Rice research: Past, present, and future. *Journal of Integrative Plant Biology* 49; 2007:729–730.

Machlis, J. *The Enjoyment of Music: An Introduction to Perceptive Listening*. New York: W. W. Norton, 1955.

Macinnis, P. *Bittersweet: The Story of Sugar*. Crows Nest, Australia: Allen & Unwin, 2002.

Malthus, T. *An Essay on the Principle of Population*. Oxford: Oxford University Press, 2000.

Manetas, Y. *Alice in the Land of Plants: Biology of Plants and Their Importance for Planet Earth*. New York: Springer, 2012.

Mangelsdorf, P. C. *Corn: Its Origin, Evolution and Improvement*. Cambridge, MA: Harvard University Press, 1974.

Mann, C. Reseeding the Green Revolution. *Science*, 277; 1997:1028–1042.

Marti, D. B. *To Improve the Soil and the Mind: Agricultural Societies, Journals, and Schools in the Northeastern States, 1791–1865*. Ann Arbor, MI: Published for the Agricultural History Society and the Department of Communications Arts, New York State College of Agriculture and Life Sciences, Cornell University by University Microfilms International, 1979.

Martin, J. H., R. P. Waldren, and D. L. Stamp. *Principles of Field Crop Production*. Upper Saddle River, NJ: Pearson/Prentice-Hall, 2006.

Maxwell, F. G. and P. R. Jennings. *Breeding Plants Resistant to Insects*. New York: John Wiley, 1980.

Mayr, E. *The Growth of Biological Thought: Diversity, Evolution, and Inheritance*. Cambridge, MA: The Belknap Press of Harvard University Press, 1982.

McDowell, G. R. *Land-Grant Universities and Extension into the 21st Century: Renegotiating or Abandoning a Social Contract*. Ames, IA: Iowa State University Press, 2001.

McKay, J. P., B. D. Hill, J. Buckley, C. H. Crowston, M. E. Wiesner-Hanks, and J. Perry. *Understanding Western Society: A Brief History*. Boston and New York: Bedford/St. Martin's, 2012.

Meltzer, M. *The Amazing Potato: A Story in Which the Incas, Conquistadors, Marie Antoinette, Thomas Jefferson, Wars, Famines, Immigrants, and French Fries All Play a Part*. New York: HarperCollins Publishers, 1992.

Molecular expressions: Cell biology and microscopy, structure and function of cells and viruses. http://micro.magnet.fsu.edu/cells/plants/nucleus.html (accessed December 5, 2014).

Montgomery, S. H. and N. I. Mundy. Microcephaly genes and the evolution of sexual dimorphism in primate brain size. *Journal of Evolutionary Biology* 26, 2013:906–911.

Moulton, F. R. (ed.), *Liebig and after Liebig*. Washington, DC: American Association for the Advancement of Science, 1942.

Neiswander, C. R. and E. T. Hibbs. Corn borer first discovered on Middle Bass Island in 1921. *Ohio Farm and Home Research,* 1954: 52–54.

Nordin, D. S. *Rich Harvest: A History of the Grange, 1867–1900*. Jackson: Mississippi State University Press, 1974.

O'Connell, S. *Sugar: The Grass that Changed the World*. London: Virgin Books, 2004.

Oleson, A. and S. C. Brown (eds), *The Pursuit of Knowledge in the Early American Republic: American Scientific and Learned Societies from Colonial Times to the Civil War*. Baltimore, MD: Johns Hopkins University Press, 1976.

Oleson, A. and J. Voss (eds), *The Organization of Knowledge in Modern America, 1860–1920*. Baltimore, MD: Johns Hopkins University Press, 1979.

Opportunities to Meet Changing Needs: Research on Food, Agriculture, and Natural Resources. College Station, TX: Texas A&M University Press, 1994.

Pauketat, T. R. *The Oxford Handbook of North American Archaeology*. Oxford: Oxford University Press, 2012.

Pearson, C. J. *Field Crop Ecosystems*. New York: Elsevier, 1992.

Pearson, L. C. *The Diversity and Evolution of Plants*. Boca Raton, FL: CRC Press, 1995.

Perkins, J. H. *Geopolitics and the Green Revolution: Wheat, Genes, and the Cold War*. New York and Oxford: Oxford University Press, 1997.

Pierre, W. H., S. R. Aldrich, and W. P. Martin (eds), *Advances in Corn Production: Principles and Practices*. 2nd edition. Ames, IA: Iowa State University Press, 1967.

Plato. *Great Dialogues of Plato*. Trans. W. H. D. Rouse. New York: New American Library, 1956.

Plato. In Hamilton, E. and H. Cairns (eds), *The Collected Dialogues of Plato including the Letters*. Princeton, NJ: Princeton University Press, 1989.

Prance, G. and M. Nesbitt (eds), *The Cultural History of Plants*. New York and London: Routledge, 2005.

Raven, P. H., R. E. Evert, and S. E. Eichhorn. *Biology of Plants*. 6th edition. San Francisco: W. H. Freeman, 2013.

Reader, J. *Potato: A History of the Propitious Esculent*. New Haven, CT: Yale University Press, 2009.

Richards, M. P. and R. W. Schmitz. Isotope evidence for the diet of the Neanderthal type specimen. *Antiquity* 82; 2008:553–559.

Ridley, M. (ed.), *Evolution*. Oxford and New York: Oxford University Press, 1997.

Rightmire, G. P. *Homo erectus* and middle pleistocene hominins: Brain size, skull form and species recognition. *Journal of Human Evolution* 65; 2013:223–252.

Rossiter, M. W. *The Emergence of Agricultural Science: Justus Liebig and the Americans, 1840–1880*. New Haven, CT: Yale University Press, 1975.

Rubatzky, V. E. and M. Yamaguchi. *World Vegetables: Principles, Production, and Nutritive Values*. New York: Chapman & Hall, 1997.

Russell, E. J. Rothamstead and its experiment station. *Agricultural History* 16; 1942:161–183.

Russell, W. A., W. D. Guthrie, and R. L. Grindeland. Breeding for resistance in maize to first and second broods of the European corn borer. *Crop Science* 14; 1974:725–729.

Salaman, R. *The History and Social Influence of the Potato*. Cambridge: Cambridge University Press, 1985.

Sanchez, A. *The Teeth of the Lion: The Story of the Beloved and Despised Dandelion*. Blacksburg, VA: McDonald and Woodward, 2006.

Schapsmeier, E. L. and F. H. Schapsmeier. *Henry A. Wallace of Iowa: The Agrarian Years, 1910–1940*. Ames, IA: Iowa State University Press, 1968.

Schiebinger, L. and C. Swan (eds), *Colonial Botany: Science, Commerce, and Politics in the Early Modern World*. Philadelphia, PA: University of Pennsylvania Press, 2007.

Schopf, T. I. M. (ed.), *Models of Paleobiology*. San Francisco, CA: Freeman, 1972.

Schulze, R. *Carolina Gold Rice: The Ebb and Flow History of a Low Country Cash Crop*. Charleston, SC: History Press, 2005.

Schumann, G. L. *Plant Diseases: Their Biology and Social Impact*. St. Paul, MN: APS Press, 1991.

Shaffer, R. Evolution, humanism and conservation. *Humanist* 72; 2012: 19.

Sharma, S. D. (ed.), *Rice: Origin, Antiquity and History*. Enfield, NH: Science Publishers, 2010.

Sharples, R. W. *Theophrastus of Eresus: Sources for His Life, Writings, Thought and Influence*. 5 vols. Leiden: E. J. Brill, 1995.

Shull, G. H. The composition of a field of maize. *Report of the American Breeder's Association* 4; 1908: 296–301.

Shull, G. H. A pure line method of corn breeding. *Report of the American Breeder's Association* 5; 1909: 51–59.

Slater, A., N. W. Scott, and M. R. Fowler. *Plant Biotechnology: The Genetic Manipulation of Plants*. 2nd edition. Oxford and New York: Oxford University Press, 2008.

Smith, C. W. *Crop Production: Evolution, History, and Technology*. New York: Wiley, 1995.

Smith, A. F. *The Tomato in America: Early History, Culture, and Cookery*. Urbana and Chicago: University of Illinois Press, 2003.

Smith, R. Agricultural experiment stations provide catalyst for 150 years of innovations in agriculture. *Southwest Farm Press* 36; 2009, 8–9.

Smith, C. W., J. Betran, and E. C. A. Runge. *Corn: Origin, History, Technology and Production*. Hoboken, NJ: Wiley, 2004.

Smith, C. W. and R. H. Dilday (eds), *Rice: Origin, History, Technology, and Production*. Hoboken, NJ: John Wiley, 2003.

Sprague, G. F. and J. W. Dudley (eds), *Corn and Corn Improvement*. Madison, WI: American Society of Agronomy, 1988.

Stern, K. *Introductory Plant Biology*. New York: McGraw-Hill, 2003.

Stewart, C. N. Jr. *Plant Biotechnology and Genetics: Principles, Techniques, and Applications*. Hoboken, NJ: Wiley, 2008.

Stocking, G. W. Eugene Du Bois and the Ape Man from Java. *Isis: Journal of the History of Science in Society* 81; 1990:785–786.

Stringer, C. Africa and the origins of modern humans. *African Archeological Review* 14; 1997:209–212.

Stringer, C. Modern human origins—Distinguishing the models. *African Archeological Review* 18; 2001:67–75.

Stringer, C. and P. Andrews. *The Complete World of Human Evolution*. London: Thames & Hudson, 2012.

Stringfield, G. H. Heterozygosis and hybrid vigor in maize. *Agronomy Journal* 42; 1950:45–52.

Stringfield, G. H. Maize inbred lines of Ohio. *Research Bulletin 831 of the Ohio Agricultural Experiment Station*, 1959: 28–42.

Suri, T. Selection and comparative advantage in technology. http://www.yale.edu/~egcenter.

Suszkiw, J. Crop plants: At the root of civilization. *Agricultural Research*, 1999:2–6.

Tattersall, I. Human origins: Out of Africa. *Proceedings of the National Academy of Sciences of the United States of America* 106; 2009:16018–16021.

Taylor, T. N., E. L. Taylor, and M. Krings. *Paleobotany: The Biology and Evolution of Fossil Plants*. 2nd edition. Amsterdam: Elsevier, 2009.

Templeton, A. R. Genetics and recent human evolution. *Evolution* 61; 2007: 1507–1519.

Theophrastus. *De Causis Plantarum*. 3 vols. Trans. B. E and G. K. K. Link. Cambridge, MA: Harvard University Press, 1990.

Thorne, C. E. The maintenance of fertility. *Circular 40 of the Ohio Agricultural Experiment Station* 1905: 1–37.

True, A. C. *History of Agricultural Education in the United States, 1785–1923*. Washington, DC: U.S. Department of Agriculture Miscellaneous Publications 36, 1926.

True, A. C. *History of Agricultural Experimentation and Research, 1607–1925, Including a History of the United States Department of Agriculture*. New York: Johnson Reprint, 1970.

Tuttle, R. H. The pitted patter of laetoli feet. *Natural History* 99; 1990: 60–65.

UCLA Fowler Museum of Cultural History, 2003.

Vacuole. http://www.princeton.edu/~achenay/tmve/wiki100k/docs/vacuole.html (accessed December 5, 2014).

Varro, M. T. *On Agriculture*. Trans. W. D. Hooper. Cambridge, MA: Harvard University Press, 1993.

Vavilov, N. I. *World Resources of Cereals, Leguminous Seed Crops and Flax, and Their Utilization in Plant Breeding*. Moscow: The Academy of Sciences of the USSR, 1957.

Wacher, J. (ed.), *The Roman World*. 3 vols. London and New York: Routledge, 1987.

Wade, N. Neanderthal women joined men in the hunt. *New York Times*, 2006. http://www.nytimes.com/2006/12/05/science/05nean.html?_r=0.

Waldbauer, G. *Insights from Insects: What Bad Bugs Can Teach Us*. Amherst, NY: Prometheus Books, 2005.

Walden, D. B. (eds), *Maize Breeding and Genetics*. New York: John Wiley, 1978.

Wallace, H. A. Public and private contributions to hybrid corn—Past and future. *Proceedings of the Tenth Annual Hybrid Corn Industry—Research Conference, 1955*. Chicago, IL: American Seed Trade Association, 1956.

Wallace, H. A. and W. L. Brown. *Corn and Its Early Fathers*. Revised edition. Ames, IA: Iowa State University, 1988.

Wanjie, A. (ed.), *The Basics of Plant Structures*. New York: Rosen Publishing, 2014.

Wayman, E. Discovering human ancestors. *Smithsonian* 42; 2012:38–39.

Weaver, T. D. Did a discrete event 200,000–100,000 years ago produce modern humans? *Journal of Human Evolution* 63; 2012:121–126.

Welch, F. J. Hybrid corn: A symbol of American agriculture. *Proceedings of Sixteenth Annual Hybrid Corn Industry—Research Conference, 1961*. Washington, DC: American Seed Trade Association, 1962.

Went, F. W. *The Plants*. New York: Time-Life Books, 1965.

White, K. D. *Roman Farming*. Ithaca, NY: Cornell University Press, 1970.

White, T. D. *Human Osteology*. 2nd edition. San Diego: Academic Press, 2000.

White, P., G. Timothy, J. Hammond, and E. James. Improving crop mineral nutrition. *Plant & Soil* 384; 2014:1–5.

Willis, K. J. and J. C. McElwain. *The Evolution of Plants*. Oxford: Oxford University Press, 2002.

Woeste, K. E., S. B. Blanche, K. A. Moldenhauer, and C. D. Nelson. Plant breeding and rural development in the United States. *Crop Science* 50; 2010:1625–1632.

Wong, K. Human or hobbit? *Scientific American* 311; 2014:28–29.

Woolfe, J. A. *Sweet Potato: An Untapped Food Resource*. Cambridge: Cambridge University Press, 1992.

Wu, F. and W. P. Butz. *The Future of Genetically Modified Crops: Lessons from the Green Revolution*. Santa Monica, CA: Rand Corporation, 2004.

Zimmer, C. Interbreeding with Neanderthals. *Discover* 34; 2013:38–44.

Index

Sucker, 8
Sugarcane, 126
   new opportunities and problems, 130–131
   in New World, 128–129
   origin and diffusion in Old World, 126–128
Sweet potato, 141
   Columbian Exchange, 141–144
   New World origins, 141

T

T-DNA, *see* Transfer DNA (T-DNA)
Terrestrial plants, rapid evolution of, 30
   cuticle, 31
   evolution of vessels, 32
   land plants evolution, 30
   spores, 30–31
Theophrastus, 105
Tissue culture, 219
Tobacco leaf, 13
Tod A, *see* Toxin A (Tod A)
Tolerance, 182
Tooth abrasion, 67
Toxin A (Tod A), 225
Transcription RNA (tRNA), 219
Transfer DNA (T-DNA), 220
Transgene, 219
tRNA, *see* Transcription RNA (tRNA)
Tropical rainforests, 40
Turgor pressure, 19

U

Unisex flower, 37
United Nations Educational, Scientific,
   and Cultural Organization
   (UNESCO), 201
United Nations Food and Agriculture
   Organization (FAO), 210, 242
University of Georgia, 165
U.S. Agency for International Development
   (USAID), 197
U.S. Department of Agriculture (USDA), 132,
   145, 157

U.S. Environmental Protection Agency (EPA),
   167, 223
U.S. National Plant Germplasm
   System, 168

V

Vascular bundles, 12
Vascular plants, 33
Vavilov, Nikolai, 74
Vein, 11, 12
Vitamin A deficiency, 251
Volcanism, 41

W

Washington, George, 155
Water, 18; *see also* Flowers; Fruits;
   Leaves; Photosynthesis;
   Seeds
   capillary action, 19
   ions transportation, 20
   osmosis, 18–19
   stomata, 19
Weeding, 111–113
Wheat, 68–69, 75, 208
Woman, 62
   edible plants, 63
   female field workers, 62
   hunter-gatherers, 63, 64
Wood, 9

X

Xyla, 12
Xylem, 12

Y

Younger Dryas, 77

Z

Zea mays, 89